# A USER'S GUIDE TO
# BUSINESS ANALYTICS

# A USER'S GUIDE TO
# BUSINESS
# ANALYTICS

## AYANENDRANATH BASU

*INDIAN STATISTICAL INSTITUTE*
*KOLKATA, INDIA*

## SRABASHI BASU

*BRIDGE SCHOOL OF MANAGEMENT*
*GURGAON, INDIA*

&

*WORLD CAMPUS INSTRUCTOR*
*THE PENNSYLVANIA STATE UNIVERSITY, USA*

**CRC Press**
Taylor & Francis Group
Boca Raton   London   New York

CRC Press is an imprint of the
Taylor & Francis Group, an **informa** business

A CHAPMAN & HALL BOOK

CRC Press
Taylor & Francis Group
6000 Broken Sound Parkway NW, Suite 300
Boca Raton, FL 33487-2742

Printed on acid-free paper
Version Date: 20160509

International Standard Book Number-13: 978-1-4665-9165-3 (Hardback)

**Visit the Taylor & Francis Web site at**
**http://www.taylorandfrancis.com**

**and the CRC Press Web site at**
**http://www.crcpress.com**

Printed and bound in the United States of America by
Edwards Brothers Malloy on sustainably sourced paper

To Professor Atindra Mohan Gun

and

To the memory of Professor Bruce G. Lindsay

*with a lifetime of gratitude*

# Contents

# *Preface*

This is a book on predictive analytics. If technology was the competitive edge for business during the later part of the 20th century, for the 21st century it is going to be knowledge. Easy availability of technology at a fraction of the cost compared to what it was in the 1990s, has made all businesses hungry for data. E-commerce, e-business, e-transactions as well as less technology–intensive methods of doing business now generate a plethora of data. Information technology (IT) and IT–enabled services (ITeS) are at the core of all major and minor players in every domain. Automation produces an ever-expanding universe of data, spiraling and re-orienting itself, and producing various patterns in the data.

Before getting into the subject matter, we want to emphasize a personal preference which will have some effect on the style followed in this book. The primary subject matter here deals with data and knowledge. In Latin, data is the plural of datum; however, this Preface is perhaps the only occasion when we will use the word datum in this book. Most style guides have now come to accept the use of the noun data with the singular verb. We do acknowledge that the use of the plural verb with data is still the correct grammatical usage. However, modern usage often treats the word data as a collective noun taking a singular verb. As practitioners, we have, over the years, become used to writing the noun data with the singular verb; this, we believe, is true for many other practitioners as well. This is the expression that we will follow throughout the book. This is just our preference, and does not represent a lack of knowledge of the rules of grammar; neither is it an effort to offend the traditionalists.

Big Data, as the expanding data conglomerate is familiarly known, is a treasure trove of information. In the past, complex data mining techniques were used to explore and manipulate the data, by either visualization or intensive mathematical methods. In the main, data mining techniques were the forte of specialists. However, this was before the advent of modern computer technology, which has made real-time capture of a huge volume of data possible. Big data has made traditional data mining methods driven by the specialists obsolete to some extent. Microscopic and labor-intensive data mining techniques are being replaced by automated software-driven methods which are being controlled and interpreted by analysts. Analytics seems to be the inevitable tsunami that is slated to inundate the field of business intelligence.

In all fields where business intelligence plays a role, competition is being driven by data analytics. Take, for example, the case of revenue management in the hospitality or aviation industries. At precisely which time point how many rooms or seats would be made available and at what price level, for how long should those levels hold or for loyal customers what should be the optimum discount offered – these decisions were previously controlled by account managers on a case by case basis. Not any more! Now the whole system is automated and an optimized revenue management suite is in place which, potentially, can update the system in real time on the basis of each sold or canceled seat or room so that a customer gets the best possible deal in booking over the web instantaneously. Automated optimized web-based systems are the rule of the game.

In all spheres of business, intelligence is being built in to extract granular level information to drive competition. Marketing departments would like to put customers in well-defined groups for targeted marketing efforts rather than blindly put everybody in the same basket and incur avoidable loss. As communication channels and media exposure are expanding at an exponential rate, more effective and tailored contents are required to reach potential customers with a higher probability of acceptance of the products and to retain them. Insight into customer preference must now be data-driven and customer relationship management must be leveraged using choices of customers which are expressed through their product purchase pattern and feedback. To steer ahead of competition, not only a very large database must be maintained, it should also be accessed in real time for knowledge development and corresponding actions.

The message here is very clear. Each and every business-related decision and action is now thoroughly rooted in data. Historical data is collected and examined from all angles for knowledge gain. This is now routed through predictive modeling for further enhancement of business. Extraction of knowledge from observed facts, therefore, cannot remain in the domain of specialized experts only. An effective business analyst needs to understand the stories spelled out in data. Whatever be her subject matter expertise, she needs to understand the analytical logic behind the recommendation made by the software because, whenever big data is involved, software is heavily depended upon to manipulate the data. Nevertheless, knowledge extraction is not merely processing and manipulation of historical data.

Data is composed of observed facts and the current status of the business. Consolidation of facts leads to information gain. Enterprise resource planning (ERP) and similar systems in a business merge and process all business-related facts stored in various integrated systems and present information at managerial fingertips. However thorough and comprehensive a report may be, it is worthless if the information cannot instigate a control over the future. Predictive modeling based on historical data builds intelligence in the procedure which orients the process against possible pitfalls. Predictive modeling is an artifact of software output and hence these outputs need to be assimilated and interpreted correctly.

This book is for those who have a rudimentary training in statistics and are faced with a large volume of business data; this book is for those who are responsible for predicting the future status of a business with a high degree of accuracy. It is not possible to discuss all possible predictive modeling and data mining methods in a single volume. Two of the most used methods, regression and time series forecasting, have been discussed extensively. Our emphasis is not on the theoretical aspects. Nonetheless, wherever appropriate, we have provided mathematical justifications in the appendix for the technically advanced reader. Skipping the technical details will not arrest the flow of the book. We have focused on various real-life situations that analytics experts are faced with as a matter of course and for which they need a definitive solution. All spheres of business and businesses in all domains are coming to realize the advantages of analytics. Hence analytics specialists are coming from all types of backgrounds and training. Reliance on software has increased many times. It is therefore mandatory that software output be interpreted accurately. At the same time, analysts should realize the pitfalls of blind dependence on software. After all, beyond a certain point, one must not let the computing system take over and dictate human analytical capability! The objective of this book is to guide analysts in making decisions based on an understanding of broad statistical rules as well as the stories told through data mining.

All the examples considered in this book have been solved using the R software exclusively. Obviously a very attractive feature of R is that it is free. That does not take away from the fact that R is also a highly capable software and is used extensively in the analytics field. Its availability in the public domain guarantees that it is accessible to everyone for sharing and exchanging programs and codes. This universality of R makes it a very suitable software for use in the analytics industry. All variable names in the R examples start with capital letters, both in the R output and in the text. Sometimes long variable names have been used to keep them self explanatory. For example, Duration.of.Credit.Month will represent the period of a loan (in months).

Who will benefit from this book and why is this book necessary? Readers will notice that we have provided a large number of references after each chapter. In doing so, we have tried to keep abreast of the literature and keep the reader aware of the same. However, we feel that there are few, if any, books which provide a substantially comprehensive description of predictive business analytics while building the statistical methods from scratch. The book is not only written from the business perspective, it also strongly emphasizes the predictive part, which is necessary to set it apart from just another book on business statistics.

We have followed the American (rather than the British) system of spelling in this book. Unlike the use of the singular verb with the noun "data", this does not represent any particular preference of the authors. We have just chosen the easiest way to go along with the default spell-check routines.

In the past, most basic textbooks on statistics and probability (or any text using statistical distributions) have contained detailed tables of probabilities and quantiles of standard probability distributions including the normal, and the sampling distributions arising from the normal (the $\chi^2$, $t$ and $F$ distributions). This has continued to fairly recent times, more by habit than by real utility. At the present time, we deem this to be unnecessary. The probabilities and the quantiles of the different distributions can now be easily obtained from any one of a host of standard statistical software or programming language such as R; they are available even in Excel, which is accessible to almost anybody with a computer. As such, these tables do not form a part of this book.

The culture of analytics must be ingrained in its core for a business to forge ahead and succeed. Decisions need to be based on hard facts, even after acknowledging the necessity for imaginative entrepreneurship. This book aims to arm industry leaders with an adequate understanding of data-based decision making from a practitioner's viewpoint.

It is our pleasure to acknowledge the assistance of many of our colleagues, friends and students who have contributed in various ways to the preparation of the manuscript. In particular we mention the support received from Dr. Abhijit Mandal, Dr. Abhik Ghosh, Dr. Kiranmoy Das, Mr. Arun Kumar Kuchibhotla, Mr. Apratim Dey and Mr. Taranga Mukherjee. They proofread the raw manuscript, helped with the preparation of the figures, supplied appropriate references and provided assistance in many other forms. All of them are past or present students of the Indian Statistical Institute, or have been associated with the Institute in the capacity of project-related personnel. Dr. Das is currently a Professor at the Institute. We are grateful to all of them.

We also gratefully acknowledge the support and cooperation received from the CRC colleagues including Ms. Aastha Sharma, Mr. Delroy Lowe, Ms. Robin Lloyd-Starkes, Mr. Alex Edwards and Mr. Gary Stallons. Mr. Shashi Kumar's expert advise and assistance has helped us overcome the hurdles of typesestting in LaTeX.

Finally, special thanks are due to a special person–our daughter Padmini. Apart from occasional proofreading and correction of typos, her silent understanding made the long and difficult stretch of manuscript writing far more bearable than it could have been.

Ayanendranath Basu
Srabashi Basu
April 2016

# 1

## *What Is Analytics?*

*Business Analytics*, or simply, *analytics*, seems to be one of the most commonly used words in the world of business today. From the top brass to the newest recruit, everybody claims to be applying analytics in their respective domains to raise the efficiency of their business operations. Unfortunately, only a handful of organizations are currently able to take real advantage of analytics to maximize return on investment or to find the lowest hanging fruits, and they are undoubtedly the leaders in the respective industries.

The popularity of analytics is on the rise along with the data resource gathered by any business. In today's world data is collected and stored, not merely in terabytes but in petabytes or in even higher orders. All real-time transactions over the net, customer preferences, customer demography and other detailed information are stored in vast data storages; resources are even pooled across all sorts of social and other network sites. Computing power has increased manyfold so that accessing and getting information out of the stored data can be easily done with only a click of the mouse.

Any organization can do that. And actually almost every organization is doing that, and doing just that! The huge data warehouses, most of the time, are underutilized, being accessed only to produce reports, without any attempt to draw on the hidden knowledge. The principal benefit of analytics is to purify the knowledge embedded in the transactional databases so that this can be applied to further business goals and increase revenue. Most organizations, though, are satisfied with stopping at the threshold of analytics. It needs to be clearly understood that managerial report generation is not core analytics.

Analytics may be considered as a three-step business tool, the first of which is report generation. This is looking at the history of business activities to get a fair idea of how a business performed in the past. This is known as *descriptive analytics*. The next step is *predictive analytics* or forecasting, where a model is built using the historical knowledge gathered. This is the essential step of analytics, where facts or information from the past are leveraged to understand the future course the business may assume. There may also be a third step, where the modeling knowledge is applied in an optimum way to exert a greater control on the future. This is called *prescriptive analytics*. Most businesses, at least in India, have not matured to the stage where they are even able to take advantage of modeling. There is no dearth of data, but only a limited amount of information is being extracted, from which it is not possible to gain maximum knowledge. A vast majority of businesses are still sitting on a pile of expensive data but are making little use of it and running the business on gut feeling.

So what really is analytics? Without going into any technicalities, analytics may be defined as an intelligent process of converting collected data and information into knowledge for improved decision making. At the core of analytics lie information retrieval, data mining technologies and statistics. These are assisted and honed to precision by domain knowledge and business intelligence.

## 1.1   Emergence and Application of Analytics

A common belief is that analytics is necessary for big data only. That is to say, only if a business is big enough, would it benefit from analytics. Before we discuss analytics, let us look into the idea of big data. Interestingly enough, big data is not for big businesses only—even a medium sized business may encounter big data in its transactions. Big data does not only indicate the massiveness of the collected information, but encompasses the variety and complexity of the total information generated. Today's technology is able to record hundreds of thousands of activities per second through diverse platforms, be it through digital pictures, sensory mechanisms, postings at social media sites, cell phone signals and purchase or transaction records. It is not possible to store such a variety of records in a simple relational database. Alternative storage and access facilities are required, possibly distributed over a good many servers. The need to extract information from such a huge mass of facts in real time accelerated the emergence of big data technology. Analytics plays a pivotal role in capturing and presenting the knowledge content therein.

The diversity of information presents new opportunities of analysis. Take, for example, fraud analytics. Banking and credit card companies have been doing fraud analytics for a long time. However, leveraging with innovative applications of text mining and network link analysis, financial institutions are looking into the likelihood of fraudulent claims and prioritizing their resources accordingly. Opportunities for customer relation managers lie in capturing social media postings regarding their products. The information that is not available via traditional survey questionnaires may actually be captured using various discussion postings.

Almost all business applications can benefit from analytics. In the insurance business, especially in non-life insurance, the premium charged is a product of a base cost and multipliers corresponding to levels of customer profiling factors. Determination of a car insurance premium, for example, is a product of a base cost and multiplicative factors corresponding to the insured person's gender, age, age and make of the car, residential area and a multitude of other relevant conditions. To determine these multiplicative factors so that profit is maximized under price elasticity, which in turn determines the conversion rate of a policy, complex analytics methods are applied.

Marketing managers have been applying analytics to a certain extent for a long time. Every airline maintains a database on their loyalty customers to understand the nature of their travel pattern, differential demands and possibility of cost saving on behalf of the organization, without having major impact on travel comfort. Prime time spots on major media channels are expensive and their costs are determined by collecting and analyzing data on viewers' preferences for various programs. Commercial organizations sending bulk emails to promote clients' products are always analyzing click-rates and conversion-to-purchase rates depending on subject headers. Needless to say, the whole e-commerce industry flourishes on analytics.

To understand big data and to leverage the information content, analytics is the technology to apply. Analytics includes data analysis through statistics, but is not limited to it. Machine learning and computer science play vital roles in data analytics and so does domain knowledge.

## 1.2 Comparison with Classical Statistical Analysis

Statistics has been used to solve industrial problems for a long time. There are many instances where statistical procedures were developed through industrial applications – Student's $t$ by Gosset is a case in point. But the major difference, as we see it, between classical statistical applications and analytics is two-fold. The first and the foremost is availability of volume and variety of data. Possibly for the first time in the history of data-driven applications, the problem centers around availability of too much data, making elimination of noise from the signal extremely difficult.

The second difference is in the solution approach to an industrial problem. Previously, the statistical approach was less focused in that, in the absence of a precise statistical solution, an approximate method would have been acceptable. But not any more! With the advent of very high computational power and development in computational theory, if a theoretical approach is not tractable for a particular problem, computational approaches, including those related to machine learning or other computer science techniques, are adopted until a solution with an acceptably high accuracy for the business problem is found. Since real-time computational power is now within our grasp, present day business tolerates a very narrow margin of error only.

Analytics is therefore very much dependent on efficient computation, intelligent coding and ability to find a logical way to solve a problem. Analytics is not magic, but applied shrewdly, it can work wonders in extracting the knowledge embedded in the data. A logical approach, based on statistics, which digs down deep into the business process, breaks the whole problem into smaller sub-problems, tries to find patterns in seemingly unrelated modules of data and finally extracts relevant information for knowledge management and has a tangible impact on business is at the core of analytics.

Analytics involves data mining. The term data mining came from computer science and machine learning. Data mining is the process which helps in the discovery of patterns among seemingly unrelated masses of data collected from many different sources. The two main streams of data mining are model building and pattern recognition. Examples of model building involve regression and predictive modeling including time series analysis; pattern recognition involves multivariate clustering, discriminant analysis, identification of principal components, factor analysis and more.

## 1.3   Theory versus Computational Power

Classical statistics leans on theoretical support; analytics borrows strength from theory but tries to adapt itself all the time to get into the core of the data. Analytics is completely data-driven and statistics is a technology to understand the fantastic behavior of data. Take, for example, the case of outliers. Generally speaking, a student of classical statistics would like to downweight the outliers so that the remaining major (and pure) part of the data is closer to the assumptions made so that the estimates and testing procedures are suitably justified. Business analytics may actually focus on the outliers. While analyzing loyalty customers for an international airline, it is the high-value customers who are targeted. Their behavior, demands and consumption propensity are closely monitored so that their value to the airlines can be increased even more. For revenue management, it is the outliers that bring value to an organization and therefore require further scrutiny.

As mentioned earlier, big data is not only voluminous in terms of records, but it may contain innumerable characteristics on each record. In classical statistics it may be shown that any null hypothesis of equality (or any other relevant hypothesis) may be eventually rejected if the sample size increases indefinitely. For a huge sample size, the traditional p-value associated with such a test may not make much sense. Hence an innovative approach is required to analyze such data.

Given the state of computational power available at our fingertips, innovation is bound to harness this capability. Analytics, especially predictive analytics, is now closer to computer science than it ever was before. Many techniques, e.g., neural networks, among others, have been integrated seamlessly into the analytical realm. Clustering and classification algorithms are heavily dependent on computational power. Simulation and resampling methods are applied intensively to justify future predictions in situations where distributional assumptions are violated.

Computational techniques have given analysts unprecedented power, but, at the same time, they have increased the chances of blindly applying any technique to any situation, without paying heed to its applicability. The analytical insight does not come with any software application; a software application enhances analytical insight. It is always dangerous to put the cart before the horse. In this age of easily available software, precisely this tendency has increased. Unless one knows how to read and interpret data, it is not possible to elicit the knowledge embedded in the data. Blind application of software on a large number of records will not necessarily provide insight into the data; rather it is possible that in the mire of information all grains of truth will be inextricably lost.

Computational power is to be enjoyed, but not at the cost of foregoing theoretical support. In fact, now that everyone, with or without a proper training in statistics or data analysis, is able to access statistical software, understanding of the basic principles is even more vital. A data analyst does not need to be bothered with the mathematical intricacies. But, without the rationale and logic behind the application procedures, an analyst will not be able to fully understand the data-generating mechanism and will not be able to further the business goal to the optimum limit.

## 1.4 Fact versus Knowledge: Report versus Prediction

What is important for a business to survive? The easiest answer is to maintain and keep increasing its revenue. The health status of a business needs to be continuously monitored through the data collected. Data will include, but is not limited to, all structured transaction data, internal systems data, sales and logistical data, etc., which are carefully recorded in the enterprise resource planning (ERP) or any other system. At the same time, external data and information are extremely vital for the survival of a business. Customer feedback surveys, call center records and pre-launch market surveys are a few of the channels to understand the competitiveness of the business and its products and the synergy generated among current and prospective buyers. Very recently there has been a drive to capture the unstructured information regarding the performance of a business. What structured customer feedback will not be able to capture may be floating around in social media postings.

All aspects of an organization are monitored and all sorts of transactional data are stored in data warehouses. Previously data was stored in stand-alone data marts, but nowadays organizations are pushing toward data warehouses where data of various dimensions and volumes can be juxtaposed and slicing and dicing of data at all levels is possible for an in-depth study. However, examination of historical data is only the first step in leveraging knowledge. Facts are retrospective while analytics is forward looking. Unless the past performance is well understood and knowledge is assimilated in the business process, data collection, storage and mining are not going to add value for an organization.

## 1.5 Actionable Insight

Thus the ultimate aim of data analytics is to gain *actionable insight* into the business process. For a business organization, all data is related to customer interaction in some form or other. Think of sales data, marketing data, customer relationship data, human resources data, transactional data, financial data — everything informs the analyst about the status of the business. Of these various data types, some will not be under control for the business. For example, if the overall economy shows signs of stagnation, a particular business will not be able to control it, but it can plan on less demand being generated. Actionable insight is the knowledge gained from mining the collected and stored data that can be directly fed into the business system and can be used to make decisions regarding the future of the business, be it short, medium or long term.

The challenge here is that actionable insight depends upon uncertainty. A fact that is certainly known to everybody is not an insight. Predictive analytics thrives on the inherent variability in business transactions. This is the reason why analytics has deep roots in probability and statistics. In addition, data mining, pattern recognition and model building help to extract knowledge about the core business process from noisy data.

## 1.6 Suggested Further Reading

The field of analytics is evolving and the literature is fairly new. Possibly the most prolific educator in this area is Tom Davenport, with several outstanding publications under his belt; see Davenport and Harris (2007), Davenport et al. (2010) and Davenport and Kim (2013). For an overview of analytics applications, analytics success stories, and how businesses can profit from analytics, these books merit special mention. They do not deal with the theory behind analytics but explain in a logical way why analytics is the way to success for any business. Among other books available on the market to get an insight into analytics, Hardoon and Shmueli (2013) and Camm et al. (2013) are good reads. Saxena and Srinivasan (2013) provide a number of business case studies.

There are many excellent books on data mining with application to business analytics, such as Linoff and Berry (2011). These books are for a more technical audience and contain advanced data mining technology. The focus of many of these books is domain specific, such as marketing, customer relationship management (CRM), etc. We have not listed such books here. Several books have been cited in appropriate chapters where data analytic applications are considered. However, no domain-specific books have been cited, since we have tried to keep this book for a general audience.

Analytics cannot be performed without the help of advanced software. There are a number of good user's guides on various software, such as SAS, SPSS, R, etc. In the next chapter we introduce R and the development in the rest of the book is illustrated solely through R applications. Reading material on R is cited at the end of Chapter 2.

# 2

## Introducing R—An Analytics Software

Business analytics is based solely on data, often massive data, and adds value to a business by extracting hidden patterns and knowledge from the data. Data itself comprises structured and unstructured formats containing text, audio, video, images and even network information. The days of simple relational databases are gone and MS Excel or similar software are not enough to mine the data. Hence an analyst must be adept at using at least one analytics software.

Among software, choices are many, such as Minitab, SAS, SPSS, Stata, R and many more. Of these, SAS is probably the most used software. It has been around since 1976 and provides valuable support to many application areas of analytics. But SAS is an expensive licensed software. It requires one-time installation and yearly license updates. R, on the other hand, is completely free and open source. The analytics world, one may say, is divided into two camps — those who swear by R and those who do not! It is not our intention to disparage SAS or any other software. But we feel that R is very powerful and can do very well indeed in a non-legacy set up. Organizations that are just starting up or have been in business for a shorter period often choose R. However, learning R might not be the easiest task in the world. We have used only R for all the examples that are discussed in this book. For many of the illustrations, important parts of the codes are also provided along with the R output. In this chapter we provide a very brief glimpse into the working of R and indications as to where one should look for help in case one gets stuck. It is not possible to provide comprehensive guidance on R in a single chapter. Indeed, in this brief note we are not even able to scratch the surface of R functionalities. But a context setting for the very first users of R should be helpful.

## 2.1 Basic System of R

A major advantage of R is that it is completely free for anybody to install, use, modify and redistribute. The copyright for the primary source code for R is held by the R Foundation (`https://www.r-project.org/foundation/`) and it is published under GNU General Public License version 2.0

(`http://www.gnu.org/licenses/gpl-2.0.html`). The primary R system can be downloaded from Comprehensive R Archive Network (CRAN, `https://cran.r-project.org/`). The CRAN site also contains many add-on packages, called R libraries, many of which are user developed and distributed for anybody to use.

R works on almost every kind of platform, and installation of R is done from a CRAN mirror directly. In addition, it is always a good idea to install the integrated development environment called RStudio. Installation of this IDE provides a window-like feature for R. One can create a shortcut for RStudio on the desktop and clicking on it will start R. The interface will look like Figure 2.1.

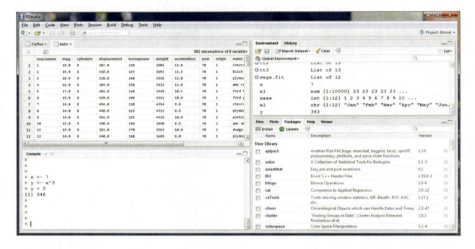

**FIGURE 2.1**
Screenshot of RStudio interface.

In the Console window (bottom left) one can directly input the R commands. It is also possible to write codes in R in the window above (top left) and submit all instructions in a batch. R accepts all commands at R prompt (>). The assignation is done by using an arrow (<-). To assign the value of 8 to a variable x and to print out the variable, the following sequence of commands is followed:

```
> x <- 8
> x
[1] 8
```

On the console, simple mathematical operations are done.

```
> y <- 4
> y
```

```
[1] 4
> x/y
[1] 2
> a <- x/y
> a
[1] 2
```

The result of a mathematical operation can be displayed on the screen or saved as a variable and accessed later. The result can also be subjected to further operations.

```
> b <- (a - 3)^3
> b
[1] -1
```

Note that ^ indicates exponentiation.

R is an object-oriented program. That means it is focused more on data structure than programming logic. Variables, data, functions, results, etc., are stored in the active memory of R in the form of objects with names which may be accessed later. The user can manipulate these objects with arithmetic or logical operators and functions. Data types are at the core of R programming. There are different classes of objects in R, such as character, numeric, logical and complex.

Examples of numeric objects are numbers in R which are treated as double precision real numbers. The special number `Inf` represents infinity and the value `NaN` represents an undefined value. R designates missing values with `NA`.

```
> 1/0
[1] Inf
> 0/0
[1] NaN
```

In R the concepts of vector, matrix, array, data frame and list are important. A vector is characterized by, among other things, its length and is created by putting numbers or characters together. To enter data in a vector the command c (combine or concatenate) is used, as illustrated in the sequence below.

```
> x <- c(1, 3, 5, 7)
> x
[1] 1 3 5 7

> y <- c(1, 4, "a", "TRUE")
> y
[1] "1"     "4"     "a"     "TRUE"
```

We will not provide an extensive description of data types, but it may be noted that the elements of the vector y have now been coerced into character type variables, as indicated by the numbers within double quotes. This is because both numeric and character objects were entered in y. However, in the following situation y stays a numeric vector, since a and b contain numeric values.

```
> y <- c(1, 4, a, b)
> y
[1]  1  4  2 -1
```

A list is a special type of vector which can contain objects of different types. Often outputs of R functions are stored in R as lists. A matrix is characterized by rows and columns. A data frame is a special type of matrix where the columns admit variable names. When a dataset is exported in R, it is automatically stored as a data frame. We can also think of data frames as special type of lists where each column is of same length.

## 2.2  Reading, Writing and Extracting Data in R

Since this is a book on analytics, we will keep our discussion on R completely confined to data-related topics. R does not read directly from MS Excel files but it can read .csv files. There are other methods to read data in, but since text data is outside our purview, read.table() and read.csv() are the two important commands to get familiar with at this point. The basic structure of both commands is read.table(file.name, header=T, sep=","), where file.name is the file name with full address; header specifies whether the file has any header and what the separator character is. While reading files it is also possible to control skipping lines from the beginning of the file, ignoring blank lines and many other customizations. In RStudio, .csv files can be directly imported by using the command "Import Dataset" from the menu on the top right panel. Similarly, write.table() and write.csv() commands are used to save data outside of R, in a specific directory under a specific file name.

```
> Datafile<- read.csv("C:/Users/sbasu/Desktop/College.csv")
> is.data.frame(Datafile)
[1] TRUE
> dim(Datafile)
[1] 30 19
```

The file College.csv is read and saved in R under the name Datafile. If it is not saved as an R object, then the file is displayed and scrolled on the console and cannot be accessed for any further operation. Datafile is a data frame with 30 rows and 19 columns. As soon as the dataset is read in R, it is automatically viewed in the top left panel of RStudio. For a small file, the entire file may be viewed. For a large file the top 1000 rows are viewed. To view the top three rows on the console, one has to explicitly extract those rows.

```
> Datafile[1:3,]
            Name Private  Apps Accept Enroll Top10perc Top25perc
1 Michigan State      No 18114  15096   6180        23        57
2  Arizona State      No 12809  10308   3761        24        49
3     Penn State      No 19315  10344   3450        48        93
  F.Undergrad P.Undergrad Outstate Room.Board Books Personal PhD
1       26640        4120    10658       3734   504      600  93
2       22593        7585     7434       4850   700     2100  88
3       28938        2025    10645       4060   512     2394  77
  Terminal S.F.Ratio perc.alumni Expend Grad.Rate
1       95      14.0           9  10520        71
2       93      18.9           5   4602        48
3       96      18.1          19   8992        63
```

The operator [ ] extracts elements from a matrix or data frame. To extract only one element with a specific cell number from a matrix, the row and the column number are given.

```
> Datafile[2,7]
[1] 49
```

The element in the second row and seventh column is extracted and the value of that element is displayed on the screen. Similarly, one or more rows can be extracted as well as one or more columns. In a data frame each column is named according to the column header and can be addressed using the $ operator. For example, `Datafile$Accept` will address the fourth column of Datafile. The same can also be done by writing `Datafile[,4]`. This will now perform as a variable in R. Note that, to extract the variable Accept, we have to write `Datafile$Accept`. The variable Accept does not exist in the global environment in R, but it exists only with reference to the data frame `Datafile`. It is also possible that two different data frames contain two variables of the same name.

Names of objects in R must start with an alphabet, but it can contain numerals. R is case sensitive. Hence x and X are two different entities in R.

## 2.3    Statistics in R

R is an analytics software and it has almost unlimited potential for statistical analysis of data. R has built-in functions for basic to sophisticated statistical functions. For example, the function "mean" returns the arithmetic average of a variable.

```
> mean(Datafile$Accept)
[1] 6793.7
```

A function can return one scalar output or a vector as output.

```
> summary(Datafile$Accept)
   Min. 1st Qu.  Median    Mean 3rd Qu.     Max.
   2165    5045    6076    6794    8225    15100
```

One of the most important features of R is R libraries. A library is a package of functions which may or may not be installed at the same time as R is installed. The packages that are installed at the same time as R are called the system library. These are the very basic features which are required for basic operations. In addition to the basic functions, there are many user-developed functions which are tested and validated by the R users community and then made available to the world. Because of this flexibility, R is quick to implement advanced analytics functionalities.

The libraries need to be downloaded and installed from the same CRAN mirror sites where R is available. Once they are installed, they stay available locally. But, unlike the systems library, which is loaded every time R starts, to use these packages one has to load them for each R session. There are overlaps among libraries and their functionalities.

R functions have the ability to save the outputs in a named list. The benefit of saving the output is that later on various parts of that output can be extracted and used for further analysis. Some of these features will be clear from subsequent chapters. There are innumerable functions in R that are used for data analysis. Trying to list them all would be foolish. Even listing common R libraries is not a good idea, especially since the list is increasing continuously.

Hence let us discuss the help facility in R. The function `help.start()` provides general help. In RStudio the page shown in Figure 2.2 opens in the bottom left panel.

For the functions available within R, `help(function name)` or `?(function name)` gives all the information regarding that function. Another useful command is `apropos`, which lists all functions containing the name of the function.

```
> apropos("mean")
 [1] ".colMeans"    ".rowMeans"    "colMeans"    "kmeans"
```

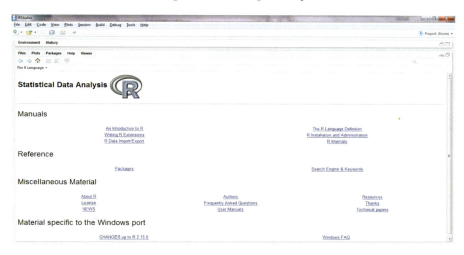

**FIGURE 2.2**
Screenshot of R Help page.

```
[5]  "mean"           "mean.Date"      "mean.default"   "mean.difftime"
[9]  "mean.POSIXct"   "mean.POSIXlt"   "rowMeans"       "weighted.mean"
```

If the function belongs to a package in R which is not loaded for the current R session, ?? will provide a list of all functions which contain that name, along with the library name which includes that functionality and a brief one-line description. For example,

```
> ??mean
```

provides a long list, a part of which is shown in the screenshot in Figure 2.3.

R has a vibrant user community and help is also available from sites like stackoverflow.com.

## 2.4   Graphics in R

R has highly developed graphics facilities. The basic command for that is plot(). The command

```
> plot(Datafile$Accept, Datafile$Enroll)
```

will produce the plot given in Figure 2.4.

The plain vanilla plot does not look impressive. But all aspects of the

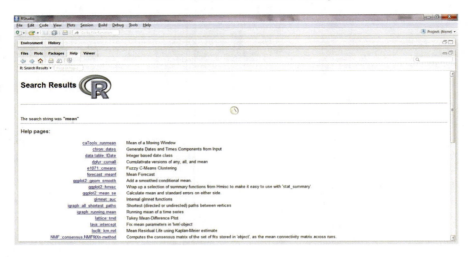

**FIGURE 2.3**
Screenshot of Help on R function list.

basic plot can be customized by specifying the $X$-axis and the $Y$-axis, proper labeling, adjusting the size and shape of the points and using color. Other than the basic plot as shown here, there are graphics libraries which produce even more sophisticated graphics. One such group of functions, belonging to the library `ggplot2`, is used extensively in Chapter 4 for data visualization.

In RStudio, all graphs appear in the Plots window in the bottom right panel. It is possible to direct graphical outputs to pdf devices. It is also possible to save graphs in other formats and copy from the Plots window. Other than simple two-dimensional plots, many statistical functionalities can also be rendered graphically. Several useful functionalities are considered and showcased in subsequent chapters, wherever relevant.

## 2.5    Further Notes about R

Explicit loops do not work well in R. Since R loads all data in the memory, taking each element of a vector (or matrix) and performing the same operation repeatedly is not an efficient way to work in R. The best way to manage memory and speed in R is to work through vectorized operations, i.e., applying functions on a vector in R which is actually applied on individual elements in R. R also has functionalities that can perform parallel processing. This

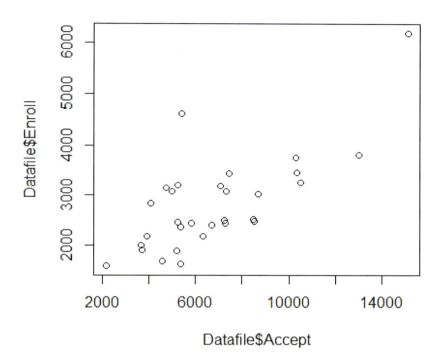

**FIGURE 2.4**
Example of a plot in R.

facility helps R to perform many operations in the cloud and thereby enhances efficiency.

Beginners in R language will benefit by using `swirl()`, which is a software package to learn R within the R console. Users of `swirl()` get immediate feedback as they are taken through self-paced lessons. The learning environment is interactive and both R and data science can be learned through `swirl()`. Just like R, `swirl()` is also free and open source. R has many datasets available within the system. Many authors also contribute codes used in their books as well as datasets to be used in R. These datasets can be used to learn and practice R coding.

## 2.6   Suggested Further Reading

There are many resources to learn R on one's own, including videos on YouTube. Many of these videos are produced by R gurus and are appropriate for R beginners. One series of more than 30 videos under the name "R Tutorials from Scratch" is a good resource. The Johns Hopkins University conducts a Coursera course on R Programming as part of Data Science Specialization. This course is conducted by Roger Peng and the course material is available (Peng (2015)). Several concepts in this material are supported by videos.

Because of R's usage, application and association with analytics and data science, there are many books on the market that deal with statistics with R; see Dalgaard (2008), Kabacoff (2011), Hothorn and Everitt (2014) and Akritas (2014). To get over the initial learning difficulty in R, Burns (2013) is a good resource.

To a large extent, R learning is done through usage. As it is an evolving language and freely developed by users all over the world, there does not exist a set of support documents published by a single source that contains all the relevant information. R documentations are available from the CRAN Mirror site, and are downloaded automatically along with each and every R library. These documentations contain an immense, and often daunting, amount of information. Among other information available for R on the net, cheat-sheets containing a list of frequently used R commands are very helpful for beginners.

# 3

## Reporting Data

### 3.1 What Is Data?

Anything that is observed or conceptualized falls under the purview of data. Often we do not even think of data as a separate entity, it is so much internalized and synthesized in our thought processes. If a commuter is asked to state the average time of a daily commute, he will throw out a number, such as 60 minutes. This is data on commuting. Ask a mother at what time she wakes up to prepare her children for school and you will learn that the time is 6 in the morning. This is data on household behavior. Similarly, if you ask a brand manager whether to mix the color pink with the scent of sandalwood in a soap, she might answer that such a combination will not have a high preference among prospective buyers! Such an opinion on the part of the brand manager as well as on a preference scale rating by a prospective buyer are all examples of data.

In a somewhat restricted view, data is something that can be measured. One can imagine billions of instances of measurements, e.g., rainfall amount, yield of wheat per hectare, heights and weights of Indian males and females, student ratings for an instructor on a 10-point scale, change in the IBM stock as reflected in the New York Stock Exchange in a day, quarterly profit of Microsoft over the last two decades and click-rate per minute in an on-line shopping portal. Until recently, data included only such measurable entities. With the advent of big data and major innovations in extraction and exploration of free text and other electronic media generated material, unstructured and previously unmeasurable contents are also coming under the purview of data. However, for this book, we will follow the traditional definition and treatment of data.

Data represents facts, or something that has actually taken place, observed and measured. Sensex breaching 20,000 on a certain day is an observed item of data. But a market analyst's projection for a rise in IBM's stock price as a result of this event may not come under the ambit of data. Data may come out of passive observation or active collection. Each data point must be rooted in a physical, demographic or behavioral phenomenon, must be unambiguous and measurable. Data is observed on each unit under study and stored in an electronic device such as a database.

Depending on how many different types of information are collected from each unit, data may be univariate, bivariate or multivariate; technically anything beyond univariate is multivariate, but we are making a conscious distinction in treating the bivariate case separately. As the names suggest, if only one type of information is collected from each unit, the data is univariate. A simple example is the set of observed heights of a group of adult Indian females. If two separate pieces of information are collected from each unit, data is bivariate, e.g., height and weight of each female in the group. If more than two types of information are collected from each unit, e.g., height, weight, nutritional intake, health status, cholesterol level, information on whether the individual is following a daily exercise regime, etc., then the data is multivariate.

A database is an organized collection of data where all information is stored in such a way as to facilitate easy access and analysis of the stored information. A database is constructed as a relational database where a unique primary key field identifies and links data from different tables. This ensures that all data from the same unit, even though stored in different tables, can be accessed and analyzed simultaneously. An extension of a relational database is a data warehouse. Creation of data warehouses is considered to be the first step toward data analytics. A data warehouse stores historical data integrated in a specific manner and is often subject oriented, i.e., data on sales only, data on demographic and transactional characteristics of loyalty customers, etc. Once entered in a data warehouse, specifics of data do not change. Data warehouses are not for looking at current business transaction data, but contain consolidated transaction data on various time points. Data warehouses are used for data analysis and dashboarding.

Data can have varied forms and structures but in one criterion they are all the same – data contains information and characteristics that separate one unit or observation from the others. Before we formally introduce various types of data, let us consider several business cases, which will be used throughout the book.

**Case Study 3.1.** (German Credit Data) Banks enhance their businesses by giving credit to loan applicants. It is in their interest to approve the loans but at the same time they would like to minimize the risk of defaulting on the loans. Banks maintain records in their data warehouses on previous loan applicants and their borrowing history; in particular, this includes information on whether they were defaulters on loan repayment in the past. Along with this crucial information, profiles of applicants are maintained which include information on the demographic and economic status of the applicants. The objective of maintaining such historical records is to use the knowledge in predicting whether a future loan applicant will be a credit risk or not. It is assumed that the demographic, behavioral and economic profile of an applicant also has an impact on the probability of being a defaulter and being a liability for the bank.

There are in all 21 variables on which data has been collected for 1000 loan applicants of which 20 variables describe an applicant's demographic and economic status and one variable gives an assessment of

the creditworthiness of the applicant. A borrower profile includes data on age, marital status, occupation, length of current employment, concurrent (other) liabilities, percentage of income spent on installment payments, savings, credit history with the bank, number of dependents, amount of loan applied for, etc. The full data can be downloaded from the URL
`http://www.stat.uni-muenchen.de/service/datenarchiv/kredit/kredit_e.html.` ◈

**Case Study 3.2.** (Wine Quality) The wine industry shows a recent growth spurt as social drinking is on the rise. The price of wine depends on a rather abstract concept of wine appreciation by wine tasters; such opinions may have a high degree of variability among the different experts. Pricing of wine depends on such a volatile factor to some extent. Another key factor in wine certification and quality assessment is physicochemical tests which are laboratory based and take into account factors like acidity, pH level, presence of sugar and other chemical properties. For the wine market, it would be of interest if the human quality of tasting can be related to the chemical properties of wine so that the certification and quality assessment and assurance process becomes more objective.

Two datasets are considered of which one dataset is on red wine and has 1599 different varieties and the other is on white wine and has 4898 varieties. All wines are produced in a particular area of Portugal. Data is collected on 12 different properties of the wines one of which is quality, based on sensory data, and the rest are on chemical properties of the wines including density, acidity, alcohol content, etc. The full data can be downloaded from the URL
`http://archive.ics.uci.edu/ml/datasets/Wine+Quality.` ◈

**Case Study 3.3.** (Coffee Brand Selection). To understand market preference for coffee brands, 2111 German households were selected and their coffee purchases across three years (January 1988–December 1990) were registered. Data was collected on the demography of the households in the panel, e.g., size of household, age of the members, their economic, social and educational status, etc., as well as on how many packets of coffee were purchased, at what price and with what frequency. Each household may have made more than one purchase during these three years, and indeed almost all of them did. There is a total of 130,986 records of purchase. The full data is downloadable from the URL
`http://www.stat.uni-muenchen.de/service/datenarchiv/kaffee/kaffee_e.html.` ◈

## 3.2 Types of Data

Data may be of many varieties. In the German Credit Data example the main variable of interest, Creditworthiness, can take only two values, 1 (creditworthy) and 0 (non-creditworthy), whereas the variable Credit Amount (loan amount asked for) has a minimum of 250 Deutsche Mark (DM) and a maximum of 18,424 DM. It is true that not all values between 250 and 18,424 DM appear in the data, but all loan amounts between the bank-specified minimum and maximum amounts are possible! In the dataset on Coffee Brand Selection, nine different brand varieties are mentioned. Consequently the variable Coffee Brand has nine values. Depending on the nature of the observations, data is classified into several categories.

Before we proceed further with classification of data, it is important to understand what a variable is. Mathematically speaking, a variable is a quantity which may take different values in different situations. It is designated by an alphabetic character and is not fully known, but completely specified through a given relation. Take, for example, the relationship $Y = 2X$, $X$ and $Y$ both being variables. There is a definite relationship between $X$ and $Y$ so that once you specify one, the other is completely known.

In statistics we introduce the concept of a random variable. A random variable is associated, in a probabilistic manner, with an attribute of the units that comprise the data to be analyzed. In the German Credit Data, there are 21 attributes associated with each individual; hence in this dataset there are 21 random variables. A formal probabilistic definition of a random variable will be given later in Chapter 6. The primary difference between a mathematical variable and a random variable is that the latter may have different values for different units in the dataset but these are not known with certainty beforehand. Consider the attribute Creditworthiness. It can take only two values, 1 and 0. But which value a particular applicant will assume is known only after the actual observation is made, and not before. A random variable is associated with the possible outcomes of an experiment and their corresponding probabilities. An experiment, like throwing a coin or casting a six-faced die, may have more than a single possible outcome. In the case of the first experiment the number of possible outcomes is two while in the case of the second experiment the number of possible outcomes is six. Hence the associated random variables may take any one of the two values in the first case and any one of the six values in the second case. However, until the outcome is noted, it is not known whether a particular toss will result in a head or in a tail. Similarly, for the German Credit Data, each loan application is an experiment with two possible outcomes, but the outcome is not known beforehand. Henceforth we will sometimes use the words data and random variable interchangeably since an observed data is a realized value of a random variable in a given scenario.

### 3.2.1   Qualitative versus Quantitative Variable

The most critical classification divides data into two types, *qualitative* and *quantitative*. As the names suggest, qualitative data deals with the description of the observations. Here the variable can assume two or more distinct labels. For example, the variable Creditability in German Credit Data has two labels, creditworthy and non-creditworthy. The variable Coffee Brand in Coffee Brand Selection Data can assume nine distinct labels corresponding to the nine different brands. A qualitative variable may also be of two distinct types. A qualitative variable is called a *nominal* variable if there is no hierarchical relationship among the different levels of the variable. An example of a nominal variable is the selected Coffee Brand. A qualitative variable is called an *ordinal* variable if the values of the random variable have a natural ordering; e.g., in the Coffee Brand Selection dataset there is a variable named Social Level of Householder with five levels—upper class, upper-middle class, middle class, lower-middle class and lower class. There is a distinct ordering among the five levels as defined here, but the ordering, or the difference between any two levels of the variable, cannot be quantified.

Any random variable taking numerical values gives rise to quantitative data. Quantitative data results either from counting or from taking measurements on relevant aspects of an observed unit. Consider, for example, the variable Amount of Credit from German Credit Data. If subject $A$ has applied for 1049 DM and subject $B$ has applied for 2799 DM, we know that subject $B$ has applied for exactly $(2799 - 1049 =)1750$ DM more than subject $A$. All variables considered in Wine Quality Data are quantitative.

### 3.2.2 Discrete versus Continuous Data

Any random variable that can take only countably many values is a *discrete random variable*, whereas a random variable that takes all values within a certain range is called a *continuous random variable*. Technicality aside, a discrete variable is one which can have a limited number of values, either descriptive levels or numerical values. All qualitative variables, nominal and ordinal, are discrete by definition. Quantitative variables may also be discrete. An example of such a variable is Quality in Wine Quality Data. Each wine is tasted by three wine tasters and ranked on a scale of 1 to 10; the value of the variable Quality for this particular wine is the middle rank of these three. This variable is ordinal in nature.

The principal difference between a discrete and a continuous variable is in the range of values the random variables can assume, and not in the actual values recorded. Due to limitations in measuring instruments, or because an extreme degree of accuracy is not required, it may seem that all values within a range are not possible and definitely are not observed. But that does not change the nature of a continuous variable or transform it into a discrete variable. Discrete and continuous variables are discussed more formally in Chapters 6 and 7.

### 3.2.3 Interval versus Ratio Type Data

Continuous variables are classified into two levels. *Interval data* is numeric, continuous and the difference between any two measurements has a definite meaning. However, for this data the concept of 0 is not absolute, but is a matter of convenience. For a *ratio variable*, 0 is an absolute number and represents the absence of the characteristic. Consider, for example, the total revenue and the net revenue of a company. Total revenue is 0 only if the company has not done any business and it is an absolute 0. Net revenue can be 0 even if the company has done a lot of business but has not been able to make a profit. Total revenue is a ratio variable while net revenue is an interval variable. In an interval variable the difference or distance between two numbers is important. The flight time of a commercial airlines flight from Kolkata to New Delhi is an interval variable since whether it is a 9 am or a 7 pm flight, the attribute a passenger is interested in is the length of the flight.

In the hierarchy of data, nominal is at the lowermost rank as it carries the least information. The highest type of data is ratio since it contains the maximum possible information. While analyzing the data, it has to be noted that procedures applicable for lower data types can be applied for higher ones, but the reverse is not true. Analysis procedures for nominal data can be applied to interval type data, but it is not recommended since such a procedure completely ignores the amount of information an interval type data carries. But the procedures developed for interval or even ratio type data cannot be applied to nominal or ordinal data. A prudent analyst should recognize each data type and only then decide on the methods applicable.

## 3.3    Data Collection and Presentation

With the advent of cloud computing and distributed data storage, it may seem that data is automatically collected, but that is hardly the case. There must be a conscious effort and planning behind capturing data, which is even more important when data is being generated at the rate of millions of bytes per nanosecond! Data collection is a process, and the veracity of data depends on the sanctity of the process. Traditionally, data was collected through survey instruments. A decennial census is an example of such a data collection process; so are customer feedback forms handed out after a meal in a restaurant and ladies going from door-to-door asking whether a particular brand of electrical appliances is being used in that household. The modern data collection process is far more sophisticated, where every click in Snapdeal or Amazon is recorded to reveal every customer's choice and ultimate purchase pattern, which items are frequently purchased together and many other demographic and economic information.

There is no single method for data collection for business – be it through electronic media or using traditional methods. We will simply indicate several options that are available and used in various realms. Data is collected either on the entire group (the population) or from a selected subset of the group (the sample). A population is defined as the largest possible set of individuals or items with strictly defined characteristics. Enumerating an entire population is known as a census. As we are all aware, every tenth year India undergoes a census where an attempt is made to enumerate each and every person living in India. In statistics, the term "population" does not necessarily refer to the collection of the residents of a region, state or country, but it has a wider implication. All persons watching a particular news channel is the viewer population for that channel. All women applying for a loan from a nationalized bank is the population of women loan applicants. All drug companies developing generic drugs form the population of generic drug developers. The characteristics that set a population apart must be rigorously defined and maintained so that there is no confusion regarding membership of the population.

A sample is a smaller subset of the population. A sample has the same set of characteristics as the population but it does not include every member of the population. In almost all practical cases, data is collected from a sample, not from the population. Populations are generally too large, and the constraint of time and resources makes it virtually impossible to enumerate the whole population most of the time. In some situations, collection of data may also be a destructive process, making complete enumeration impossible; to estimate the lifetime of a light bulb, one needs to burn the bulb out.

A sample is assumed to represent the population. This is a strong assumption and is generally possible only if the sample is randomly chosen, i.e., if each unit in the population has an equal chance of being included in the sample. There are a number of different mechanisms to ensure that the sample selected is a random sample. At present we will not go into detail about sample selection. We will not even make any explicit reference to whether data is coming from a sample or from a population. We will act on the given data to extract relevant information.

Data that is collected from various aspects of a business has only one purpose, and that is to gain knowledge about the business. Data constitutes pieces of information on the history and the current status of a business which, when manipulated intelligently, will provide knowledge about the business, which in turn will lend the business a competitive edge. The very first step of understanding the numbers is to interpret them properly. Proper collection, collation, organization and presentation facilitate information gathering. Data is the driving force behind analytics. Every piece of information is valuable and must be considered.

Nevertheless, vast quantities of observations on a large number of variables need to be properly organized to extract information from them. Broadly speaking, there are two methods to summarize data: visual summarization and numerical summarization. Both have their advantages and disadvantages and applied jointly they will extract the maximum information from raw data. To understand thousands of rows of data in a limited time, there is no alternative to visual representation. The objective of visualization is to reveal the hidden pattern through simple charts and diagrams. Visual representation of data is the first step toward exploration and formulation of analytical relationships among the variables. In a whirl of complex and voluminous data, visualization in one, two or three dimensions helps data analysts to sift through the data in a logical manner and understand the data dynamics. Data visualization forms a part of *Exploratory Data Analysis* (EDA). It is an integral component of reporting and preparation of dashboards to understand the current status of a business. It is instrumental in identifying patterns and relationships among groups of variables. Visualization techniques depend on the type of variables. Techniques available for nominal variables are generally not suitable for visualizing continuous variables and vice versa. Data often contains complex information. It is easy to internalize complex information through the visual mode. Graphs, charts and other visual representations provide quick and focused summarization.

**Case Study 3.3.** (Continued) Effective visualization must tie in with business questions. More than 130,000 coffee purchase records are available for 2111 households between January 1988 and December 1990. Each household may have made one or more coffee purchases over these three years. For each household the time between two successive purchases is given. Other variables contain the demographic profiles of each household, their choice of coffee brands, number of packets and cost of each packet of coffee purchased. The key business questions may be the following.

1. Which brands of coffee are purchased?

    (a) Are they purchased in equal proportions?

    (b) Is any particular brand preferred over the others?

    (c) What are the prices of the different brands of coffee?

    (d) How frequently does a household buy coffee?

    (e) How many packets of coffee are bought at a time?

2. What are the factors that have an impact on a particular household's coffee purchase pattern?

    (a) Does brand preference depend on income?

    (b) Does coffee consumption depend on household size?

    (c) Does purchase depend on a consumer's age?

The very first step in any analysis is to know all the relevant attributes individually, which is also known as univariate analysis. Univariate analysis includes enumerating all possible values of the variables, examining frequencies of the values and checking if any of the variables is discrete, whether one or more levels of the variable can be combined for a more meaningful analysis, etc. Even when the ultimate goal is building sophisticated models for prediction, univariate analysis is a must. Without this seemingly naive step, the structure of data is not clearly understood, which may later lead to invalid models.

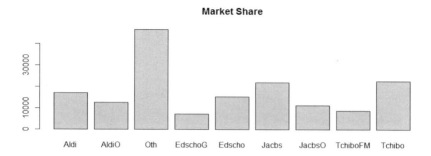

**FIGURE 3.1**
Bar chart showing market share of coffee brands.

Bar charts and pie charts are among the most common graphical modes of data display. Both these graphs present the same information but in different formats. Bar charts present frequencies (counts) of different groups. They can also present relative frequencies (proportions). However, pie charts are the preferred form of presenting relative frequencies. The total pie must necessarily be 100%. A particular segment (wedge) of the pie represents the proportion of that group in the entire sample. Depending on the requirement, either a bar chart or a pie chart may be used. Binary, nominal or ordinal discrete data may be represented through bar charts and pie charts. Figure 3.1 displays the market share of different brands of coffee. The R commands to produce the bar chart are given below.

```
> barplot(cnt, names.arg=names, main="Market Share")
```

A more important point from a business perspective is to know about the revenue share of various coffee brands. However, revenue share may not be readily available from the data since the unit price of coffee is given over a range. A decision needs to be taken regarding how best to estimate revenue share. The rule used here is

- If the range given is closed (e.g., 6.5–8.5), then its midpoint (7.5) may be taken to represent the entire range.

- If the range is open (e.g., < 6.5 or > 8.5), a convenient representative price is used (e.g., 6 and 9, respectively).

Naturally, revenue share depends on the representative price.

Since all brands together make up 100% of market and revenue share, for a quick description, the slices of the pie in Figure 3.2 can be compared. The codes for this figure are given below.

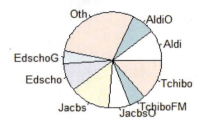

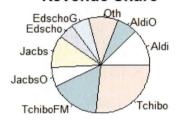

**FIGURE 3.2**
Pie chart showing market and revenue share.

```
> par(mfrow=c(2,1), mar=c(0,1,1,1)+0.1, oma=c(1,0,0,0)+0.1 )
> pie(Table1$Purchase, labels=c("AldiO","Oth","JacbsO","Tchibo",
  "Jacbs","TchiboFM","Edscho","EdschoG","Aldi"), main="Market Share")
> pie(Table2$Price.Share, labels=c("AldiO","Oth","JacbsO","Tchibo",
  "Jacbs","TchiboFM","Edscho","EdschoG","Aldi"), main="Revenue Share")
```

From the figure, the following points are clear. About 30% of purchases
are of Andere Kaffeemarken (Other Brands, Oth). To understand brand popularity, this group may possibly be ignored since this does not correspond to
any single coffee brand. Next highest popularity is enjoyed by Tchibo Other
(Tchibo) and Jacobs Kronung (Jacbs). Considering all varieties of Tchibo
and Jacobs, these two German brands are almost equally popular. Next in
popularity is Aldi, which happens to be a supermarket brand.

Another important point to note is that market share and revenue share
of brands are quite different, as is evident from the two pie charts. Andere
Kaffeemarken has only a 10% share of revenue, while it enjoys 30% of market
share. Even though Tchibo brand with all its variants has about 20% of market

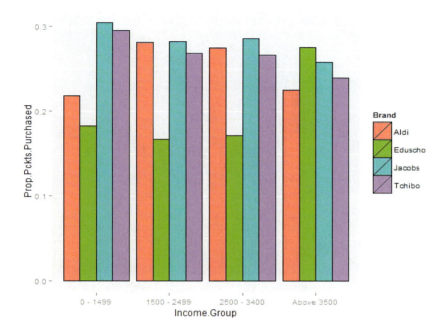

**FIGURE 3.3**
Grouped bar chart for differential brand selection.

share, it has the highest revenue share (about 43%). For Jacobs and Eduscho brand variants, share of market and share of revenue are similar.

Another key business question is to understand the influencers of coffee purchase. Possible important predictors are monthly household income and price of coffee per packet. Univariate analysis is important but it can provide only a limited insight into the complex interactions among all the attributes. It is commonly believed that education and social status are dependent, and so are education and income as well as income and social status. Brand managers may be interested in knowing the coffee purchase pattern among different social sectors to modulate their sales strategy accordingly. The brand preference of the four major brands across the different income groups is displayed in Figure 3.3. It is clear that the popularity of the brands Jacobs and Tchibo is about the same in all income groups. Eduscho is most popular among the highest income bracket. In the two middle income groups Jacobs, Tchibo and Aldi are almost equally preferable. Jacobs and Tchibo enjoy popularity across all income groups while Aldi and Eduscho brands display differential popularity. The codes to produce the figure are given below.

```
> library(ggplot2)
> ggplot(Table3, aes(Income.Group, Prop.Pckts.Purchased,
  fill = Brand)) + geom_bar(stat="identity", position = "dodge")
```

◇

Bar charts and pie charts are appropriate for visualizing the frequency distribution of discrete variables. These cannot be used to represent continuous variables. Continuous variables take innumerable values within a range, and special diagrams, such as frequency histograms, are required for proper representation. *Histograms* are bar charts in which the bars generally have the same width and are contiguous. The bars represent groups of values of the variable. Each bar is represented by its midpoint (or class boundaries). The height of each bar indicates the frequency or relative frequency of that group. The classes are constructed so that each data value falls into one class and one class only, and the class frequency is the number of observations in the class. There is a relationship among class width, data range and the number of desired classes. Often class width is defined as the ratio of the range and the number of desired classes. Class widths are also taken according to analytical requirements. Histograms describe the shape of the distribution. Class widths have a major impact on the appearance of the histogram.

The shape of a distribution provides an overall picture of the various values a random variable can assume. The first and foremost of the common shapes is the symmetric bell-shaped distribution where most of the values are concentrated around the middle of the range. A bell-shaped symmetric random variable has many desirable analytical properties. The two other common shapes are positively skewed and negatively skewed distributions. In the former, values are concentrated toward the lower end of the range of the random variable; large values occur with lower frequency but smaller values occur with high frequency. A random variable with a negatively skewed distribution takes lower values with smaller frequency. Of the three, this is the least common shape of a distribution. All the three shapes are unimodal in that they have a single peak (i.e., mode). See Figure 3.4 for an illustration of the three shapes.

A histogram is a discretized version of the distribution of a random variable. The class width of a histogram controls the shape of a histogram to a large extent. If the classes are wider and contain more data points, histograms are flatter. Alternatively, if the classes are narrower and contain fewer data points, histograms are more spiky. To get a good idea of the proper shape of the distribution, it is important to determine an appropriate class width. Note that class width and number of classes are inversely related. There is no hard and fast rule of what would be an optimum number of classes.

However, noting that the class width (bin width) and the starting point of the histogram impact the shape of the histogram, several recommendations have been put forward in the literature. Two common rules of thumb for determining the number of classes are Sturges' Rule and the Rice Rule. Sturges' Rule says that the data range should be split into $k$ equally spaced classes

## Examples of Distributions

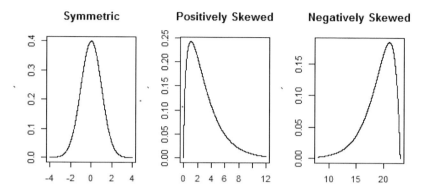

**FIGURE 3.4**
Symmetric, positively and negatively skewed distributions.

where
$$k = [1 + \log_2 n],$$

$n$ is the sample size and $[\ ]$ denotes the ceiling operator or the smallest integer larger than the expression. Hence for a sample of size 1000, $\log_2 n = 9.9$ and $k = 11$. An alternative rule recommends choosing $k = 2\sqrt[3]{n}$. Following this rule, the number of class intervals will be 20 when the sample size is 1000. These empirical rules only take into account the size of the sample but not the actual spread of the data. However, when all is said and done, practice and applicability determine the number of classes.

**Case Study 3.4.** (Coffee Brand Selection) Figure 3.5 shows a histogram for the variable Number of Days (between coffee purchase), the values of which are shown on the $X$-axis. The first class has limits $[1, 5]$; the second class has limits $[6, 10]$ and so on. The frequency of the first class is 30,000 (approximately, as appears from the graph). This implies that approximately 30,000 coffee purchases are between 1 and 5 days (of the previous purchase); 45,000 coffee purchases are between 6 and 10 days and so on. Note that the total number of coffee purchases recorded is 130,986 and the total number of households is 2111; each household may have made repeated purchases during these two years. The total frequency is 130,986 in this case and the height of all the bars should add up to this number. The width of the first few classes is equal and it is equal to 5 (days). However, the last class contains all observations greater than 120. It is often a good idea to draw histograms with the last one or two classes wider than the rest so that the full range of the data is covered but the classes in the tail of the histogram are not too sparse. The shape of the

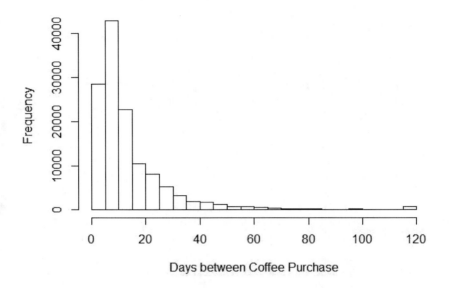

**FIGURE 3.5**
Distribution of number of days between coffee purchases.

histogram is positively skewed, indicating that coffee is purchased frequently. Most of the purchases are made at least once every three weeks. There are only a few instances when coffee is purchased once in a month or even more infrequently, but such cases are very rare. Figure 3.5 effectively summarizes the information contained in the dataset. The codes are given below.

```
> with(Table4, hist(Days_between_Purchase, xlab="Days between
  Coffee Purchase", main=""))
```

The empirical rules mentioned above would have a number of classes anywhere between 18 and 100. However, since all of those classes would have equal width, the lower few would contain almost all the observations and there would be a very long tail. Overall such a histogram will be less informative than the one provided.

One important limitation of the histogram is that it does not show individual purchase patterns. Once the data is summarized, individual observations lose their identity. In this case, information is over repeated purchases, but such individual identity cannot be preserved in such a graphical summary. An alternative way of presenting an individual pattern is through a time series graph where time is shown on the $X$-axis and the variable value is presented on the $Y$-axis.

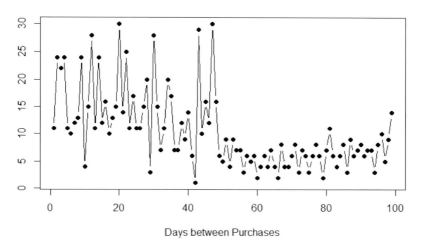

**FIGURE 3.6**
Time series data on coffee purchase for a sample household.

One particular household's purchase pattern is shown in Figure 3.6. The codes to produce this figure are given below. Over three years this household has made 99 purchases, as shown on the horizontal axis. On the vertical axis the number of days between two consecutive purchases is recorded. For example, the first purchase shown is after 11 days, whereas the second purchase shown is after 24 days. The pattern of purchases shows a distinct change in the level around the 50th time point. Toward the later half of the series coffee is purchased much more frequently.

```
> plot.ts(Purchase[,2], type="b", pch=19, xlab="Days between
  Purchases",ylab="",main="Purchase Pattern for ID No 1", cex=0.5)
```

So far we have discussed only a few of the more common diagrams. In addition to these, stem and leaf diagrams, dot plots, etc., are also used when sample size is small to moderate. However, to visualize voluminous data they are not much in demand. New methods of visualization based on Business Intelligence software are in vogue to draw out hidden patterns in the data. Along with R, Tableau, Qlik View, Python, etc., are used extensively for data visualization. A few of the more modern and investigative visual techniques will be discussed in Chapter 4.

---

## 3.4 Reporting Current Status

Business health monitoring today requires manipulation of enormous datasets. Millions of cases, represented by rows of data matrix, and possibly hundreds of attributes on each of these cases, represented by columns of the matrix, need to be manipulated and mined for information. However appealing the visualization techniques may be, graphs alone are not enough to provide insight into the status of a business. Numerical summarization of variables, overall and by groups, is required to get succinct answers to key business questions.

**Case Study 3.5.** (German Credit Data) The data here represents demographic, economic and other conditions of 1000 loan applicants to a German bank. The business question for the bank is to maximize the number of loans given out and minimize the number of defaulters. To tackle the overall problem, it needs to be broken down into smaller, and easier, subproblems. Some of the important questions the loan manager would like to get answers to are

1. Who are the loan applicants?

   (a) What are the demographic profiles of the loan applicants?
   (b) What are the economic profiles of the loan applicants?

2. What is the purpose of the loan?

3. What is the amount of the loan requested?

4. What is the duration of loan?

5. What is the collateral provided by the applicant and what is its value?

Let us take up these key questions one by one. There are 1000 loan applicants with various combinations of demographic profiles. The variables constituting demographic profile may include sex and marital status, age and the number of dependents. The very first challenge an analyst faces is how to summarize the data through numbers. This approach will help to understand certain facts about the data which are not possible to glean from histograms or any other graph. Some of the examples are given below.

6. What is the typical amount of credit applied for by demographic categories?

7. What is the extent of the variability of credit applied for by demographic categories?

There are a few major distinctions between a visualization-based approach to data description and numerical summarization. Visualization shows an overall picture whereas numerical summarization provides a set of numbers as representative of the data. Graphs are good to describe the shape of the data whereas numerical summarization is appropriate for comparing two or more groups. Visualization must be considered to identify the outliers or extreme values in the data; numerical summarization focuses on the central part of the data. Graphical analysis is inappropriate for statistical inference; numerical summary values obtained from the sample are instrumental to understand the underlying population.

Based on sex and marital status, all 1000 loan applicants are divided into four categories. The relative frequency histograms of the loan amount for the four groups separately is given in Figure 3.7. See Table 3.1 for a description of the four groups. It is quite clear that the four groups have different distributions of the loan amount variable , but to facilitate quick comparison numerical measurements are necessary. Figure 3.7 is produced by the following code.

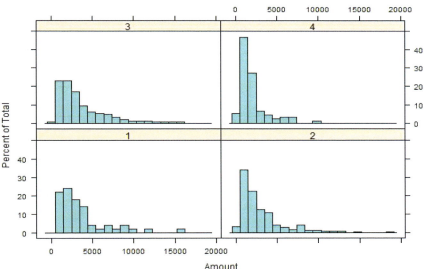

**FIGURE 3.7**
Relative frequency histogram of loan amount.

```
> library(lattice)
> histogram(~Credit.Amount|factor(Sex.Marital.Status),nint=20,
  data=German.Credit,main="Relative Frequency Histogram
  of Loan Amount",xlab="Amount")
```

### 3.4.1 Measures of Central Tendency

If you are to report only one representative value of loan amount for all 1000 applicants in Case Study 3.5, which value would you choose? Which representative value would you report for each of the above four demographic categories? Further, if, as a loan disbursement manager, you would like to know the typical amount of loans (in monetary units) applied for, to plan for next year, how should you go about it? The one single value of the variable that may be taken to represent the whole distribution is known as a measure of *central tendency*. The name comes from the fact that generally most values tend to cluster around the center. The two most commonly used measures of central tendency for continuous variables are the *mean* (or the arithmetic average) and the *median* or the central value when the data is arranged in the increasing (or decreasing) order. Suppose a sample of size $n$ is observed and the values of the variable of interest are $X_1, X_2, \ldots, X_n$. Then the mean of this set is denoted by $\bar{X}$ and is computed as $\bar{X} = \frac{1}{n} \sum_{i=1}^{n} X_i$. The median of a sample of size $n$ is denoted by $\widetilde{X}$ and is the middlemost observation of the ranked values. If $n$ is an odd number, then the median is the middlemost observed value. If $n$ is an even number, the sample median is the average of the two middlemost observed values. Formally,

$$\widetilde{X} = \begin{cases} X_{\left(\frac{n+1}{2}\right)} & \text{if } n \text{ is odd,} \\ \frac{1}{2}\left(X_{\left(\frac{n}{2}\right)} + X_{\left(\frac{n}{2}+1\right)}\right) & \text{if } n \text{ is even.} \end{cases}$$

Conventionally $X_{(i)}$ denotes the $i$-th ranked observation in the dataset. $X_{(1)}, X_{(2)}, \ldots, X_{(n)}$ denote the sample observations ranked from smallest to largest.

Both the mean and the median of a sample are useful to look at and their relative positions provide important insight into the data. For a symmetric distribution, bell-shaped or otherwise, the mean and the median are identical. If a distribution is right (positively) skewed, the mean is greater than the median; for a left (negatively) skewed distribution the mean is less than the median. Note that the mean is based on the actual values of the observations, whereas the median is based on their relative positions. Hence the mean is much more influenced by even a small number of extreme observations. If there are outliers in the data, they have a larger impact on the value of the mean compared to the median. Hence the median is said to be a more robust measure of central tendency, and, if there is a strong belief that the data contains outliers, the median is the recommended measure of central tendency to be used. See Figure 3.8 for the relative positions of the mean and the median for different types of distributions.

Another common measure of central tendency is the *mode*, or the most frequent observation. However, for a continuous variable, this cannot be a useful measure for obvious reasons. For a categorical variable with a moderate to large number of categories, the modal class is the most frequent category. However, in such a case the concept of a single numerical value of the mode may not exist.

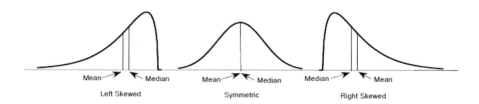

**FIGURE 3.8**
The mean and the median for distributions with different shapes.

Even though the mean may be heavily impacted by a small number of extreme observations, it is the most commonly used measure of central tendency. Apart from the fact that the mean is intuitively easier to grasp, there are several other reasons in its favor. For example, means from different groups can easily be combined to find the overall mean for the whole sample, but no such combination is possible in case of the median. If there are $K$ groups and $\bar{X}_k$ is the group mean and $n_k$ is the total count in the $k$-th group, $k = 1, 2, \ldots, K$, then the overall mean for the whole sample is

$$\bar{X} = \frac{(n_1\bar{X}_1 + n_2\bar{X}_2 + \cdots + n_k\bar{X}_k)}{n_1 + n_2 + \cdots + n_k}.$$

An alternative measure of central tendency which may be taken as a compromise between the mean and the median is the *trimmed mean*, where the $100p\%$ trimmed mean is the arithmetic average of the middle $100(1 - 2p)\%$ observations. Unless otherwise mentioned, a 10% trimmed mean is calculated, i.e., the arithmetic average is calculated based on the middle 80% of observations. One of the drawbacks of the trimmed mean is that it is not based on the entire sample. As the trimming proportion increases, the number of observations contributing to the trimmed mean is reduced.

**Case Study 3.6.** (German Credit Data) The histograms in Figure 3.7 indicate that within each of the four demographic strata the distribution of

**TABLE 3.1**

The central tendencies for each of the four demographic groups in the German Credit Data

| Group | Count | Mean | Median | 10% Trimmed Mean |
|-------|-------|------|--------|------------------|
| 1. Male Divorced | 50 | 3684.70 | 2793.5 | 3085.78 |
| 2. Male Single | 310 | 2877.78 | 1959.0 | 2370.48 |
| 3. Male Married/Widowed | 548 | 3661.88 | 2666.0 | 3146.47 |
| 4. Female | 92 | 2045.54 | 1476.0 | 1693.96 |
| Overall | 1000 | 3271.25 | 2319.5 | 2754.59 |

the loan amount is positively skewed. In all four groups the mean is larger than the median (see Table 3.1). Note that the overall mean $3271.25 (= (50 \times 3684.7 + 310 \times 2877.78 + 548 \times 3661.88 + 92 \times 2045.54)/1000)$ can be found from group means.                                                      ◈

### 3.4.2   Measures of Dispersion

In addition to a measure of central tendency, a measure of *dispersion* is required for the numerical description of a dataset. Consider, for example, two samples of size 5 with observations −0.5, −0.25, 0, 0.25 and 0.5 in the first sample and −10, 6, 3, 1, 0 in the second sample. Both have the same mean (zero) but their spreads are different. Dispersion measures the spread of a sample. The simplest measure of dispersion is the *range*, which is the difference between the largest and the smallest observations. But this measure depends only on two observations from the sample and ignores the concentration of the data.

The most commonly used measure of dispersion is the *variance* of the observations, which is measured as the sum of squared deviations of each individual observation from the average, divided by $n - 1$, where $n$ is the sample size. Note that the sum of the deviations of the observations from the mean is identically zero; hence it cannot be used as a measure of dispersion. Formally

$$Var(X) = \frac{1}{n-1}\sum_{i=1}^{n}(X_i - \bar{X})^2 = s^2$$

$$s = \sqrt{Var(X)}$$

where the *standard deviation* $s$ of $X$ is the positive square root of the variance. Typically, the pair (mean, standard deviation) is used for the description of a dataset under consideration. Another popular measure of dispersion is the *mean absolute deviation about the median* or $MAD$ which is defined as

$$MAD(X) = \frac{1}{n} \sum_{i=1}^{n} |X_i - \widetilde{X}|.$$

To further describe the shape of the distribution, *quantiles* are used. Quantiles are a set of numbers which divide the data into several bins so that the proportions in the bins are equal. The most important quantiles are the *quartiles*. The three quartiles divide the data into four equal parts, and are denoted by $Q_1$ (first quartile), $Q_2$ (second quartile) and $Q_3$ (third quartile), respectively. If $n$ is the total number of observations, then $Q_1$ is the $(n + 1)/4$-th ranked observation and 25% of the observations are less than or equal to $Q_1$. The measure $Q_2$ is the $(n + 1)/2$-th ranked observation and 50% of the data are less than or equal to $Q_2$. The measure $Q_3$ is the $3(n + 1)/4$-th ranked observation and 75% of the data are less than or equal to $Q_3$. Note that $Q_2$ is the median of the data. Note also that $Q_1$, $Q_2$ and $Q_3$ are not necessarily equidistant on the range of the observations, but they split the data into four equal parts.

Relative positions of the quartiles are determined by the shape of the distribution. If $Q_2 - Q_1 = Q_3 - Q_2$, then the distribution is symmetric, since the data concentration on both sides of the median is similar. If, on the other hand, $Q_2 - Q_1 < Q_3 - Q_2$, then the distribution is positively skewed because there is more data concentration on the left hand side of the distribution, indicating that numerically smaller values occur more frequently, but there are a few very large observations. Following the same argument, $Q_2 - Q_1 > Q_3 - Q_2$ indicates a negatively skewed distribution. The difference between $Q_3$ and $Q_1$ is known as the *interquartile range* or IQR. The IQR contains the middle 50% of the data and provides a good idea of the spread of the distribution without the outlying observations. In fact, a comparison of the spread in the middle 50% and the full range is useful to detect outliers.

Why is it important to identify outliers? These are the values which, if the distributional assumptions are correct, are unlikely to be observed. To understand their implication, the relation between the population and a sample thereof needs to be revisited. Instead of being ignored or deleted from the data, these values should be further scrutinized to study the nature and cause of unexpected patterns in the data.

Analytics, especially predictive analytics, is forward looking. Based on past observations, models are built to predict what is likely to happen in the future. Past observations are representatives of the data generation mechanism, and the objective of model building is to understand and replicate the same for the future. Randomness in the observations needs to be accounted for in order to get at the data generation mechanism. This gives rise to the concept of a probability distribution, which is discussed at greater length in Chapters 6 and 7.

If the distribution is symmetric and bell-shaped, i.e., the middle part of the distribution contains most of the data, the mean, the median and the mode of the distribution are identical (or close). There also exists an empirical rule which gives some indication of how the data are clustered around the mean. According to the empirical rule, the interval $\bar{X} \pm s$ contains approximately 68.26% of the observations, the interval $\bar{X} \pm 2s$ contains approximately 95.44% of the observations and the interval $\bar{X} \pm 3s$ contains approximately 99.72% of the observations. Hence it is very rare for an observation to lie at a distance of more than $3s$ from the sample mean (on either side). If observations are found to lie at a distance larger than that, there is enough reason to believe that they may not have originated from this particular data-generating mechanism.

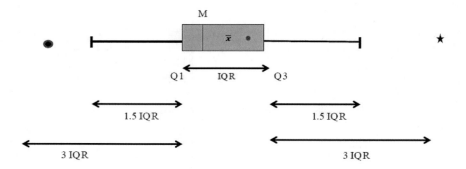

**FIGURE 3.9**
Box and whisker plot. The quantity M represents the median of the distribution.

The empirical rule is a strong rule whenever bell-shaped distributions are concerned. For other distributions another tool, known as the boxplot, is used to identify the outliers in the sample. The boxplot utilizes the IQR and constructs two sets of boundaries at $1.5 \times \text{IQR}$ and $3 \times \text{IQR}$ distance from $Q_1$ and $Q_3$ on the outer side. The box in the middle represents the spread of the middle 50% of the data. If there are any points outside $[Q_1 - 1.5 \times \text{IQR}, Q_3 + 1.5 \times \text{IQR}]$, those points may be considered outliers. To be even more conservative, one may decide to consider as outliers only those points which are outside the $[Q_1 - 3 \times \text{IQR}, Q_3 + 3 \times \text{IQR}]$ limits. This is done if, using the shorter limits, a higher proportion of data is identified as outliers. Boxplots (sometimes called box and whisker plots) are very important visual tools to identify outliers. See Figure 3.9.

The following case study helps to illustrate these points.

**Case Study 3.7.** (Wine Quality) Several chemical properties of wines from a particular area of Portugal are measured with an objective of making wine quality certification free of volatility due to human tasting. Three chemical properties of red wine are considered for illustration (see Figure 3.10). The variable Density, as given in the Wine Quality Data, has a symmetric distribution. The numerical summary values of this variable are presented below.

| Variable | Mean | SD | Min | $Q_1$ | Median | $Q_3$ | Max | IQR |
|---|---|---|---|---|---|---|---|---|
| Density | 0.997 | 0.002 | 0.990 | 0.996 | 0.997 | 0.998 | 1.004 | 0.002 |

The histogram and boxplot of the variable Density are given in the top panel of Figure 3.10. The mean (represented by the filled read oval) and the median (represented by the filled blue oval) are practically identical; so are the standard deviation and the inter-quartile range. Recall that these are sample statistics. Even if in the population the mean and the median are identical, one can hardly expect it to happen in a sample. There are only a few values outside the $[Q_1 - 1.5 \times \text{IQR}, Q_3 + 1.5 \times \text{IQR}]$ limits and they seem to be symmetrically distributed. The quantities $Q_1$ and $Q_3$, represented by empty blue ovals, are placed at equal distance from the median on both sides. Within these two limits 50% of the data is contained. Red empty ovals represent the $\bar{X} \pm 3s$ interval, which should contain, by the empirical rule, 99.72% of the data. Blue crosses represent the interval $[Q_1 - 1.5 \times \text{IQR}, Q_3 + 1.5 \times \text{IQR}]$.

Consider now the variable Fixed Acidity in red wine. The numerical summary is presented below.

| Variable | Mean | SD | Min | $Q_1$ | Median | $Q_3$ | Max | IQR |
|---|---|---|---|---|---|---|---|---|
| Fixed Acidity | 8.32 | 1.74 | 4.6 | 7.1 | 7.9 | 9.2 | 15.9 | 2.1 |

The mean is slightly larger than the median and the median is closer to $Q_1$ compared to $Q_3$, indicating moderate positive skewness. The histogram and the boxplot are given in the middle panel of Figure 3.10. Outliers are present only on the right hand side, again indicating positive skewness. Note also that the lower boundary $\bar{X} - 3s$ does not show up on the histogram, as this is completely outside the range of the data.

These effects are more pronounced in case of the distribution of the variable Residual Sugar in red wine, as seen in the bottom panel of Figure 3.10. The numerical summary is given below and indicates the presence of a pronounced positive skewness. Note that the median is closer to $Q_1$ compared to $Q_3$, while the mean is very close to the latter.

| Variable | Mean | SD | Min | $Q_1$ | Median | $Q_3$ | Max | IQR |
|---|---|---|---|---|---|---|---|---|
| Residual Sugar | 2.54 | 1.41 | 0.9 | 1.9 | 2.2 | 2.6 | 15.5 | 0.7 |

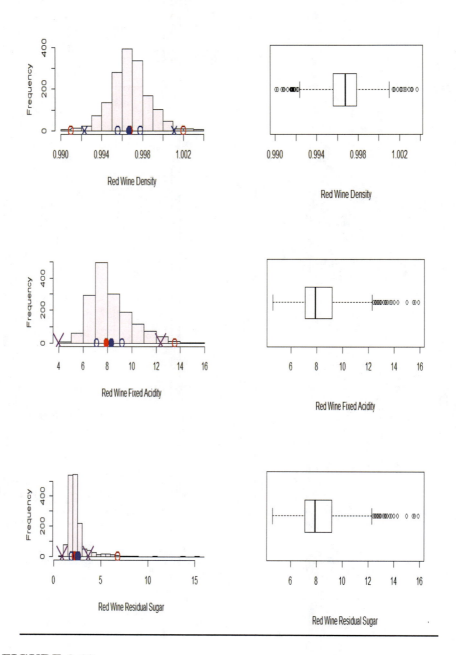

**FIGURE 3.10**
Relative positions of mean, median and outliers for three red wine variables for the Wine Quality Data.

As mentioned before, univariate analysis is an important first step in data analytics. Numerical and visual summarizations put together help analysts to describe basic features of the attributes. Even when the number of variables is large, automated comparisons of the mean and the three quartiles, standard deviation and the IQR are possible. Flags can be raised depending on gross departures from expectation, thus setting up the platform for a deep dive initiative.

### 3.4.3 Measures of Association

So far the behavior of one variable at a time has been discussed. However, when several variables are observed on each unit, it is likely that these variables will be associated. Association or dependence means that, on an average, the increased value of one variable will lead to a systematic change (increase or decrease) in the value of the other. It is not necessary that, for each unit of observation, the pair of values will increase or decrease simultaneously. The most common visual representation of association when both variables are continuous is the scatterplot, a simple two-dimensional plot of the variables. The arrangement of the points gives an indication of the nature of their dependence.

**Case Study 3.8.** (Wine Quality) In Figure 3.11 observations on several pairs of variables on red wine characteristics are plotted to illustrate the different possible relationships that exist between the variables. Note that these are by no means exhaustive types of relationships that may exist between a pair of variables.

Consider the plot of Citric Acid versus Alcohol in red wine. Since the plot forms a rectangular structure, there does not seem to exist any major dependence between these two characteristics. A very similar story is told by the plot of Chlorides versus Alcohol. Moreover, it also indicates that the variability in Chlorides is very small compared to that of Alcohol. Hence the points form a column rather than a rectangle. The plot also indicates the existence of a few very large values of Chlorides.

Compare to this set the other two plots of pH versus Density and Fixed Acidity versus Density. In the former there is a tendency of the points to go downward from the upper left corner to the lower right corner. This happens when, in general, larger pH values are associated with lower Density values in red wine. Also note that not all wines having a high pH value have low Density. Hence from the scatterplot we may conclude that there is a negative dependence between pH and Density in red wines. Alternatively, higher values of Fixed Acidity are associated with higher values of Density and the tendency of this plot is for the concentration of points to move from the lower left corner to the upper right corner. One may conclude that, in general, as the Fixed Acidity content in red wine increases, so does its Density and these two characteristics are positively associated.                                    ◈

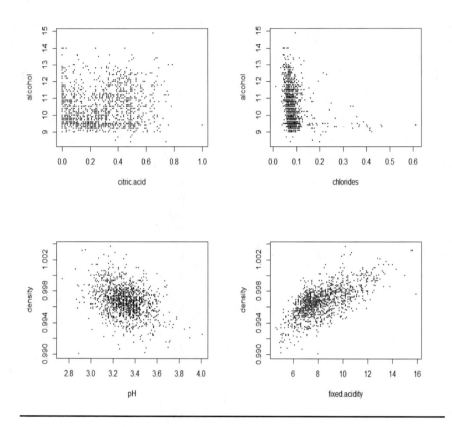

**FIGURE 3.11**
Scatterplot of red wine characteristics.

Scatterplots show the existence and direction of the relationship between two variables but do not fully quantify the strength of the relationship. For quantification of the strength of dependence, the correlation coefficient is used. We first define the covariance measure; this measure takes into account the joint variability of the paired values from the center of the data. Denote a pair of variables by $(X, Y)$ so that the observed pair of values on the $i$-th unit is denoted by $(x_i, y_i)$, $i = 1, 2, \ldots, n$. The covariance between $X$ and $Y$ is defined as

$$Cov(X, Y) = \frac{1}{n-1} \sum_{i=1}^{n} (x_i - \bar{x})(y_i - \bar{y}).$$

If, in general, larger values of $Y$ are associated with smaller values of $X$, covariance has a negative sign. If, on the other hand, larger values of $Y$ are associated with larger values of $X$, the covariance will have a positive sign. The correlation coefficient between $X$ and $Y$ is defined as

$$r = Corr(X, Y) = (Cov(X, Y))/(s_x s_y),$$

where $s_x$ and $s_y$ are the standard deviations of $X$ and $Y$, respectively. Since standard deviation is a positive quantity, the sign of the correlation coefficient is identical to that of the covariance. The R function to compute correlation is `cor()`. The correlation coefficient $r$ is independent of changes in the scale of measurement of the variables. See Chapter 6 for a more theoretical discussion of the properties of the correlation coefficient.

The correlation coefficient measures the strength of the linear relationship between $X$ and $Y$. If the dependence between $X$ and $Y$ is not linear, then $r$ is not the appropriate measure to use. The value of $r$ is always between $-1$ and $+1$. If $X$ and $Y$ have a perfect linear relationship, then $r = +1$ or $-1$, depending on the direction of relationship. In a practical situation though, a value of $\pm 1$ is not expected. As the magnitude of the correlation coefficient gets closer to 1, the relationship between $X$ and $Y$ gets closer to perfect linearity. A correlation coefficient equal to 0, or close to 0, indicates $X$ and $Y$ are not linearly related (or only weakly related). The correlation of any variable with itself is always 1. Similarly, the correlation of any variable with a constant is identically 0. Empirically, a correlation less than 40% in magnitude is taken as evidence of weak association; a correlation between 40% and 70% indicates moderate association, while a correlation above 70% indicates strong association.

**Case Study 3.8.** (Continued) The matrix in Table 3.2 exhibits the pairwise correlation coefficients for all the variables related to chemical attributes of red wines. The R command to produce all pairwise correlations in a matrix format for the data frame Red.Wine is `cor(Red.Wine)`. With 11 variables in the data frame, R produces an $11 \times 11$ matrix. The main diagonal contains the value 1, as this is the correlation of each variable with itself. The correlation coefficient $r$ is symmetric, hence the lower (or the upper) triangular part of the matrix provides complete information.

In the table, the 11 variables are coded as follows. FA: Fixed Acidity; VA: Volatile Acidity; CA: Citric Acid; RS: Residual Sugar; Chl: Chlorides; FSD: Free Sulfur Dioxide; TSD: Total Sulfur Dioxide; Dens: Density; pH: pH; Sulph: Sulphates; Alc: Alcohol.

The correlation coefficient between Citric Acid and Alcohol is 0.11 whereas the correlation between Chlorides and Alcohol is $-0.22$. Since both these values are less than the empirical threshold of 0.40, they indicate no major linear dependence exists between these pairs. The scatterplots given in Figure 3.11 also testify to this fact. The correlation between pH and Density is $-0.34$, indicating a moderate negative dependence. The correlation between Fixed Acidity and Density is 0.67, which indicates moderately strong positive dependence. Moreover, note that the sign of the former correlation is negative and the latter positive. The scatterplots also show the same directions of dependence.

**TABLE 3.2**

Pairwise correlation coefficients of Red Wine characteristics

|        | FA     | VA     | CA     | RS    | Chl   | FSD   | TSD   | Dens   | pH     | Sulph |
|--------|--------|--------|--------|-------|-------|-------|-------|--------|--------|-------|
| VA     | −0.26  |        |        |       |       |       |       |        |        |       |
| CA     | **0.67**  | **−0.55**  |        |       |       |       |       |        |        |       |
| RS     | 0.11   | 0.00   | 0.14   |       |       |       |       |        |        |       |
| Chl    | 0.09   | 0.06   | 0.20   | 0.06  |       |       |       |        |        |       |
| FSD    | −0.15  | −0.01  | −0.06  | 0.19  | 0.01  |       |       |        |        |       |
| TSD    | −0.11  | 0.08   | 0.04   | 0.20  | 0.05  | **0.67**  |       |        |        |       |
| Dens   | **0.67**  | 0.02   | 0.36   | 0.36  | 0.20  | −0.02 | 0.07  |        |        |       |
| pH     | 0.68   | 0.23   | **−0.54**  | −0.09 | −0.27 | 0.07  | −0.07 | −0.34  |        |       |
| Sulph  | 0.18   | −0.26  | 0.31   | 0.01  | 0.37  | 0.05  | 0.04  | 0.15   | −0.20  |       |
| Alc    | −0.06  | −0.20  | 0.11   | 0.04  | −0.22 | −0.07 | −0.21 | **−0.50**  | −0.21  | 0.09  |

In fact, the correlation matrix in Table 3.2 shows that most of the chemical characteristics of the red wines do not show a high degree of association among themselves. Most of the correlations are small in magnitude. Only a few pairs of variables show moderately high association, either positive or negative; they are highlighted in the table. Note that the magnitude of linear dependence is quantified by the absolute value of the correlation coefficient whereas the sign indicates the direction of dependence. The usefulness of the correlation coefficient will become clearer in the regression model building applications considered in Chapter 9. ◈

## 3.5 Measures of Association for Categorical Variables

The concept of association between two variables is valid for all variable types, but the correlation coefficient is ideally defined for continuous variables only. Other measures of association are necessary for categorical variables.

If both variables are nominal, statistical procedures exist to test whether there is any association between the variables. Consider a situation in retail analytics where a brand store has to take a decision on shipping formal shirts of different colors to different parts of a country. The brand manager may need to investigate whether there exists any association between geographic region and choice of colors. A measure of association may be defined by comparing color distribution across geographic regions. If there is no association, then the value of this quantity is expected to be small. Thus, in the case of nominal variables, it is possible to quantify association, but the direction of the association does not make sense. We will consider tests of hypothesis to determine the presence or absence of association in Chapter 8. For the moment we look at measures of association between two variables which are ordinal.

**Case Study 3.9.** (Coffee Brand Selection) Table 3.3 displays the number of coffee packets of different prices bought by customers belonging to different income levels.

**TABLE 3.3**
Contingency table for household income and coffee price

| Income Level | Price per Packet 1 | 2 | 3 | Grand Total |
|---|---|---|---|---|
| 1 | 7509 | 16,914 | 9057 | 33,480 |
| 2 | 8293 | 17,736 | 8514 | 34,543 |
| 3 | 9130 | 21,298 | 10,129 | 40,557 |
| 4 | 4705 | 11,426 | 6275 | 22,406 |
| Grand Total | 29,637 | 67,374 | 33,975 | 130,986 |

Naturally one would like to see whether household income has any impact on the price level of coffee purchased. In the following we define measures that will quantify such associations. ◈

Before the actual measure of association is defined, let us understand what are concordant and discordant pairs. Concordant pairs are those pairs of individuals where one individual scores higher on both ordered variables than the other individual. Similarly discordant pairs are those pairs of individuals where one individual scores higher on one ordered variable and the second individual scores higher on the other. Suppose $C$ denotes the number of concordant pairs in a dataset and $D$ denotes the number of discordant pairs. If the association between two variables is positive, then $C$ is greater than $D$. If the variables are negatively associated, then $C$ is expected to be less than $D$ and if the variables are not associated, then $C$ and $D$ are expected to be close. The difference between $C$ and $D$ can be exploited to define measures of association between ordinal variables. Two measures are used often. Goodman and Kruskal's gamma is defined as

$$\gamma = (C - D)/(C + D).$$

Kendall's tau is defined as

$$\tau = \frac{(C - D)}{0.5\sqrt{(n^2 - \sum r_i^2)(n^2 - \sum c_j^2)}},$$

where $n$ is the sample size, $r_i$ is the sum of the observations in the $i$-th row, and $c_j$ is the sum of the observations in the $j$-th column of the contingency table of counts. For both measures the numerator decides the direction of the association; the denominators are normalizing factors.

**Case Study 3.9.** (Continued) Following are the R codes to determine the association between coffee price and income level (**#** denotes comments in R).

```
> library(vdcExtra)
> GKgamma(Kafe)
gamma          : 0.011
std. error    : 0.004
CI            : 0.004 0.018
> Kafeg<- GKgamma(Kafe)
> names(Kafeg)          # Shows the contents of the output Kafeg
[1] "gamma"    "C"       "D"        "sigma"     "CIlevel"
[6] "CI"
> Kafeg$C               # Number of concordant pairs
[1] 1981091382
> Kafeg$D               # Number of discordant pairs
[1] 1937081383
```

The total number of concordant pairs is higher than the total number of discordant pairs. There seems to be a significant positive association between income level and coffee price. Kendall's tau is not directly available in R for more than two vectors, but it can be calculated by using the formulation shown above using the available $C$ and $D$ values.                    ◈

## 3.6   Suggested Further Reading

There are innumerable books on descriptive statistics. Since before the era of analytics, business managers used statistics and data analysis in their respective domains. Market research and sample surveys are examples of statistical applications in business. There are many references for managerial level statistics, such as Levin and Rubin (2011), Black (2012), Mann (2013), Anderson et al. (2014) and Lind et al. (2014). Many of these books contain exercises and partial solutions. All books at this level illustrate points through case studies and data analysis. The exercises contain various types of data for understanding applications. In addition to these, descriptive statistics textbooks used at the university undergraduate level are good resources to understand fundamental concepts. Readers will also find alternative graphs, charts and measures of central tendencies and dispersions, which are not frequently used or have limited applications in big data analytics. A few standard books are Freedman et al. (2007), Moore et al. (2009) and Ott and Longnecker (2010). For a rather mathematical treatment of the concepts discussed in this chapter, Stuart and Ord (2010) is a classic resource.

# 4

## *Statistical Graphics and Visual Analytics*

Graphs, charts, diagrams and figures of various types have been in use for presentation of raw and processed data for a long time. They have been accepted as a part and parcel of descriptive statistics for data summarization, as discussed in Chapter 3. There we have presented a few common examples of visual summarization. Historically, graphics were used for presentation purposes. Only recently they are also being used for exploration, which has opened up a whole new focus area, more familiarly known as visual analytics. To make informed decisions in business as well as in the fields of science, in government and public policy making, visual analytics is essential. However, it does not seem as though a lot of attention has been given to developing any theory of graphics. Several guidelines for practitioners have been propagated, but again most of that involves only common sense.

Presentation graphics and exploration graphics differ in their forms and objectives. Presentation graphics are static and often they are used to substantiate a finding. Exploration graphics must be dynamic. Before one starts to explore, there is no preconceived notion of what to look for and where, or the best method to uncover the hidden patterns in the data. Exploration graphics are expected to leave a trail of the process to show how and why a particular section of the data is identified to provide actionable insight.

Visual analytics may be defined as the science of analytical reasoning facilitated by interactive visual interfaces. The main difference between graphs, charts and other visual summarization considered in the previous chapter and visual analytics is that the former are used to showcase already known information whereas the latter is used for data mining and pattern recognition. Improvement in data collection and data storage devices has led to information overload. The principle behind visualization of data is to utilize the capacity of the human brain for recognition of patterns. Visual analytics has been a part of astronomy, chemical and biological sciences for a long time. Only recently has it become part of business data analytics.

Visual analytics is driven by data density, data availability and computational prowess. It is also a multi-disciplinary research area, bringing together visualization science, data mining, mathematical and statistical models, data management, user interface techniques and human perception and cognition. The theory behind data visualization has been under development for the past decade or so. The user interface is being supplied by various business intelligence tools like Tableau, Python, Qlik View, SAS Visual Analytics and

R. Many of these tools are highly interactive and are able to combine various facets of information. The main value addition of visual analytics is on-demand analysis of data which can be churned out interactively and instantaneously. Data visualization can quickly indicate interesting patterns in the data and identify areas for further analysis. Separate models may be developed depending on isolated data patterns to understand departure from the main concentration. Proper visualization facilitates macro – as well as micro – level pattern identification with a minimum amount of turn-around time. Interactivity in visual analytics facilitates exploration of data to achieve insight.

Visual analytics provides multiple views of multiple slices of data from multiple angles. Ideally, it should be able to provide all views from raw data to data abstraction. Large amounts of data encompassing multiple attributes need to be presented in small spaces. Human eyes cannot comprehend more than three dimensions at the same time. Innovative mechanisms are necessary to represent multi-dimensional data. Similarities and anomalies among different segments of data and relationships among different variables must come out of visualization exercises. Exploration of data will lead to information extraction and distillation.

The visualization framework recognizes three levels of analysis – micro, meso and macro. The micro, or the individual level, consists of small datasets with less than 100 records. An example is a network of Facebook or Twitter interactions among groups of friends. At the meso level the dataset contains between 100 and 10,000 records. An example is the daily usage pattern by the inpatients and outpatients of a healthcare facility. Finally, the highest level of analysis is the macro level where a large number of observations is considered. Similarly, visualization can also be planned according to whether it is topical, temporal, geospatial or of some other type.

The major turning point in visual analytics is the integration of dynamic character, simultaneous analysis at the macro and micro levels as well as automatic use of color. This book does not propose to veer into visual analytics in the proper sense. But in this chapter we introduce interesting plots and diagrams which are at the boundary of visual analytics and expand the realm of descriptive statistics into the analytical horizon. To that end we introduce several case studies. The nuances of analytical visualization will be illustrated through these case studies. All the graphics in this chapter have been produced using the R library `ggplot2`. It constitutes a powerful set of graphic tools which are flexible and can be adapted to include various dimensions of a dataset.

## 4.1   Univariate and Bivariate Visualization

Anyone dealing with any amount of data is familiar with bar charts, pie charts, etc. These are all examples of univariate visualization. Similarly, scatterplots are examples of bivariate visualization. In the former set only one variable is studied graphically, while in the second one each point is placed according to its value on two attributes. These charts and graphs are equipped to handle a considerable volume of data and a reasonable number of levels of a categorical attribute. But, with a large volume of data, such simplistic figures are not enough to bring out the patterns in the data.

**Case Study 4.1.** (United States Census Data)  The data relating to United States Census was taken from the UC Irvine Machine Learning data repository `http://archive.ics.uci.edu/ml/datasets/Census+Income`. The complete data has 48,842 instances divided into a training set and a test set. We have used 30,162 complete cases. The attributes associated with each observation include Age, Workclass, Fnlwgt (a computed variable), Education, Education.Year (years of education), Marital.Status, Occupation, Race, Sex, Hours.per.Week (work hours per week) and Native.Country. These attribute names refer to the actual variable names used in R; most of them are self explanatory. The goal of the exercise is to determine whether these attributes are enough to predict Income. It may not be out of context to note here that the error rate of prediction as shown in the literature is quite high with many of the standard prediction methods. That indicates that the data has very high level of variability. ◈

### 4.1.1   Univariate Visualization

The very first step in exploring a continuous variable is to draw histograms and boxplots. However, when the number of observations is very large (as in Case Study 4.1 above), multiple histograms may be necessary, as illustrated below.

**Case Study 4.1.** (Continued) Since prediction of income is the ultimate goal, it is important that possible contributors to income be investigated separately as well as simultaneously. For the purpose of illustration, some of the possible contributors to income are studied in detail, such as Hours.per.Week, Age and Education.Year (see Figure 4.1).

The following code was used to produce the histogram of Age in Figure 4.1. The other histograms can be similarly produced.

```
> qplot(Age, data=USCensusData, geom="histogram", binwidth=5,
  main="US Census Data")
```

However, overall histograms do not impart a lot of information. It is well

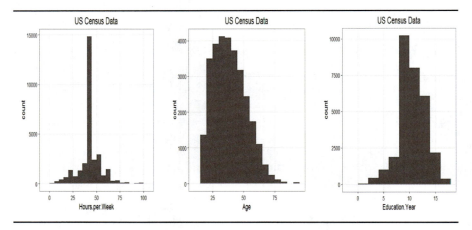

**FIGURE 4.1**
Examples of histograms.

known that Sex and Race do play a part in determining annual income. At the same time, these two socioeconomic variables may be confounded with other predictors, e.g., Hours.per.Week and Education.Year. The stacked histogram is one way to compare the distribution of Age and Education.Year (see Figure 4.2) over the covariate Race.

The following code was used to produce the stacked histogram of Age by Race. The other stacked histogram can be similarly produced.

```
> qplot(Age, data=USCensusData, fill=Race, geom="histogram",
  binwidth=5)
```

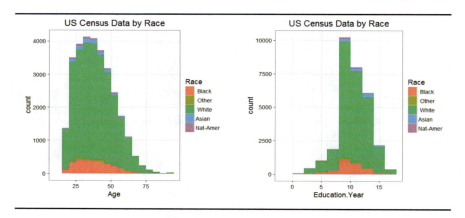

**FIGURE 4.2**
Examples of stacked histograms.

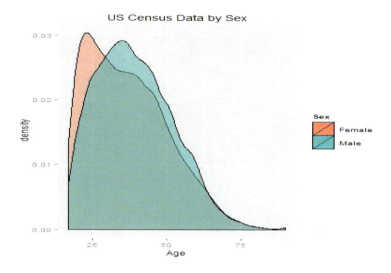

**FIGURE 4.3**
Example of overlaid density curves.

The best manner of visualization depends on a number of characteristics, including counts in different groups. The problem with Figure 4.2 is that whites being an overwhelming majority, almost all other groups are being masked. Stacked histograms are more useful as visual tools when the groups have comparable sizes.

If comparison is made on the basis of Sex, the density curves based on the age distribution of the US population show that females have a lower modal value compared to males. The female population is younger but the spreads of the two populations are about the same (see Figure 4.3).

The code to produce Figure 4.3 is given below.

```
> qplot(Age, data=USCensusData, geom="density", fill=Sex,
  binwidth=5, main="US Census Data by Sex", alpha=0.02)
  + scale_alpha(guide = 'none')
```

When the number of groups to be compared is large, it is a better strategy to compare the conditional distributions. For a formal definition of conditional distribution see Chapters 6 and 7. Since it is extremely unlikely that, in a sample with multiple groups, all the groups will have comparable sizes, histograms for conditional distributions must have density, i.e., relative frequency, instead of count in the $Y$-axis. This facilitates comparison, but does not provide any insight into the size of the strata. In the current sample, whites dominate for both males and females and the next largest group is black. This is very clear from the stacked histograms. The discrepancy is so large that any further model building exercise may not contain any influence of the other races, however much those races are different from the two largest groups. On the other hand, the sex distribution is equitable across the races.

The age distributions across all race and sex combinations are similar (see Figure 4.4). It is true that the histograms for white for both males and females are smooth compared other races, but that is possibly an artifact of their sizes. The most striking difference among these histograms corresponds to Education.Year for the Asian population. The modal class for both males and females is shifted right, indicating a higher education level for this group compared to the others. It could be argued that a younger population may have fewer number of years of education, but that is not true in this case, as the age distribution for Asian is not different from the others. The R code for this figure is given below.

```
> qplot(Age,..density.., data=USCensusData, geom="histogram",
  binwidth=5)+facet_grid(Race~Sex)
```

Another effective means of univariate visualization of data is the boxplot, whose width may be constructed as a function of sample size. The skewness of education among males and females of different races is very evident. The most interesting observation is that in many instances the distribution of Education.Year is negatively skewed (see Figures 4.5 and 4.6). However, what is not clear is the weights of individual points in Figure 4.6, i.e., whether each point represents one observation or more. This confusion arises because Education.Year variable can take only integer values and many observations may be coincident. The R code for Figure 4.6 is given below. The Race.Sex variable is a composite variable created over all combinations of Race and Sex.

```
> qplot(Race.Sex, Education.Year, data=USCensusData,
  geom="boxplot",varwidth=T)
```

The jitter plot provides a better visualization in this situation (see Figure 4.7). Through controlling the opacity of the points, the concentration of points in an area is recognized. It is also possible to control the opacity of the points. In the current plot the opacity is 1/1000, i.e., to get a completely opaque point 1000 points need to be coincidental. Instead of just the values of the outlier points, this technique provides an estimate of the number of points having identical values. The sparseness or otherwise of the distribution of the points provide the same information on the sample size in each stratum. Here each point is subject to a small random shift, so that coincident points are differentiated visually. The codes for this plot are as given below.

```
> qplot(Race.Sex, Hours.per.Week, data=USCensusData,
  geom="jitter",varwidth=T, alpha=1/1000)
```

This visualization clearly depicts the concentrations of education levels. For blacks, males and females both have about same relative frequency at

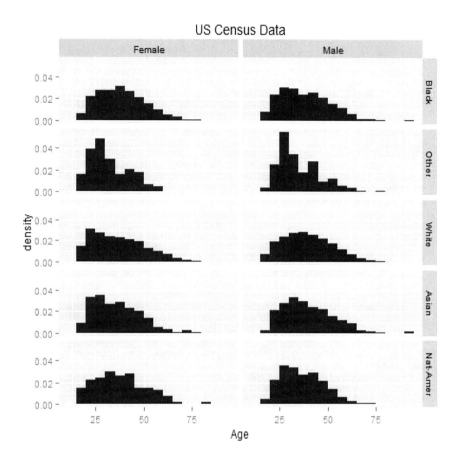

**FIGURE 4.4**
Conditional histogram: Example 1.

some years in high school and interestingly enough at one year of college. Compared to that group, white males have almost equal concentration at all levels from high school years to the highest number of years of education. For Asian males, and females also to a certain extent, comparatively denser concentration is at higher number of years of education. ◈

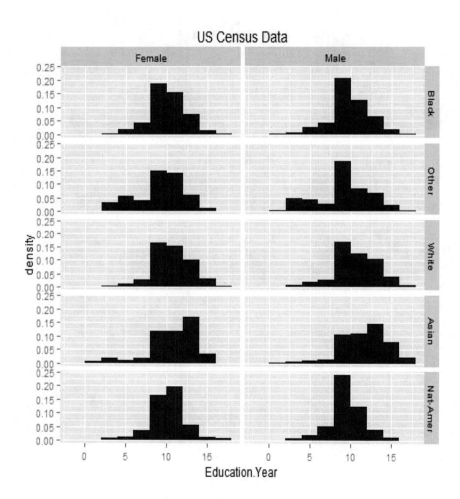

**FIGURE 4.5**
Conditional histogram: Example 2.

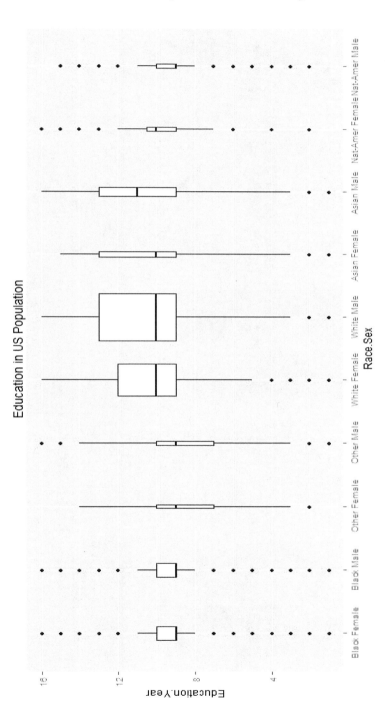

**FIGURE 4.6**
Conditional boxplots.

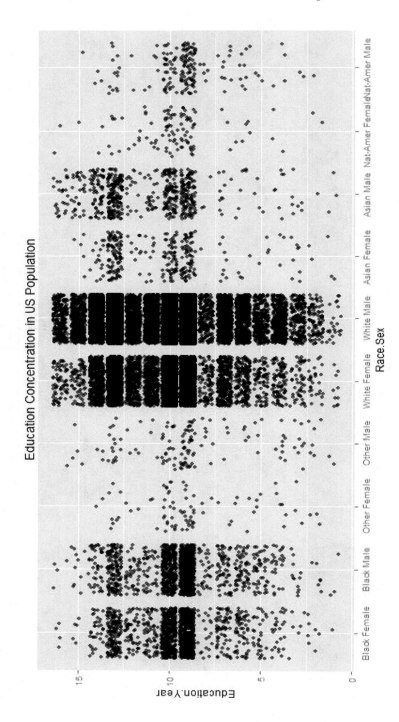

**FIGURE 4.7**
Conditional distribution.

## 4.1.2 Bivariate Visualization

By no means the graphs discussed in the previous section are the only possible visualizations for one dimensional data; they are among the most commonly used. However, challenges become tougher when several attributes need to be considered simultaneously. Note that visual techniques need to be aligned with business questions. Visualization also deals with pattern recognition in the data, i.e., whether an overall pattern would be the same for smaller subsections of the data.

**Case Study 4.2.** (United States Census Data) For this dataset, describing the joint behavior of the variables Age and Education.Year may throw some light on whether educational level is increasing in recent times. In other words whether it is true that younger population is more educated compared to older generation. This knowledge is important to gather as it is linked to the development index of a country and sets a context for business development. In order to assess that the data needs to be controlled for age, since very young population would not have completed their education.Another important information that can be gleaned from the data is whether joint distribution of Age and Education show different pattern across type of employment (Workclass). Since only a very small proportion of population will be in workforce after age 70, the data is truncated at that age. The joint behavior of Age and Education.Year is illustrated through contour plots. Contour plots join points of equal probability. Within the contour lines concentration of bivariate distribution is the same. One may think of the contour lines as horizontal slices of a 3-dimensional bivariate density. Contour plots are concentric; if they are perfect circles then the random variables are independent. The more oval shaped they are, the farther they are from independence. Different patterns indicate difference in the dependence of Education.Year on Age (see Figures 4.8 – 4.10). Two concentric figures indicate bimodality in the data. The codes for Figure 4.8 are as given below. The codes for Figures 4.9 and 4.10 are similar.

```
> d0 <- ggplot(OlderWhBlAs, aes(Age, Education.Year),
  main="Contour Plots: US Population")
> d0 +  stat_density2d()+facet_grid(Race~Sex)
  +ggtitle("Contour Plots: US Population")
```

Joint distribution of Age and Education.Year across Race and Sex reveals that for blacks, years of education is concentrated around 9-10 years of schooling and this pattern is consistent across all ages and gender. However a small proportion of blacks do have college education. Note that this group is less than 50 years old, indicating in the younger age group higher education shows some penetration. In the white population spread of education is more and the distribution seems distinctly bimodal. Among Asians the spread is even wider.

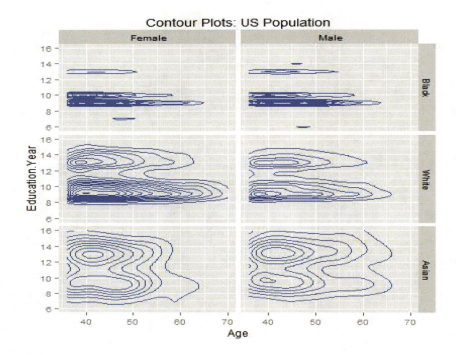

**FIGURE 4.8**
Contour plot: Example 1.

A related question may be whether the joint distribution of Age and Education.Year is the same across combinations of Race and Sex as well as across the levels of Occupation. The contour plots provided in Figures 4.8–4.10 provide more in-depth understanding of US Census data.                                         ◈

To isolate patterns in the data it is important to investigate data from all aspects. This brings us to the problem of dealing with multiple dimensions.

## 4.2 Multivariate Visualization

In the previous sections multiple dimensions of data have not been considered from a true multivariate viewpoint. It is not easy to include more than two dimensions in a single graph. So far conditional bivariate visualizations have been considered, where each bivariate plot is conditioned on given values of

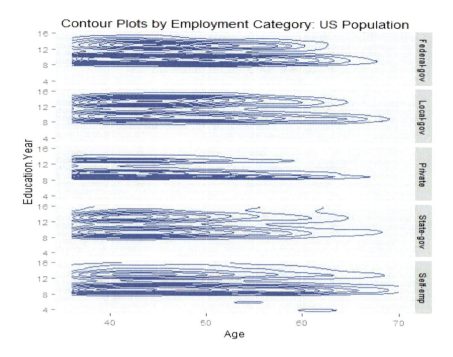

**FIGURE 4.9**
Contour plot: Example 2.

qualitative attributes. Contour plots of Age and Education.Year are drawn at all combinations of Occupation, Race and Sex. However joint representation of all characteristics has not been considered so far. One natural manner in which several characteristics of each observation can be shown is by using color, shape and size of the plotted points, in addition to the attributes represented by the rectangular axes.

**Case Study 4.3.** (Auto Data) Within R's own data library Auto Data contains 392 observations on different makes and models of cars with a number of characteristics of the cars including its horsepower, miles per gallon, origin or country of make, time to acceleration, year of make as well as other variables. A fuller description of this data is given in the Chapter 9 where this data is studied in detail. To study all features of the automobiles simultaneously it would be necessary to incorporate as many features in a visualization as possible. The number of data points are very small compared to the US Census data. Conditioning on one or more characteristics may not be fruitful as only a handful of observations may be available at that level. It is sensible to

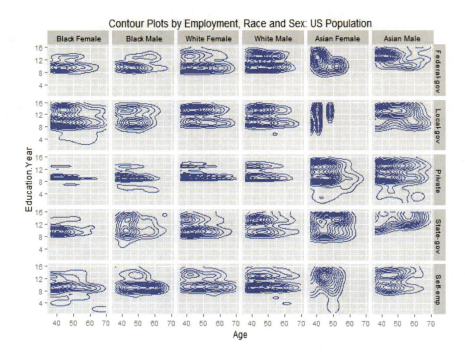

**FIGURE 4.10**
Contour plot: Example 3.

condition on categorical variables, but not on continuous ones, unless they are bucketized.

In Figure 4.11 a number of cars are plotted according to their Horsepower and MPG (miles per gallon) performance. Three other characteristics of the data are being represented through shape of the points, size of the points as well as on a color scale. The cars have three places of origin (1: US, 2:Europe, 3:Japan), are made between the years 70 and 82 and have time to acceleration on a scale between 8 and 24. Year is categorized into four levels (and named Year_cat) with A being the earliest part of the year and D being the latest part. Acceleration is considered a continuous variable. The categorical variable Year_cat is represented using four distinct colors whereas Acceleration being a continuous variable is being represented by means of increasing size of the points. Three distinct shapes denote three places of origin of the cars. Following are the R codes for Figure 4.11.

```
> qplot(Horsepower, MPG, data=AutoCat,color = Year_cat,
  shape=Origin, size=Acceleration)+ ggtitle("Horsepower
  and MPG: Auto Data")
```

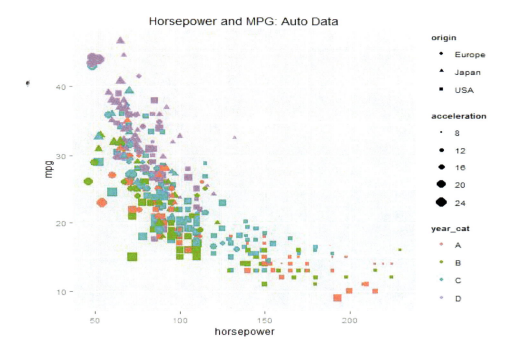

**FIGURE 4.11**
Scatterplot: Example 1.

From the scatterplot in Figure 4.11 it seems that there is a shift in the cars' features. Cars made during the earlier part of this 12 year period have a wider range of Horsepower and MPG combination. Some of those cars have very high Horsepower and low MPG performance whereas a few are at the other end of the spectrum. Cars made recently have better MPG at all levels of Horsepower. Japanese cars are more recent and mostly have advantageous position with respect to MPG at a given level of Horsepower.

However attractive and informative such a scatterplot may be, there is a limitation of dimensions that may be presented through a single plot, especially if the dimensions are continuous. Among the expressions used, color and size are suitable for continuous variables but shape is not. Too many features on a single graph do not provide clarity of perception either.

An alternative graphical presentation is a matrix of pairwise scatterplots along with univariate visualization. That is not the same as looking at all aspects of the observations simultaneously, but it provides pairwise comparison of several continuous attributes together. When all is said and done, 3D graphics are best when used interactively, but for static presentation, higher dimensional graphs are not very informative. The R codes for Figure 4.12 are given below.

```
> attach(Auto)
> X <- cbind(MPG, Horsepower, Displacement, Weight, Acceleration)
> scatterplotMatrix(X, diagonal=c("boxplot"),
  groups=AutoCat$origin, by.groups=T, reg.line=F, smoother=F,
  pch=19, cex=0.8)
> mtext ("Scatterplot Matrix of Car Attributes: By Origin",
  side=3, line = 3)
```

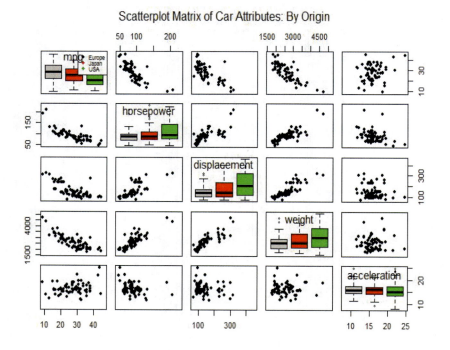

**FIGURE 4.12**
Pairwise scatterplots.

The scatterplot matrix in Figure 4.12 is arranged so that each individual plot has one variable on the $X$-axis and another on $Y$-axis. For all blocks on

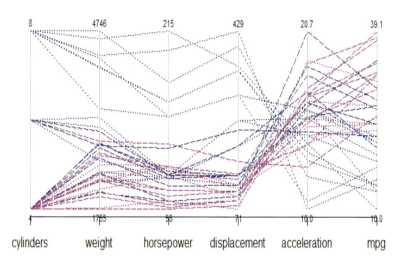

**FIGURE 4.13**
Example of profile plot for cars.

the first row, MPG appears on the $Y$-axis whereas for all the blocks on the first column MPG appears on the $X$-axis. For all the blocks on the second row Horsepower appears on the $Y$-axis whereas for all blocks on the second column, Horsepower appears on the $X$-axis. Hence for the scatterplot on the first row and second column, Horsepower constitutes $X$-axis and MPG constitutes $Y$-axis. The scatterplot on the third row and fourth column has Weight on $X$-axis and Displacement on $Y$-axis. Note that the arrangement is symmetric. The blocks on the main diagonal show boxplots of corresponding variables in groups defined by origin.

An innovative way to present multiple dimensions in the same figure is to use a parallel coordinate system. Each dimension is presented by one co-ordinate and instead of plotting coordinates at right angle to one another, each coordinate is placed side-by-side. The advantage is that, several different continuous and discrete variables can be handled within a parallel coordinate system, but if the number of observations is too large, the profiles do not separate out one from the other and patterns may be missed.

In Figure 4.13 different attributes of the cars have been considered. Only

35 observations have been considered and all of them are different varieties of Toyota or Ford, categorized into two groups: produced before 1975 and produced in or after 1975. The older models are represented by dotted lines whereas the newer cars are represented by dashed lines. The Fords are represented by blue color and Toyotas are represented by pink color.

The codes for Figure 4.13 are given below.

```
> library(MASS)
> # Colors by condition:
> car.colors<-ifelse(test = Comp1$Make=="Ford", yes = "blue",
  no = "magenta")
> # Line type by condition:
> car.lty<-ifelse(test = Comp1$year < 75, yes = "dotted",
  no = "longdash")
> parcoord(Y, col = car.colors, lty = car.lty, var.label=T)
> mtext ("Profile Plot of Toyota and Ford Cars", line = 2)
```

The differences among the four groups are very clear from the figure. Early Ford models had 8 cylinders, were heavy and had high horsepower and displacement. Naturally they had low MPG but needed less time to accelerate. No Toyota belonged to this category. All Toyota cars are built after 1975, have 4 cylinders (with one exception only) and on MPG performance belongs to the upper half of the distribution. Since only 35 cars are compared in the profile plot, each car can be followed over all the attributes. However had the number of observations been higher, the distinction among the profiles would have been lost and the plot would not be informative.        ◈

Another interesting system for plot for comparison of individual units simultaneously on a large number of attributes is the Chernoff Face. Human faces are drawn based on values of the attributes. Parts of a face such as eyes, nose, mouth etc, their position and shape vary according to values of the attributes. R can incorporate up to 15 different variables in drawing the faces. If the number of variable is less than 15, variables are recycled. Each individual face is then arranged in a matrix format.

**Case Study 4.4.** (College Data) A dataset on 777 colleges and universities is available in ISLR package in R. A total of 18 attributes, including number of applicants, number accepted, number enrolled, expenses on books, student-faculty ratio etc, are collated for each of the colleges. For comparison purpose we have selected 15 largest state universities/colleges and 15 largest private universities/colleges to see if state universities are similar among themselves and are different from the private universities. The results are presented in Figure 4.14. The R codes for this figure are given below.

```
> Map(faces(CollegeSmall[,3:19],  scale=T, face.type=0,
  labels=CollegeSmall[,1])
```

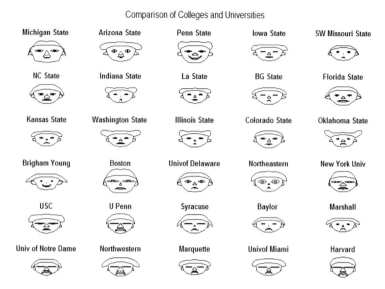

**FIGURE 4.14**
Chernoff face example.

The first 15 faces correspond to state universities while the last 15 faces correspond to private universities. It is expected that state universities will behave in a similar fashion and they will be different from the private universities. To some extent that conjecture seems to be true, but not always. State Universities generally enjoy large number of applications and allow large number of acceptances, but so do a few private universities. A comparison of Michigan State, Arizona State and Penn State and University of Delaware and University of Boston attest to the fact. Consider another variable S.F.Ratio. Most of the universities do not show a high variance in that attribute except for University of Miami, U Penn and New York University. For these universities S.F.Ratio is very small.

For comparison of a small number of observations on 15 attributes or less, the Chernoff face is an attractive technique. However, to what extent two items are similar, depends on the interpretation.

Another plot for multivariate comparison is the Star Plot (Figure 4.15) where stars are drawn according to rules as defined on the characteristics. Each axis represents one attribute and the solid lines represent each item's

**TABLE 4.1**
Attributes and facial characteristics for Chernoff faces

| Attribute of Drawing | height of face | width of face | structure of face | height of mouth | width of mouth |
|---|---|---|---|---|---|
| Attribute of Educational Institute | Apps | Accept | Enroll | Top10perc | Top25perc |
| Attribute of Drawing | smiling | height of eyes | width of eyes | height of hair | width of hair |
| Attribute of Educational Institute | F.Undergrad | P.Undergrad | Outstate | Room.Board | Books |
| Attribute of Drawing | style of hair | height of nose | width of nose | width of ear | height of ear |
| Attribute of Educational Institute | Personal | PhD | Terminal | S.F.Ratio | perc.alumni |

value on that attribute. All attributes of the observations are possible to be represented; however for the sake of clarity on the graph we have chosen to include only 10 attributes.                                                              ◈

## 4.3   Mapping Techniques

There are many other multidimensional techniques for pattern recognition. We will discuss two of them. The first one is called the heatmap and uses parallel coordinate systems, but instead of representing the observations quantitatively, they are color coded. Heatmaps have been in use for a long time in representing topographical patterns and weather data. A heatmap is structured as a matrix of rows and columns where each row represents an observation or a unit in a dataset while each column represents a variable. Usually lighter colors represent lower values and darker colors represent higher values. Following each row it is possible to understand the behavior of each unit, and following each column it is possible to understand how each observation is performing on that parameter.

Figure 4.16 is a heatmap which presents a comparison of 40 US universities and colleges on multiple parameters. The variables of interest are represented on the $X$-axis and the names of the academic institutions are given opposite each row. The legends based on colors are also given. From the blue end of spectrum toward the red end, the numerical values of variables are increasing. The R code for the same is given below.

Startplot of University Data

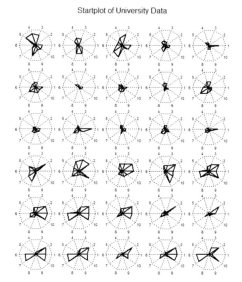

**FIGURE 4.15**
An example of a star plot.

```
> library(pheatmap)
> pheatmap(Coll2[1:40,], cluster_cols=F,
  title="Heatmap of Colleges")
```

The heatmap reveals that, on a few of the variables considered, the institutions are not too different among themselves, e.g., Grad.Rate, PhD and Books. On the other hand, F.Undergrad shows the maximum variability. The variables Accept, Expend and Outstation are also different in their values over the 40 observations.

The heatmap includes a dendrogram or clustering indicator of the observations on the left hand side of the diagram. Dendrograms will be discussed in detail in Chapter 11 along with hierarchical clustering algorithms. In short, the dendrogram shows which of the observations are more similar to each other. For example, Florida State University and the University of Georgia are similar and hence they combine to form a small cluster. The University of Southern California (USC) and New York University (NYU) are also similar and form another small cluster. At every step the smaller clusters combine to form larger clusters and finally all observations are combined to form one cluster. The lengths (or heights, depending on whether the dendrogram is horizontal or vertical) of the arms of the dendrograms indicate how similar or dissimilar are the individual items that form a cluster.

Heatmaps are more attractive than profile plots on a parallel coordinate

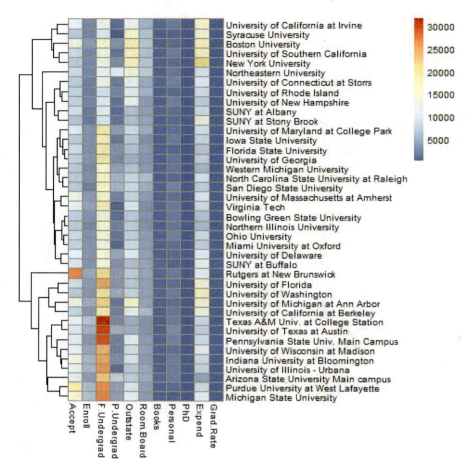

**FIGURE 4.16**
A heatmap of US universities.

system, even though each has its positive points. Heatmaps display all objects in the dataset with equal clarity; objects are not hidden behind or obscured by other objects. If there are even a moderately large number of observations, parallel coordinates look cluttered and it becomes almost impossible to glean information from that diagram. Each measure in a heatmap, however, resides in its own cell in the matrix, so there is never any occlusion. On the other hand, multivariate profiles exhibited as a series of colors (such as in a single row in our sample heatmap) are not as easy to perceive and remember as the single pattern of ups and downs formed by lines in a parallel coordinates display.

Another similar display of multivariate objects is called a Treemap. Treemaps are rectangular visualizations of a large number of observations, often displaying hierarchical relations among variables. Treemaps utilize areas of rectangles to represent numerical values of a variable. To indicate hierarchy, large rectangles are filled with smaller rectangles. Each rectangle in a treemap represents a node in a tree with area proportional to the value of the node. The space-filling approach produces a layout in which node values remain comprehensible at a high data density.

Figure 4.17 describes a set of 50 academic institutions, colleges and universities in the United States in terms of the two parameters Applications and Acceptances. This set has a substantial overlap with the set of institutions used in the heatmap, but is not a superset of the same. The filling in is done using density, i.e., the ratio of Acceptance and Application. The darker the color of fill, the higher the ratio. The short forms of the names of the colleges and universities are indicated in the rectangles. The relevant R codes are given below.

```
> library(treemap)
> treemap(College50, index="X.1", vSize="Apps", vColor="Accept",
  type="dens")
```

It is clear from Figure 4.17 that there is a huge difference between acceptance ratio among the institutions. As expected, Harvard, Yale, Columbia, MIT, Princeton, etc., accept only a tiny fraction of the applicants whereas RPI, Baylor and similar institutions accept majority of applicants. There is a difference between BU and Lehigh in the number of Applications but both of them accept a similar proportion of applicants.

Theoretically, it is possible to have several levels of hierarchy in the variable ordering. But cognition of comparison among observations decreases with higher levels of hierarchy.

## 4.4   Scopes and Challenges of Visualization

Visualization is a relatively new branch that is rapidly gaining a foothold in the analytics space. The scope and challenges of this branch of analytics are many. The most important advantage of visual analytics is its interactivity. It is most useful on the screen, and forms part of a dashboard with user control. Visualization is most effective for massive, inconsistent and messy data in its raw form. Extraction of potentially important information from large volumes of data is the objective of data visualization. Information overload is an opportunity for visualization to make major strides in the analytics field. Decision makers must be empowered to examine massive volume of multifaceted multidimensional and multisource data in a time-critical manner and effectively interact with data. Visual analytics is an iterative process which can be applied at all stages of analysis.

As we have seen before, there are different ways to visualize the same data, and, depending on the objective of the analysis and the nature of the data, there will very rarely be a hard and fast way to go forward. There are three major objectives of visualization: presentation, exploratory analysis and confirmatory analysis. Of these three, the first one, presentation, is the easiest one and has been in use for a long time. Visual summarization of data and descriptive statistics fall under this category. Here the presentation and the techniques are fixed beforehand. The aim is to effectively communicate the results of an analysis or simply present the raw and processed data. Confirmatory visual analysis starts from a hypothesized model and either confirms or rejects the hypothesis or the model. Regression diagnostics would be an example of confirmatory visualization. Exploratory visualization is the most difficult task. Proper tools and deep understanding of the domain are required for a scientific approach to data mining.

The challenges to business data visualization are many. Visualization must not be looked at as a stand-alone process. Rather, it is an integral part of the data analytics and decision making. Proper feeding of the most recent information into the system, monitoring the data inflow and continuously updating the dashboard to leverage the decision-making process in real time are some of the challenges faced. Other more important challenges are intellectual – how best to represent the data so that all the hidden patterns, correlations and causations are mined through the data.

## 4.5   Suggested Further Reading

Visual analytics and infographics are coming up as a special stream of analytics. To get a glimpse of the vast possibility in this area and to get acquainted with some of the innovations, Keim et al. (2010) is a comprehensive resource. In this edited volume cognitive aspects and scopes and challenges of data-driven visual analytics are discussed. Another interesting compilation of articles on the theory of visualization, computing and information gathering techniques, complexity and geometry of big data, technology transfer, architecture and display mechanisms is Dill et al. (2012). Borner and Polley (2014) discuss various mechanisms of infographics and recommend best practices for graphics applicable to various data types.

In addition to these, many of the guide books on R discuss graphical tools for visual analytics. Wickham (2009) in particular talks about ggplot2, the state of the art graphics tool in R. By no means is this the only available visualization tool in R. The other graphics functionalities in R are well illustrated in their respective documentations.

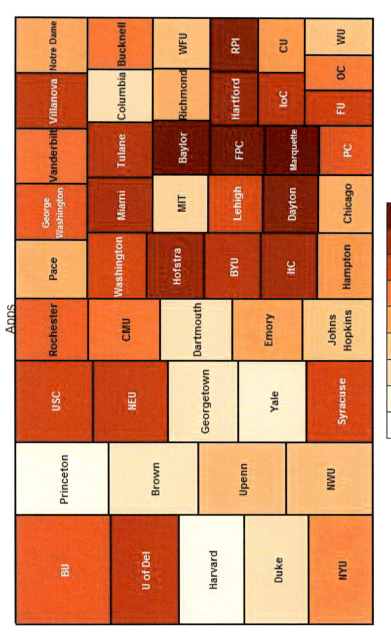

**FIGURE 4.17**
A treemap of 50 US universities.

# 5

## *Probability*

Outcomes of future events are uncertain. Decision making in this uncertain scenario assumes extreme importance in today's business world. The survival or failure of business firms primarily depends on the quality of decisions by the managers. This has become particularly important in the present age where we are swamped by a wealth of information all around us. The intuition of an individual, however experienced and smart the person may be, can be no substitute for the advanced data analytic tools that we have at our disposal at the present time. As the outcomes related to all the factors and variables under consideration cannot be known with certainty, the *theory of probability* – which quantifies the chance of the occurrence of an event – has an extremely important role in the process of informed decision making. Indeed, the theory of probability has its origin in one of the oldest forms of business and financial activity of the world – gambling – where the more analytical players wanted to systematically study the different possibilities rather than leave everything purely to chance. Intelligent decision making is enhanced and the associated business risks are minimized when the theory of probability is suitably applied to reduce the level of uncertainty. The decision to determine the level of the inventory may be aided by a probabilistic assessment of the behavior of the buyer, the strategy of launching a new product may be enhanced by a probabilistic study of the benefits of advertising and marketing, and the decision of which stocks to hold may be supported by a probabilistic time series analysis of past data. Millions of other examples may be cited. The bottom line is that a decision maker with a probabilistic understanding of the situation is far better equipped to handle the real issues compared to one who does not have this capability.

In Chapter 1 we alluded to actionable insight. It is nothing but the ability to take a justifiable decision – justification based on data collected on past business performance and the story told by the data. Uncertainty is at the root of insight – if there is no uncertainty, no insight is required. Every business will then act similarly and the outcomes will be identical; there will also be no possibility of any competitive advantage. As uncertainty is a reality, the business which is able to harness the uncertainty in the best possible manner and use it advantageously is going to be the winner. The concept of probability is deeply mathematical and a certain amount of theory cannot be avoided while introducing it. We have tried to focus on the applications rather than the proofs. The entire analytics machinery and the assimilation of data are rooted in probability. The importance of understanding probability will be clearer in subsequent chapters.

Any discourse on probability must start with a basic description of *set theory*, which is the topic of the next section. Familiarity with set theory is vital for a comprehensive development of the notion of probability.

## 5.1 Basic Set Theory

A *set* is a collection of distinct objects with some common property. The objects that make up a set are usually called the elements of the set.

We will refer to sets by upper case Roman letters like $A$, $B$, $C$, etc. Sometimes sequences of sets will be represented by subscripted letters such as $A_1, A_2, \ldots$. If the object $a$ belongs to the set $A$, it is symbolically represented as $a \in A$.

The number of elements in the set $A$ will be denoted by $N(A)$. This number is also referred to as the *cardinality* of the set.

There are many ways in which the totality of the elements of a set may be represented. The simplest is to list all the elements of the set, unless, of course, the set is so large that this is not feasible. Thus the representation $A = \{a_1, a_2, a_3, a_4\}$ means that the set $A$ is made up of the four distinct elements $a_1, a_2, a_3$ and $a_4$. We often use the symbol $\omega$ to represent a generic element of a set.

The collection of all possible elements in the population of interest is called the *universal set*. It is generally referred to as $\Omega$. All the individual elements that are within the purview of our study are members of the universal set.

A set $A$ is a *subset* of another set $B$ if every element of $A$ is also in $B$; this is expressed as $A \subset B$. In this case, whenever $\omega \in A$, it automatically implies that $\omega \in B$ also. Every set is a subset of the universal set $\Omega$.

A set which has no elements is called the *empty set*. It is generally denoted by $\varnothing$. Clearly, $\varnothing \subset A$ for any other set $A$; that is, the empty set is a subset of every other set.

The set $A^c$ is called the *complement* of the set $A$ if it consists of exactly such elements which are not in $A$. Thus every element of the universal set $\Omega$ must be a member of either $A$ or $A^c$, but not both. Clearly $A$ is the complement of $A^c$. Note that $\varnothing$ and $\Omega$ are complements of each other.

In the following, we list a few very important set operations.

1. The *union* of two sets $A$ and $B$ is the set of all elements that are either in $A$, or in $B$, or both. The resulting set is denoted by $A \cup B$. It may be represented as

$$A \cup B = \{\omega | \omega \in A \text{ and/or } \omega \in B\}. \tag{5.1}$$

In the above representation we have described a set by indicating the properties that a generic element of the set must have, rather than by enumerating all the elements of the set. The representation in Equation (5.1) indicates that an element of $A \cup B$ belongs to $A$ and/or $B$. Note that the set $A \cup B$ contains exactly those elements which belong to at least one of $A$ and $B$.

2. On the other hand, the *intersection* of any two sets $A$ and $B$ is the set that is formed by all the objects that are in both $A$ and $B$. This set is denoted by $A \cap B$, and may be represented as

$$A \cap B = \{\omega | \omega \in A \text{ and } \omega \in B\}.$$

If the sets $A$ and $B$ do not have any element in common, the intersection of these sets is empty, i.e., $A \cap B = \varnothing$. Such sets are called *disjoint* sets.

The union (or intersection) of a set with itself does not change it. Thus $A \cup A = A \cap A = A$. Also note that $A \cup A^c = \Omega$, while $A \cap A^c = \varnothing$.

The ideas of union and intersection can evidently be extended to more than two sets. Thus, given sets $A_1, A_2, \ldots, A_k$, the union of the sets, denoted $\cup_{i=1}^{k} A_i$, is the collection of all such elements which belong to at least one of the $A_i$s. On the other hand, $\cap_{i=1}^{k} A_i$ is the intersection of all these $k$ sets and consists of all such elements which belong to each of the $A_i$s. Both of these concepts can be extended to infinitely many sets.

A set of rules which are of great importance in set theoretic calculations is given by the *laws of De Morgan*. For two sets $A$ and $B$, this result states that $(A \cup B)^c = A^c \cap B^c$, and $(A \cap B)^c = A^c \cup B^c$. Once again these concepts can be easily extended to more than two sets.

Set theoretic illustrations are very often well understood through *Venn diagram* representations. These representations are very convenient ways of describing different set operations. Much of what we have described so far can be neatly described through easy to understand graphs. We give a sample of possible cases in the following.

- The upper left hand panel of Figure 5.1 shows the intersection of two sets $A$ and $B$. By definition, this is the set of the elements that are common to both sets. In terms of the Venn diagram, this is the region of overlap between the sets $A$ and $B$, indicated by the shaded region.

- The upper right panel of Figure 5.1 presents a Venn diagram which depicts the union of the sets $A$ and $B$. Notice that there is a region of overlap between the two sets. In putting the sets $A$ and $B$ together in forming the union, the overlapping part is counted twice. However, a set is a collection of distinct objects, and to form the union of $A$ and $B$, elements common to both sets must only be counted once, and not twice. Therefore, given any two sets $A$ and $B$, the number of elements in the union of the sets and the intersection of the sets satisfies the relation

$$N(A \cup B) = N(A) + N(B) - N(A \cap B). \tag{5.2}$$

- The shaded region in the lower left panel of Figure 5.1 represents the set of elements that are in $A$, but not in $B$. This is often denoted as $A - B$. In terms of the notations and concepts developed so far in this section, the following relationship is true.

$$A - B = A \cap B^c = A - (A \cap B).$$

- The shaded region in the lower right panel of Figure 5.1 is $(A \cup B)^c$. This region is also seen to be the intersection of $A^c$ and $B^c$. This gives an illustration of De Morgan's law in this particular case.

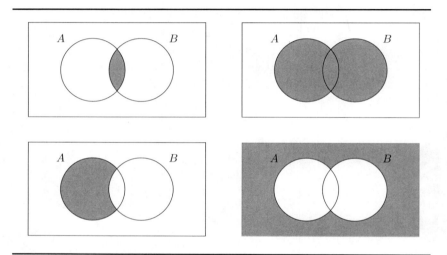

**FIGURE 5.1**
Venn diagrams.

## 5.2 Classical Definition of Probability

In practice we use the term *probability* in many different contexts in our day to day life. Not all of them come under the purview of the theory of probability that is to be formally studied here. In terms of its scientific scope, the word probability is used in connection with *experiments* that can be performed infinitely often under essentially the same conditions.

In the above spirit, let us describe the relevant set up. In our terminology, a *statistical experiment*, or a *random experiment*, will refer to any act whose outcome is unpredictable. This term might include those experiments which would come under the conventional meaning of the term, such as the experiments performed inside a biology, physics, chemistry or other scientific laboratories; the probabilistic use of the term, however, will not be limited to the above and will also include experiments like tossing a coin or rolling a die, or even noting the brand of coffee a customer is buying in the supermarket.

Any outcome of a statistical experiment is an *event*. When you are tossing a coin, for example, the possible events are $H$ and $T$, where they represent

$$H : \text{Getting a head when the coin is tossed,}$$

and

$$T : \text{Getting a tail when the coin is tossed.}$$

Similarly, when we are rolling an ordinary (six-faced) die, one can think of six possible events corresponding to each of the six faces coming out on top when the die is rolled.

Events can be *simple* or *composite*. Simple events cannot be further decomposed into simpler events. A composite event, on the other hand, is made up of more than one simple event. Consider the experiment of rolling a die. Define the events

$$A : \text{Getting the face with the number 1}$$

and

$$B : \text{Getting a face with an even number.}$$

Clearly, the first event is a simple event, while the second one is composite, being made up of three simple events.

While there are no specific rules, we often use upper case Roman letters, like $A, B, C$, etc., to refer to the possible events of a statistical experiment.

The set of all possible simple events in a statistical experiment is the *sample space* associated with the experiment, which we will denote by $\mathcal{S}$. In set theoretic notation, this is the analog of the universal set $\Omega$. An event $A$ therefore will be viewed as a set, representing the simple events that are favorable to it.

If the experiment involves tossing a coin, the sample space, under obvious notation, is $\mathcal{S} = \{H, T\}$.

If the experiment involves rolling a die, the sample space is, again under obvious notation,
$$\mathcal{S} = \{\text{"1", "2", "3", "4", "5", "6"}\}.$$

If the experiment is defined as selecting a coffee brand in a German supermarket (see Case Study 3.3), and if we consider only the brands (not their variants), then

$$\mathcal{S} = \{\text{"Aldi", "Eduscho", "Jacobs", "Tchibo", "Others"}\}.$$

At this point let us also define the following terms:

- *Mutually exclusive (disjoint) events*: Two events $A$ and $B$ are called mutually exclusive if they cannot occur together. For example, $H$ and $T$ are mutually exclusive when the experiment consists of tossing a coin. In the case of German Credit Data (see Case Study 3.1), labeling a loan applicant to be creditworthy or non-creditworthy are two mutually exclusive events. In a set theoretic sense, the sets of simple events favorable to $A$ and $B$ are disjoint. A collection of events is called a mutually exclusive collection if any two events in that collection are mutually exclusive. Mathematically, a collection $A_1, A_2, \ldots, A_k$ of events is mutually exclusive if, for any two events $A_m$ and $A_n$ belonging to this collection, we have $A_m \cap A_n = \varnothing$ whenever $m \neq n$. A mutually exclusive collection may contain infinitely many events.

- *Exhaustive events*: A collection of events is exhaustive if at least one of them must occur. Thus, if the experiment consists of rolling a die, one of the two events $A$ and $B$ must occur, where

$$A : \text{ Getting a face with an even number,}$$

and
$$B : \text{ Getting a face with an odd number.}$$

In general, a collection $A_1, A_2, \ldots, A_k$ of events is exhaustive if $\cup_{i=1}^{k} A_i = \mathcal{S}$. An exhaustive collection might contain infinitely many events.

- *Complementary events*: Two events $A$ and $B$ are complementary (or complements of each other) if they are both mutually exclusive and exhaustive, meaning $A \cup B = \mathcal{S}$ and $A \cap B = \varnothing$. In such a case we denote the event $B$ as $A^c$. The set $B$ consists of exactly those simple events which are in $\mathcal{S}$ but not in $A$.

- *Partition*: A collection of events $A_1, A_2, \ldots, A_k$ forms a partition of the sample space if the collection is exhaustive and the events in the collection are mutually exclusive. The simplest partition is provided by two events that are complements of each other. The events of buying coffees of different brands (see Case Study 3.3) partition the corresponding sample space.

We denote by $Pr(A)$ the probability of an event $A$. It is restricted to lie in the interval $[0, 1]$, with the interpretation that $Pr(A) = 1$ represents a sure event, while $Pr(A) = 0$ implies that the event $A$ will never occur.

With this background, we are now ready to present the *Classical Definition of Probability*. Assume that we have performed a statistical experiment for which

1. The total number of possible simple events, $N$, is finite, and

2. Each simple event is equally likely.

Then the probability of any event $A$ is defined as

$$Pr(A) = N(A)/N(\mathcal{S}) \qquad (5.3)$$

where $N(\mathcal{S}) = N$ is the total number of simple events and $N(A)$ is the number of simple events favorable to $A$.

Consider, for instance, the experiment of tossing a fair coin. The classical definition will indicate that the probability of getting a head for this experiment is $1/2$. Let us discuss this definition further with this particular example in mind. While the probability of getting a head is precisely defined by the classical definition, there is still a philosophical abstraction about it. Tossing the coin ten times does not necessarily lead to exactly five heads. After all, it is this uncertainty which is the basis of the theory of probability. However, as it turns out, within the fold of this uncertainty, the outcomes of the random experiments begin to show some regular behavior when the number of experiments keeps getting larger and larger. So, while the number of heads in ten tosses may not be exactly equal to five, the proportion of heads in $n$ tosses of a fair coin will, on the whole, keep getting closer and closer to $1/2$ as $n$ keeps increasing indefinitely. In this sense the term probability is referring to the *long run relative frequency* of the event of getting a head, and the classical definition essentially states that this long run relative frequency can be eventually made as close to the ratio given in Equation (5.3) as desired.

**Example 5.1.** Suppose that a lot of ten manufactured items contains two defective items. One item is chosen at random from the lot. What is the probability that the chosen item is defective?

Since there are ten items and two of them are defective, the required probability is $2/10 = 1/5$. Once again this may be interpreted as the proportion of times one would end up selecting a defective item if the same experiment is repeated infinitely often under essentially the same conditions. ‖

*Remark* 5.1. The term *at random* or *randomly* will be encountered very frequently in the subsequent description of probability. When a selection is made at random (or when the selection is made randomly), it means that all the elements are given an equal chance of being selected. ⊕

In a typical probability problem, finding the total number of simple events $N$ will usually be easy. The more difficult problem will be to find the numerator $N(A)$. Finding the numerator will often involve using different counting rules.

**Example 5.2.** Suppose that 50 people live in a particular neighborhood of a town. Out of them, 25 read the morning newspaper, 15 read the evening newspaper, and 5 read both. What is the probability that a randomly chosen person from this neighborhood does not read either of these newspapers?

Let $A$ denote the event that the randomly chosen person reads the morning newspaper and $B$ denote the event that the person reads the evening newspaper. From Equation (5.2),

$$N(A \cup B) = N(A) + N(B) - N(A \cap B) = 25 + 15 - 5 = 35.$$

Thus 35 people read at least one of the two newspapers (morning and evening). As the size of the neighborhood is 50, there are 15 people who do not read either newspaper. Thus, by the classical definition, the probability that the randomly chosen person will not be reading either newspaper is 15/50 = 0.3. ∥

Although it is clearly very simplistic, the classical definition of probability provides us with a common sense approach to finding the probability of a given event. The criticisms of the method are obvious, and they primarily center around the assumptions necessary for the application of the classical definition. We will list these criticisms in the appendix. However, in spite of these criticisms, there is a very large number of important probability problems of practical relevance that may be answered by this seemingly naive and simple definition using standard counting techniques.

## 5.3   Counting Rules

### 5.3.1   Multiplicative Rule (Basic Principle) of Counting

As indicated earlier, finding the total number of simple events favorable to $A$, which will often involve complicated restrictions, may not always be easy. In most cases the number will be too large to be enumerated completely. In such situations, often simple counting rules may be employed to systematically determine the total number of outcomes that are favorable to the event in question. The multiplicative rule of counting is one of the most basic tools in this respect. This rule and its generalizations are extremely useful in solving a large class of probability problems. The usefulness of these rules stems from the fact that, on many occasions, a statistical experiment is performed in different stages, and the entire experiment is a combination of these different parts.

**Definition 5.1.** The *multiplicative rule* (or the *basic principle*) of counting states that, if task 1 can be performed in $m$ ways, and task 2 can be performed in $n$ ways, together they can be performed in $m \times n$ ways.    ◈

**Example 5.3.** A soap manufacturer produces soaps in three different shapes (rectangular, round and oval) and in five different colors (white, cream, pink, violet and green). How many different shape-color combinations does the manufacturer produce? The multiplicative rule of counting says that, since there are three shapes and five colors, $3 \times 5 = 15$ shape-color combinations are possible in this case.    ∥

**Definition 5.2.** The above principle can easily be extended to the *generalized multiplicative rule* (*generalized basic principle*) of counting, which says that, if task 1 can be performed in $n_1$ ways, task 2 in $n_2$ ways, ..., task $k$ in $n_k$ ways, then the $k$ tasks can be performed together in $n_1 \times n_2 \times \cdots \times n_k$ ways. ◈

**Example 5.4.** Continuing with the soap manufacturer example (Example 5.3), suppose that the manufacturer, apart from producing the soaps in three different shapes and five different colors, also uses three different fragrances (rose, sandalwood and lavender). How many different varieties of soaps does the manufacturer produce? From the generalized multiplicative rule of counting, the answer is $3 \times 5 \times 3 = 45$. ∥

## 5.3.2 Permutations

Counting the number of favorable simple events in a probability problem often involves the determination of the number of distinct arrangements of a set of objects (where the elements of the set themselves may or may not be distinct). An ordered arrangement of this type is called a permutation. To begin, let us consider $n$ distinct objects, and our interest is in finding the probability that a randomly chosen arrangement satisfies a certain constraint.

**Example 5.5.** Consider the three distinct letters $a, b$ and $c$, and consider all possible permutations of these letters. Choose one of those permutations randomly. What is the probability that the permutation will have the letter $a$ in the first position?

The constraint imposed in this case is that there must be an $a$ in the first position. As this is a problem with only three elements (letters), we can answer this by complete enumeration. There are only six possible permutations of these three letters, given by $abc$, $acb$, $bac$, $bca$, $cab$ and $cba$. Since two out of these six arrangements start with an $a$, the required answer (from the classical definition of probability) is $2/6 = 1/3$. Notice that, since the permutation is chosen randomly, each of the six outcomes is equally likely, as required by the classical definition. ∥

In general, however, counting the total number of permutations or the number of permutations satisfying a constraint will require further development of counting rules. We want to start by systematically counting the total number of permutations of $n$ distinct objects taken $r$ at a time. The notation and formula for these are as follows:

$$^{n}P_{r} = \frac{n!}{(n-r)!}, \tag{5.4}$$

where $n! = 1 \times 2 \times \cdots \times n$. A systematic development of this will be presented in the appendix. By definition, $0! = 1$. When $n = r$, the expression in Equation (5.4) reduces to $n!$.

Consider, for example, the permutation of three distinct objects taken two at a time. According to the formula, the total number of permutations in this case is given by $3!/(3-2)! = 6$. As an illustration, note that, if the objects are the letters $a, b$ and $c$, these six permutations are $ab, ba, ac, ca, bc$ and $cb$.

Sometimes we are interested in permutations of objects that are not all distinct. In general, if the group of $n$ objects contains $r_1$ objects of the first type, $r_2$ objects of the second type, $\ldots$, $r_k$ objects of the $k$-th type, $\sum_{i=1}^{k} r_i = n$, the total number of permutations of these $n!$ objects is

$$\frac{n!}{r_1! r_2! \ldots r_k!}. \tag{5.5}$$

For example, if we have the objects (letters) $a$, $a$ and $c$, the distinct permutations are now *aac*, *aca* and *caa*. The total now is only three, as opposed to the six we would have had if three distinct letters ($a$, $b$ and $c$) were used. From the last formula, the total number of permutations for this case is $3!/(2! \times 1!) = 3$, which we already know. Note that $1! = 1$. An intuitive development of the formula in Equation (5.5) is given in the appendix.

**Example 5.6.** How many distinct permutations can be formed out of the letters of the word MASSACHUSETTS? Notice that there are 13 letters, but not all of them are distinct. The frequency of the letter S is four, that of A and T are two each, while there is one each of M, C, H, U and E. Noting that $1! = 1$, the required number of permutations is $13!/(4! \times 2! \times 2!) = 64,864,800$. Similarly, the number of permutations of the letters of the word MISSISSIPPI is $11!/(4! \times 4! \times 2!) = 34,650$.

$\parallel$

### 5.3.2.1 Combinations

Unlike permutations, where the order matters, sometimes we are only interested in the actual collection of objects without any reference to the order. Combinations represent unordered sets. We have seen that, for three letters $a$, $b$ and $c$, the total number of ordered arrangements is six; the arrangements themselves are $abc, acb, bac, bca, cab$ and $cba$. However, all of them represent the same set of numbers when the order is unimportant, and therefore all of them are identical when we are considering them as combinations.

We want to be able to systematically count the total number of combinations of $n$ objects taken $r$ at a time. The notation and formula for these are as follows:

$$^{n}C_r = \binom{n}{r} = \frac{n!}{r!(n-r)!}. \tag{5.6}$$

This formula may be easily derived and understood in the following way. When we find the number of permutations of $n$ objects taken $r$ at a time, the answer, as noted in Equation (5.4), is $n!/(n-r)!$. However, each collection (combination) of $r$ objects can lead to $r!$ distinct permutations, so when the size of the collection is $r$, there are $r!$ times as many permutations as combinations. Thus

$$^{n}P_r = r!(^{n}C_r).$$

This observation, coupled with Equation (5.4), immediately leads to Equation (5.6). Also notice that

$$^{n}C_r = {}^{n}C_{n-r}.$$

The last equation can be easily proved mathematically. Intuitively, it may be justified by noting that, every time we select $r$ objects from a group of $n$, exactly $(n-r)$ are left.

With this background, and using the multiplicative rule of counting, let us solve some simple probability problems. Our default notation for the number of combinations of $n$ objects taken $r$ at a time will be

$$\binom{n}{r},$$

read as "$n$ choose $r$."

**Example 5.7.** Suppose that there are two variants of the coffee brand Tchibo and three variants of the brand Jakobs on a supermarket shelf. A staff arranges the five items randomly. What is the probability that the variants of the same brand will be side by side?

To answer this, note that since there are five distinct items, in all there are $5! = 120$ ways in which they can be arranged. Since the items are arranged at random, all these 120 arrangements are equally likely. We should be able to solve the problem using the classical definition of probability provided we can count the number of ways that are favorable to the event in question, i.e., the variants of the same brand are together. Let us first consider the case where the Tchibo variants are placed on the left and the Jacobs variants are on the right. But the two Tchibo variants may be permuted within themselves in $2! = 2$ ways, and the Jakobs variants can be permuted within themselves in $3! = 6$ ways. Using the multiplicative rule of counting, then, there are $2 \times 6 = 12$ arrangements such that the Tchibo variants are on the left and the Jakobs variants are on the right. However, one could get the same number of arrangements by putting the Jakobs variants on the left and Tchibo on the right. Thus the total number of simple events favorable to the event in question is $12 + 12 = 24$. Thus the required probability is $24/120 = 1/5$.

Let us complicate this question further. Suppose a supermarket shelf has two variants of Tchibo, three variants of Jacobs, four variants of Eduscho and three variants of Aldi. If all the items are randomly arranged, what is the probability that the variants of the same brand will be side by side?

The answer is $4! \times (2! \times 3! \times 4! \times 3!)/12!$. Can you argue why? ‖

**Example 5.8.** Suppose there are ten members on the board of directors of a company. A committee of size four has to be formed from among the board members. In how many ways can this be done?

In this case the committee members are non-distinct in terms of their roles in the committee. The problem therefore is the same as that of selecting an unordered set of four objects from a total of ten objects. This is the number of combinations of four objects taken from a group of ten, and so the answer is

$$\binom{10}{4} = \frac{10!}{4!6!} = 210.$$

‖

**Example 5.9.** Contrast the previous example with the following. Four members are to be chosen from the board such that each of them fills a designated post. The posts are those of the CEO, the CFO, the CIO and the Managing Director. This problem involves more than finding a committee of four members, since in this case each of the four posts is distinct. Therefore we need to find the number of permutations of ten objects taken four at a time, and the answer to this problem is

$$\frac{10!}{6!} = 5040.$$

One can also get this number by first selecting the committee by 10!/(4!6!) ways, and then permuting the four committee members over the four positions, which can be done in 4! ways. This gives the answer 10!/(4!6!) × 4! = 10!/6! = 5040, as it should be. The same number could be reached by first choosing a CEO (in ten ways), then choosing the CFO from the remaining members (in nine ways), and then choosing the CIO and the Managing Director in eight and seven ways, respectively. When these numbers are combined by the multiplicative rule, we get 10 × 9 × 8 × 7 = 5040. ‖

**Example 5.10.** Consider a usual deck of 52 playing cards. Suppose five cards are dealt at random to a player.

(a) What is the probability that there are two kings, two queens and one jack in this five-card hand?

From the classical definition, the answer to this part is

$$\frac{\binom{4}{2}\binom{4}{2}\binom{4}{1}}{\binom{52}{5}}.$$

The above is a simple matter of counting the total number of events and the total number of favorable events.

(b) What is the probability that there are two cards of one denomination, two other cards of another denomination, and a scattered fifth card (not belonging to any of the two previous denominations) in this five-card hand?

In this case the answer is

$$\frac{\binom{13}{2}\binom{11}{1}\binom{4}{2}\binom{4}{2}\binom{4}{1}}{\binom{52}{5}}.$$

Can you figure out the arguments for the last expression? In $\binom{13}{2}$ ways we select the two denominations from which to have two cards each, and then there are $\binom{11}{1}$ ways to select the third denomination from which to select the remaining card. The rest is as in part (a). Check that one could also write the numerator of part (b) as

$$\binom{13}{1}\binom{12}{2}\binom{4}{2}\binom{4}{2}\binom{4}{1},$$

or as

$$\binom{13}{2}\binom{4}{2}\binom{4}{2}\binom{44}{1}.$$

∥

## 5.4 Axiomatic Definition of Probability

While the classical definition is very useful, it is, as we have already noted, also very restrictive in many ways. Several criticisms of the classical definition are presented in the appendix. It is clear that we need a structure which is more general than what is provided by the classical definition. For this purpose we now present the axiomatic definition of probability based on three probability axioms (sometimes called the Kolmogorov axioms). Unlike the classical definition (which still remains our basic tool for calculating probabilities in situations where the relevant conditions are met), the axiomatic definition does not directly define the probability of an event per se, but it leads to the development of a set of probability rules which support a general framework of probability.

According to the axiomatic definition, a probability function $Pr(\cdot)$ will satisfy the following axioms.

1. $Pr(A) \geq 0$ for any event $A$;

2. $Pr(\mathcal{S}) = 1$; and

3. If $A_1, A_2 \ldots$, are mutually exclusive events,

$$Pr(\bigcup_{i=1}^{\infty} A_i) = \sum_{i=1}^{\infty} Pr(A_i).$$

The above axioms help us to set up rules which we describe below through a sequence of theorems. The proofs (which may be found in any standard textbook of probability and statistics) are not of any special interest to us so we just state the results, omitting the proofs. Most of the results are fairly intuitive.

**Theorem 5.1.** *For any probability function* $Pr(\cdot)$ *and any sample space* $\mathcal{S}$, $Pr(\varnothing) = 0$.

This is as one would expect since an event which has no simple events favoring it should have a probability of zero.

**Theorem 5.2.** *For a finite sequence of mutually exclusive events* $A_1, \ldots, A_k$, *we get*

$$Pr(\bigcup_{i=1}^{k} A_i) = \sum_{i=1}^{k} Pr(A_i).$$

This is the finite sequence version of Axiom 3, and is an extremely powerful tool in our probability calculations. If the events $A_1, \ldots, A_k$ are also exhaustive, the above yields $\sum_{i=1}^{k} Pr(A_i) = 1$.

For any event $A$, we have $A \cup A^c = \mathcal{S}$. At the same time, $A$ and $A^c$ are mutually exclusive. Thus Axiom 2 and Theorem 5.2 lead to the following corollary.

**Corollary 5.3.** *For any event* $A$, $Pr(A^c) = 1 - Pr(A)$.

The following important result, which we have already made use of informally, can also be counted in this list of rules.

**Theorem 5.4.** *For any two events* $A$ *and* $B$, *we have*

$$Pr(A \cup B) = Pr(A) + Pr(B) - Pr(A \cap B).$$

**Example 5.11.** A large number of people worldwide are living without medical insurance. Hypothetical data based on the enumeration of all the adult residents of a certain city yields the following classification.

| Age Category | Health Insurance Yes | No | Total |
|:---:|:---:|:---:|:---:|
| 18 to 34 | 1800 | 600 | 2400 |
| 35 and Above | 2200 | 400 | 2600 |
| Total | 4000 | 1000 | 5000 |

An adult person is randomly chosen from the city.

(a) What is the probability that he/she has insurance coverage?

As there are 5000 people out of whom 4000 have insurance coverage, the required probability is 4000/5000 = 4/5 = 0.8.

(b) What is the probability that the randomly chosen person is 34 years old or younger?

There are 2400 people who are in the required age category, so that the required probability is 2400/5000 = 12/25 = 0.48.

(c) What is the probability that the randomly chosen person is 34 years old or younger, and has insurance coverage.

There are 1800 such people in the total population of 5000, so that the answer is 1800/5000 = 9/25 = 0.36.

(d) Suppose it is known that the randomly chosen person has insurance coverage. What is the probability that the person is 34 years old or younger?

We will defer the answer to this question for now, and revisit this example after we discuss the topic of conditional probability in the next section.

$\parallel$

**Example 5.12.** The following is a hypothetical distribution of the number of physicians in a small province of India classified over age and gender.

| | Age | | | | Total |
|---|---|---|---|---|---|
| | < 35 | 35-44 | 45-54 | > 55 | |
| Male | 1600 | 2500 | 3000 | 1400 | 8500 |
| Female | 400 | 1000 | 600 | 500 | 2500 |
| Total | 2000 | 3500 | 3600 | 1900 | 11,000 |

One physician is chosen at random from this group of 11,000 physicians.

(a) What is the probability that the physician is a male?

Since there are 11,000 physicians of which 8500 are male, the required probability is 8500/11,000 = 17/22.

(b) What is the probability that the physician is aged less than 35?

Counting the total number of cases, this probability is seen to be 2000/11,000 = 2/11.

(c) What is the probability that the physician is a male in the age category 35-44?

Here the favorable number of cases is 2500, so that the required probability is 2500/11,000 = 5/22.

(d) What is the probability that the physician is a female, and is aged between 35 and 54 years?

There are 1600 female physicians in this age bracket, so that the required probability is 1600/11,000 = 8/55.

(e) Suppose it is known that the randomly chosen individual is a female. What is the probability that she is aged between 35 and 54 years?

We will defer the answer to this question to the next section.

‖

## 5.5   Conditional Probability and Independence

To properly motivate the ideas of conditional probability, we pose the following questions.

Question 1: Consider the set up of Example 5.11. Suppose we randomly choose one of the two categories "Has insurance coverage" and "Does not have insurance coverage." Subsequently, we select one individual from the chosen category. What is the probability that the age of the individual is 35 years or more?

Question 2: Consider the set up of Example 5.12. Suppose one of the four age categories is chosen at random. Subsequently, a physician is randomly selected from the chosen age category. What is the probability that the selected physician is a female?

Question 3: A card is drawn randomly from a well shuffled deck of 52 cards. Without replacing the card drawn, a second card is drawn from the remaining 51 cards. What is the probability that the second card is an ace?

Question 4: Urn I contains three red and four green balls, while Urn II contains five red and nine green balls. An urn is chosen at random, and a ball is then randomly selected from the chosen urn. What is the probability that it is green?

The above questions are interesting in that the probabilities of the required events can be easily determined if the outcomes of the first stage of the experiments are known. In Question 1, for example, if $A$ denotes the event that the category "Has insurance coverage" is selected and $B$ denotes the event that the chosen individual is aged 35 years or more, then the probability of $B$ can be easily determined if it is known that the event $A$ has occurred (or not occurred). The probability that $B$ occurs given that $A$ has already occurred is denoted by $Pr(B|A)$. This is called the *conditional probability* of $B$ given $A$.

In Question 1, we have $Pr(B|A) = 2200/4000 = 11/20$, and $Pr(B|A^c) = 400/1000 = 2/5$; these results easily follow from an inspection of the reduced sample space after the outcome of the first stage of the experiment is known.

In Question 2, if the event $A$ denotes that the selected age category is "< 35," and if $B$ denotes that the chosen physician is a female, then $Pr(B|A) = 400/2000 = 1/5$; also $Pr(B|A^c) = 2100/9000 = 7/30$.

In Question 3, if $A$ denotes the event that the card obtained in the first drawing is an ace, and $B$ denotes the event that the card obtained in the second drawing is an ace, then the probability $Pr(B|A) = 3/51$, while $Pr(B|A^c) = 4/51$.

Similarly, if $A$ denotes the event of choosing Urn I in Question 4, and $B$ denotes the probability of selecting a green ball from the chosen urn, $Pr(B|A) = 4/7$ and $Pr(B|A^c) = 9/14$.

In each of the above cases, the occurrence or non-occurrence of $A$ makes a difference to the probability of the occurrence of $B$. Informally, that means there is some dependence on the conditioning event $A$ and the future event $B$.

It is not strictly necessary that the conditioning event should occur prior to the event whose probability we are trying to determine. Indeed, the conditioning event may occur simultaneously with the event in question, or even at a later time (as we will see in the examples involving Bayes' theorem). For the purpose of initial illustration, however, it helps to visualize the conditioning event as being a part of the outcome at a previous stage of the experiment.

We can now also attempt to answer the questions 5.11 (d) and 5.12 (e). These questions can now be understood as conditional probability questions. The reader can easily verify that the answers to these two questions are $1800/4000 = 9/20$ and $1600/2500 = 16/25$, respectively.

When we are determining the conditional probability of $B$ given $A$, the information that $A$ has occurred is already available, so we know that the simple event that was generated by the experiment belongs to the event $A$. This effectively reduces the sample space to the number of simple events that are favorable to $A$. When we ask, in addition, whether $B$ has also occurred, we are essentially asking whether the simple event that has occurred is from the set of events favorable to $A \cap B$. This argument leads to the conditional probability formula

$$Pr(B|A) = \frac{Pr(A \cap B)}{Pr(A)}. \tag{5.7}$$

A graphical representation of this is given in Figure 5.2.

Obviously, this requires that $Pr(A) > 0$. Notice that the above equation leads to the general sequential rule for probabilities through the relation

$$Pr(A \cap B) = Pr(A)Pr(B|A). \tag{5.8}$$

The conditional probability function is also a regular probability function in that it satisfies the three axioms of probability.

**Example 5.13.** Given $Pr(A) = 0.7$, $Pr(B) = 0.5$ and $Pr([A \cup B]^c) = 0.1$, find $Pr(A|B)$.

From the definition, $Pr(A|B) = Pr(A \cap B)/Pr(B)$. Since $Pr(B)$ is known, all we have to do is determine $Pr(A \cap B)$. Now $Pr(A \cup B) = 1 - Pr([A \cup B]^c) = 1 - 0.1 = 0.9$. Also

$$\begin{aligned} Pr(A \cup B) &= Pr(A) + Pr(B) - Pr(A \cap B) \\ 0.9 &= 0.7 + 0.5 - Pr(A \cap B) \end{aligned}$$

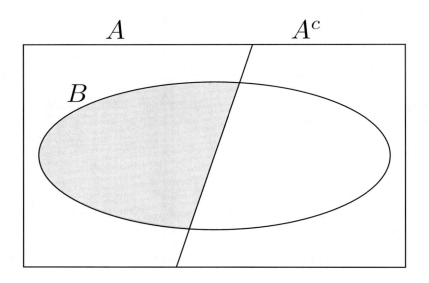

**FIGURE 5.2**
Conditional probability.

which gives $Pr(A \cap B) = 0.3$. Thus

$$Pr(A|B) = \frac{Pr(A \cap B)}{Pr(B)} = \frac{0.3}{0.5} = 0.6.$$

In Question 3, the two cards were chosen without replacement (i.e., before the second card was drawn, the first card was not replaced). Note that, when sampling is with replacement, the probability that the second card is an ace is $4/52$ regardless of whether the first card drawn was an ace or not (assuming all cards are drawn at random). The knowledge that event $A$ has occurred has no impact on the probability of $B$ in this case. When $Pr(B) = Pr(B|A)$ (as in the sampling with replacement case above), it means that the unconditional and conditional probabilities of $B$ are the same. In such cases we say that $A$ and $B$ are independent. Formally, two events $A$ and $B$ are independent if any of the following relations hold

(a) $Pr(A|B) = Pr(A)$.

(b) $Pr(B|A) = Pr(B)$.

(c) $Pr(A \cap B) = Pr(A)Pr(B)$.

The above relations are equivalent, and any of them can be proved from any of the others. However, we generally have a slight preference for relation (c), since it is symmetric in the arguments $A$ and $B$. Note also that the relation (c) is a special case (under independence) of the general sequential rule in Equation (5.8). Also, relation (c) can accommodate the cases where $Pr(A)$ or $Pr(B)$ or both are equal to zero without any problem.

### 5.5.1 Multiplicative Rule of Probability

Let us revisit Examples 5.11 (c) and 5.12 (d). Example 5.11 (c) asks for the probability of the event that the randomly chosen person is 34 years old or younger and has insurance coverage. We have already answered it by counting by the number of people who belong to the specified group. Alternatively, we can solve this by using the rules of conditional probability. Let $A$ be the event that the chosen person is 34 years or younger, and let $B$ be the event that the person has insurance coverage. We are looking for the probability $Pr(A \cap B)$. From the table in Example 5.11 we get $Pr(A) = 2400/5000$, and also $Pr(B|A) = 1800/2400$. By the sequential rule in Equation (5.8), the required probability is $Pr(A \cap B) = Pr(A)Pr(B|A) = 1800/5000 = 9/25$. Similarly, if, in Example 5.12 (d), the event that the physician is female is denoted by $A$, and the event that the physician is between 35 and 54 years of age is denoted by $B$, then answer to Example 5.12 (d) is $(2500/11,000) \times (1600/2500) = 8/55$.

Let us look at another example.

**Example 5.14.** Suppose that we have an urn which contains four green and three red balls. We draw two balls at random. What is the probability that both balls are green?

From what we have learned in the previous sections, the answer to this is

$$\frac{\binom{4}{2}\binom{3}{0}}{\binom{7}{2}} = \frac{2}{7}.$$

But now let us treat the same problem in a sequential manner. We define the events

$$
\begin{array}{ll}
A: & \text{The first ball drawn is green, and} \\
B: & \text{The second ball drawn is green.}
\end{array}
$$

Then $Pr(A) = 4/7$, while $Pr(B|A) = 3/6 = 1/2$. The required probability is $Pr(A \cap B)$, and it can be obtained sequentially as $Pr(A \cap B) = Pr(A)Pr(B|A) = \frac{4}{7} \times \frac{1}{2} = \frac{2}{7}$, as we already know it should be. ‖

We will refer to the relation $Pr(A \cap B) = Pr(A)Pr(B|A)$, already displayed in Equation (5.8), as the *multiplicative rule of probability*. Obviously, it can be extended to more than two events. Since conditioning can work both ways, we also have $Pr(A \cap B) = Pr(B)Pr(A|B)$.

## 5.5.2   Theorem of Total Probability

We have so far discussed the concept of conditional probability with the illustration of a two-stage experiment in mind. We have considered the probability of an event in the second stage with the knowledge that, in the first stage, another event has occurred (or not occurred). Knowledge of the status of the event at the first stage reduces the effective sample space and allows one to determine the relevant probability for the second stage event (or the probabilities of the joint occurrences of the first stage and the second stage probabilities) without much difficulty.

At times, however, knowledge about the status of the first stage event will be absent, but yet we will need to find the probability for the second stage event. Let us go back to our initial experiment of drawing two cards at random, without replacement, from a well shuffled deck of 52 playing cards. If we know which card was observed in the first draw, we can easily find the probability of drawing an ace in the second draw. But how difficult would that be when the information about the event in the first draw is absent? The theorem of total probability is our tool to handle problems such as these.

Unlike the case where the first stage event has a known outcome, now we have to look at the collection of all possible simple events which are favorable to the second stage event regardless of the outcome of the first stage event. Essentially, this requires going through all possible disjoint routes that the outcome of the first stage event could have taken and adding these probabilities to recover the probability in question.

**Theorem 5.5.** *(Theorem of Total Probability) Let $A_1, A_2, \ldots, A_k$ form a partition of the sample space. Let $B$ be any other event. Then*

$$Pr(B) = \sum_{j=1}^{k} Pr(A_j)Pr(B|A_j).$$

We provide a short proof of the above theorem in the appendix. The general idea is that the event $B$ is the composition of several mutually exclusive parts based on the intersection of $B$ with the elements of the partition. Putting these together and using the multiplicative rule of probability leads to the required result.

The theorem of total probability gives us the probability of a second stage event without knowing what happened in the first stage. This is the unconditional probability of the second stage event, as opposed to the conditional probability that one would have obtained if one were to find the probability of the second stage event after the information about what happened in the first stage was made available. For example, when two cards are dealt at random, successively, from a well shuffled deck of cards, the theorem of total probability can help us find the probability of having an ace in the second draw, without knowing what happened in the first draw (as opposed to finding the probability of an ace in the second draw after knowing that the first draw also resulted in an ace).

**Example 5.15.** Two cards are dealt to a player successively from a well shuffled deck of 52 regular playing cards. What is the probability that the second card is an ace?

Let us define

| | |
|---|---|
| $A_1$: | The first card is an ace, |
| $A_2$: | The first card is not an ace, |
| $B$: | The second card is an ace. |

Notice that the events $A_1$ and $A_2$ form a partition of the sample space in the first stage. One of them must occur, but they cannot occur together. To find the probability that the second card is an ace, we have to find $Pr(B)$. From the theorem of total probability,

$$Pr(B) = \sum_{i=1}^{2} Pr(A_i)Pr(B|A_i)$$

$$= (4/52) \times (3/51) + (48/52) \times (4/51) = \frac{204}{51 \times 52} = \frac{1}{13}.$$

Notice that this is the same as the probability of getting an ace in the first draw, as one would expect. When no information is available, there is no reason to expect that there will be a higher probability of an ace in the first draw over the second, otherwise all card players would want to get the first card. Here we are talking about unconditional probabilities. On the other hand, the probability of getting an ace in the second draw if you knew that the first draw also generated an ace is $Pr(B|A_1) = 3/51$. Here we are talking about the conditional probability of getting an ace in the second draw with the additional information of the result in the first draw. Calculating this probability does not need the use of the theorem of total probability.  ‖

**Example 5.16.** Suppose that Urn I contains four green and three red balls. Also suppose that Urn II contains nine green and five red balls. One urn is chosen at random, and one ball is drawn from the chosen urn. What is the probability that the selected ball is green?

At the first stage, either Urn I or Urn II could have been chosen. Thus $A_1$: Urn I is chosen, and $A_2$: Urn II is chosen form a partition of the sample space at the first stage. Also let $B$: The ball drawn from the chosen urn is green. From the theorem of total probability,

$$Pr(B) = Pr(A_1)Pr(B|A_1) + Pr(A_2)Pr(B|A_2) = \frac{1}{2} \times \frac{4}{7} + \frac{1}{2} \times \frac{9}{14} = \frac{17}{28}.$$

Note that this is different from $13/21$, as one would get if one were to simply take the ratio of the total number of green balls over the two urns and divide it by the total number of balls. The answer depends not only on the number of balls, but also on how they are distributed between the urns. ‖

## 5.6   Bayes' Theorem

In the last section we proved the theorem of total probability, which allows us to find the probability of a second stage event without knowing what happened in the first stage. The Bayes' theorem complements the former theorem in the sense that it allows us to determine the probability of a first stage event (i.e., $A_i$), given that an event in the second stage (i.e., $B$) has occurred. The proof of the Bayes' theorem is a simple consequence of the theorem of total probability and the fact that

$$Pr(A \cap B) = Pr(A)Pr(B|A) = Pr(B)Pr(A|B).$$

**Theorem 5.6.** *(Bayes' Theorem). Let $A_1, A_2, \ldots, A_k$ form a partition of the sample space. Let $B$ be any other event such that $Pr(B) > 0$. Then, for any $i = 1, 2, \ldots, k$,*

$$
\begin{aligned}
Pr(A_i|B) &= \frac{Pr(A_i)Pr(B|A_i)}{Pr(B)} \\
&= \frac{Pr(A_i)Pr(B|A_i)}{\sum_{j=1}^{k} Pr(A_j)Pr(B|A_j)}.
\end{aligned}
\tag{5.9}
$$

The unconditional probability $Pr(A_i)$, which gives the probability of $A_i$ without any reference to $B$, is sometimes referred to as the prior probability of $A_i$. The conditional probability $Pr(A_i|B)$, which is the revised probability of $A_i$ when the additional information about the occurrence of $B$ has been made available, is called the posterior probability of $A_i$ (given $B$).

In the following we will work out several interesting data examples where we will illustrate the use of the Bayes' theorem. Some of these defy intuition to some extent when one is exposed to these concepts and examples for the first time.

**Example 5.17.** In a hospital 10% of the patients are critical when they are admitted, 40% of the patients are serious when they are admitted and 50% of the patients are stable when they are admitted. Among the critical patients, 20% eventually die in the hospital; among the serious patients, 10% eventually die in the hospital; among the stable patients, 1% eventually die in the hospital. A patient is randomly chosen from the hospital.

(a) What is the probability that the patient will eventually die in the hospital? [This is the unconditional probability of the second stage event.]

(b) Given that the patient eventually dies in the hospital, what is the probability that he/she was critical when admitted? [This is the conditional probability of the first stage event, given the outcome of the second stage event.]

Let us define the events $A_1$, $A_2$ and $A_3$ as

| | |
|---|---|
| $A_1$: | Patient is critical when admitted, |
| $A_2$: | Patient is serious when admitted, |
| $A_3$: | Patient is stable when admitted. |

Also let

$B$ : Patient eventually dies in the hospital.

In part (a) we have to find $Pr(B)$. From the theorem of total probability, $Pr(B) = \sum_{j=1}^{k} Pr(A_j)Pr(B|A_j) = 0.1 \times 0.2 + 0.4 \times 0.1 + 0.5 \times 0.01 = 0.065$. In part (b) we have to find $Pr(A_1|B)$. By the Bayes' theorem,

$$
\begin{aligned}
Pr(A_1|B) &= \frac{Pr(A_1)Pr(B|A_1)}{\sum_{j=1}^{k} Pr(A_j)Pr(B|A_j)} \\
&= \frac{0.1 \times 0.2}{0.065} \\
&= 0.3077.
\end{aligned}
$$

$Pr(A_1)$ is the prior probability of $A_1$, while $Pr(A_1|B)$ is the posterior probability of $A_1$. While a priori there is a 10% chance that the patient is critical, the knowledge that the patient has died raises that to almost 31%.  ‖

**Example 5.18.** Suppose that 0.05% of the population has HIV. A test for HIV screening gives a correct diagnosis on an infected person with probability 0.99 and gives a correct diagnosis on an uninfected person with probability 0.97. A randomly chosen person is screened for HIV using the above test.

(a) What is the probability that the person will test positive?

(b) What is the probability that the person has HIV given the person tests positive?

Let us define $A_1$: Person is infected with HIV, and $A_2$: Person is not infected with HIV. Also let $B$: Person tests positive. In part (a) we have to find $Pr(B)$. From the theorem of total probability,

$$
\begin{aligned}
Pr(B) &= \sum_{i=1}^{2} Pr(A_i)Pr(B|A_i) \\
&= 0.0005 \times 0.99 + 0.9995 \times 0.03 \\
&= 0.03048.
\end{aligned}
$$

Thus there is a greater than 3% chance that a randomly chosen person will test positive, even though the virus is so rare in the population. In part (b) we have to find $Pr(A_1|B)$. By the Bayes' theorem,

$$Pr(A_1|B) = \frac{Pr(A_1)Pr(B|A_1)}{\sum_{j=1}^{2} Pr(A_j)Pr(B|A_j)}$$
$$= \frac{0.0005 \times 0.99}{0.03048}$$
$$= 0.01624.$$

Thus, even if the person tests positive, the posterior probability is only 0.01624 (less than 2%) that the person has HIV. Again, a result which would surprise most people. Although the tests are seemingly quite accurate, the errors accumulate in the population to result in a moderately large number.  ‖

**Example 5.19.** Two cards are dealt to a player, successively, from a well shuffled deck of 52 regular playing cards.

(a) What is the probability that the second card is an ace?

(b) What is the probability that the first card is an ace given the second card is an ace?

Let us define $A_1$ : The first card is an ace, and $A_2$ : The first card is not an ace. Also let $B$ : The second card is an ace. In part (a) we have to find $Pr(B)$. In Example 5.15 we have already found this to be 1/13, using the theorem of total probability. In part (b) we have to find $Pr(A_1|B)$. By the Bayes' theorem,

$$Pr(A_1|B) = \frac{Pr(A_1)Pr(B|A_1)}{\sum_{j=1}^{2} Pr(A_j)Pr(B|A_j)}$$
$$= \frac{(4/52) \times (3/51)}{1/13}$$
$$= \frac{3}{51}.$$

‖

Although we have not specifically pursued it in the present chapter, the Bayes' theorem is also useful in the discrimination problem where the experimenter has to discriminate between a finite number of classes based on the observed values of a variable of interest. When this is discussed in later chapters we will see that the selection of the optimal rule for this problem under some general conditions requires the comparison of the posterior probabilities of the different methods, and the corresponding classification rule is often referred to as the Bayes rule or the Bayes classifier. This is a problem of immense importance in applied statistics, machine learning, data mining and big data analytics.

## 5.7 A Comprehensive Example

**Case Study 5.1.** To understand the applicability of basic probability, let us consider the Auto Data (available in the ISLR package in R) containing 392 observations on different makes and models of cars with a number of characteristics of the cars. Let us also assume that this is a random sample of ALL cars made in the USA, Europe and Japan between years 1970 and 1982, both inclusive. We have already looked at various characteristics of this data and noted a general shift from low MPG cars to high MPG cars (see Figure 4.11). To quantify the shift in more concrete terms and quantify other changes in car manufacturing as well, let us try to seek answers to the following questions.

(a) What is the probability that a randomly selected car from the Auto Data is manufactured in the USA?

There are 392 cars in all, and 245 of them are manufactured in the USA, so the required probability is $245/392 = 0.625$.

(b) What is the probability that a car built before 1975 has MPG above 20?

Here we need the conditional probability of the car having MPG above 20, given that the car is built before 1975. The Auto Data contains 150 cars built before 1975, of which 50 have MPG above 20. Thus the required answer is $50/150 = 0.333$.

(c) What is the probability that a car built in the USA before 1975 has MPG above 20?

This question also asks for a conditional probability. There are 102 cars built in the USA before 1975, out of which 17 cars satisfy the given criterion. Thus the required answer is $17/102 = 0.1666$.

(d) What is the probability that a car built in 1975 or later will have MPG above 20?

In this dataset 242 cars were built in 1975 or later. Out of these 174 satisfy the given criterion. The required answer is $174/242 = 0.719$. This is a substantially higher number compared to that obtained in item (b) above. This result is consistent with the observed general shift from low MPG cars to high MPG cars.

(e) Given a car is built in 1975 or later and has MPG above 20, what is the probability that it is a Japanese car?

Here the conditioning event is that the car is built in 1975 or later and has MPG above 20. There are 174 such cars in the dataset out of which 57 are Japanese cars. The required answer is $57/174 = 0.3276$.

Now consider the data on cars from the Indian automotive industry perspective. In India, the market share of Indian cars is 55%, while, market share of US, Japanese and European cars is 5%, 30% and 10%, respectively. If 70% of Indian cars show 30 MPG performance, 10% of US cars, 80% of Japanese cars and 50% of European cars show 30 or above MPG, then

(f) What is the probability that a randomly selected car in India will have MPG 30 or more?

Using the theorem of total probability the required answer is $0.55 \times 0.7 + 0.05 \times 0.1 + 0.3 \times 0.8 + 0.1 \times 0.5 = 0.68$.

(g) What is the probability that a selected car with less than 30 MPG performance will be an Indian car?

The probability that the selected car will have less than 30 MPG performance is $1 - 0.68 = 0.32$. Thus the required probability, using the Bayes' theorem, is $0.55 \times 0.3/0.32 = 0.5156$.

## 5.8 Appendix

### 5.8.1 Criticisms of the Classical Definition

There is no doubt that the assumptions needed for the classical definition are very simplistic, and, not infrequently, one would encounter situations where they do not hold. We note our criticisms of the classical definition below.

1. The classical definition requires that the total number of simple events be finite. This is clearly a restrictive condition and is one of the major criticisms of the classical definition.

2. The definition also demands that the simple events all be equally likely, which limits the domain of its application. One cannot, for example, use it directly to find the probability of a head when tossing a biased coin or the probability of getting the face "1" when rolling a biased die.

3. It is not difficult to realize, even for somebody who is learning probability for the first time, that the above two points are obvious criticisms of the classical definition. The third criticism requires deeper thought. Careful reflection reveals that the dependence on *equally likely events* is problematic. What are equally likely events? They are events that have the same probability when the experiment is performed. Thus one is already using the concept of probability, even before defining it! When we are rolling a *fair* coin, it is already assumed that the probability of a

head (or tail) is 1/2. In that sense the classical definition of probability moves in a circle.

### 5.8.2 The Permutation Formula

Here we give a justification of the permutation formula in Equation (5.4). Note that, if there are $n$ distinct objects and if one were to arrange all of them in order, there would be $n$ ways to fill the first position of the arrangement, $(n-1)$ ways to do that for the second position, and so on. Finally, there would be just one way to fill the last position. Using the generalized multiplicative rule of counting, the total number of arrangements then will be $n \times (n-1) \times \cdots \times 2 \times 1 = n!$. Thus $n!$ ($n$ factorial) such arrangements (permutations) are possible when all the $n$ objects are considered. When only $r$ objects are considered ($r < n$), the choice ends with the $r$-th object, and the total number of ways is $n \times (n-1) \times \cdots \times (n-r+1) = n!/(n-r)!$. Since $0! = 1$, the expression in Equation (5.4) reduces to $n!$ when $r = n$, as it should.

### 5.8.3 Permutation with Non-Distinct Objects

We are interested in permutations of $n$ objects that are not all distinct. In this case the total number of distinct permutations is no longer $n!$, but is smaller than that. For example, if we have the objects $a$, $a$ and $c$, the distinct permutations are now $aac$, $aca$ and $caa$. The total now is only three, as opposed to the six we would have had if three distinct objects ($a$, $b$ and $c$) were used. In general, if $r$ of the $n$ objects are of one type (and the others are all distinct), the total number of distinct permutations turns out to be $n!/r!$. The number of permutations is now reduced by $r!$; each permutation now represents a block of $r!$ permutations which would have been distinct permutations if the $r$ objects were themselves distinct, but under the given set up are not. This number $n!/r!$ gets further reduced in the same manner if there is a second group of non-distinct objects and so on. In general, if the group of $n$ objects contains $r_1$ objects of the first type, $r_2$ objects of the second type, ..., $r_k$ objects of the $k$-th type, $\sum_{i=1}^{k} r_i = n$, the total number of permutations of these $n$ objects is

$$\frac{n!}{r_1! r_2! \ldots r_k!}.$$

### 5.8.4 Proof of the Theorem of Total Probability

*Proof.* Note that $\bigcup_{j=1}^{k} A_j = \mathcal{S}$, and

$$B = (A_1 \cap B) \cup (A_2 \cap B) \cup \ldots \cup (A_k \cap B). \qquad (5.10)$$

This is clearly exhibited in Figure 5.3, where we can see that $B$ is composed of the union of its different subparts belonging to the different segments of the partition provided by $A_1, A_2, \ldots, A_k$.

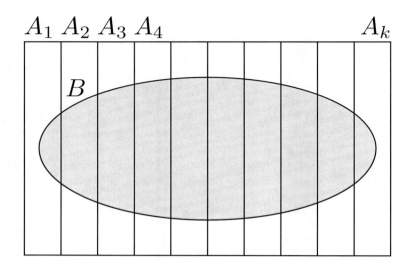

**FIGURE 5.3**
Description of the theorem of total probability.

Since the events on the right hand side of Equation (5.10) are all mutually exclusive, one can write

$$\begin{aligned} Pr(B) &= Pr[(A_1 \cap B) \cup (A_2 \cap B) \cup \ldots \cup (A_k \cap B)] \\ &= Pr(A_1 \cap B) + Pr(A_2 \cap B) + \cdots + Pr(A_k \cap B). \end{aligned}$$

By the multiplicative rule of probability, the right hand side of the last equation becomes

$$Pr(A_1)Pr(B|A_1) + Pr(A_2)Pr(B|A_2) + \cdots + Pr(A_k)Pr(B|A_k),$$

which proves the result. □

### 5.8.5    Proof of Bayes' Theorem

*Proof.* Note that $Pr(A_i)Pr(B|A_i) = Pr(B)Pr(A_i|B) = Pr(A_i \cap B)$. Thus

$$\begin{aligned} Pr(A_i|B) &= Pr(A_i)Pr(B|A_i)/Pr(B), \\ &= \frac{Pr(A_i)Pr(B|A_i)}{\sum_{j=1}^{k} Pr(A_j)Pr(B|A_j)}. \end{aligned}$$

The last step follows from the theorem of total probability. □

## 5.9   Suggested Further Reading

Unlike the previous four chapters, where the material is recent and so are the references, the tool of probability is a much older technique, and there is no scarcity of books (including good books) on probability. In such a situation any recommendation regarding what is appropriate for learners at this stage comes down to the personal preference of the authors. We feel that Ross (2010) is an excellent resource which should be at the top of this list. The book by DeGroot and Schervish (2012) provides an excellent description, not only of probability, but also of the distribution theory and statistics part. Walpole et al. (2010) is another text which is suitable for this stage and appears to be well liked. Tijms (2004) and Rozanov (1978) are two other books which may be useful for basic probability learners with some knowledge of mathematics. Akritas (2014) provides a good treatment supported by R illustrations.

The old theoretical classic is the Feller (1968) monograph, which still commands respect, although the level is far beyond what is intended in the present book.

Some of these references will be repeated in future sections, since many of the above texts discuss various topics of statistics together with probability.

# 6

## Random Variables and Probability Distributions

So far we have described the probability of events which are elements of a sample space in connection with a statistical experiment. In statistics, however, whenever we deal with probabilities, it is in relation to the possible values that a *random variable* can assume. So what is a random variable?

To explain this concept, let us start with a simple probability experiment. Suppose we have a fair coin and we toss this coin three times in succession. Denoting the possible events of *Head* and *Tail* in each toss by $H$ and $T$, the sample space of the entire experiment may be enumerated as

$$S = \{HHH, HHT, HTH, HTT, THH, THT, TTH, TTT\}.$$

As the coin is a fair one, each of these eight simple events has a probability of $1/8$.

Now, instead of finding the probabilities of each of the simple events in the sample space $S$ separately, let us view the problem in a different way. Let us define a new variable $X$, which represents the number of heads in each of the simple events, and suppose our questions are now rephrased in terms of values of the variable $X$. For example, we may want to know the value of the probability $Pr(X = 2)$. Since, according to our construction, each of the simple events $\{HHT\}, \{HTH\}$ and $\{THH\}$ corresponds to the event $(X = 2)$, the required probability in this case is $1/8 + 1/8 + 1/8 = 3/8$; similarly, $Pr(X = 0) = 1/8$, $Pr(X = 1) = 3/8$ and $Pr(X = 3) = 1/8$.

Note that the variable $X$ associates a real value with every element of the sample space. In the above example, the association induced by $X$ can be represented as follows:

| | | | | | |
|---|---|---|---|---|---|
| $HHH$ | $\rightarrow$ | 3 | $HHT$ | $\rightarrow$ | 2 |
| $HTH$ | $\rightarrow$ | 2 | $HTT$ | $\rightarrow$ | 1 |
| $THH$ | $\rightarrow$ | 2 | $THT$ | $\rightarrow$ | 1 |
| $TTH$ | $\rightarrow$ | 1 | $TTT$ | $\rightarrow$ | 0. |

This $X$ is a *random variable*. Formally, a random variable is an association which matches every element in the sample space with a real value. In statistics, once we have defined a random variable, all our probability questions will be phrased in terms of the random variable, so much so that we will tend to forget that there is a sample space operating in the background. Thus we will no longer ask for the probability of the event $\{HHH\}$, but will ask for the probability of the event $(X = 3)$ and so on. The primary reason for such a construction is that, in mathematics, it is much easier to deal with numbers than with the possibly abstract elements of the sample spaces. An experiment which naturally yields numbers as its outcome automatically defines a random variable on the corresponding sample space. For example, if the experiment consists of choosing an adult individual randomly and measuring his/her height (in appropriate units), then this height is a random variable. In terms of common day to day examples, the number of cars sold by a car dealer in a month, the number of clients coming to an ATM machine in one hour, the number of defective items in a lot of fixed size, etc., may count as legitimate random variables. Of course, random variables can be of many different kinds and these are but a few simple examples.

**Example 6.1.** Suppose an IT (Information Technology) company has four employees working on Project I and six employees working on Project II. Eight employees are randomly selected, *without replacement*, from this pool of ten employees. Let $X$ represent the number of employees working on Project I among these eight. This $X$ is a random variable. We want to describe the probability structure of this random variable.

For a given fixed value $x$, the probability $Pr(X = x)$ can be easily calculated in the same manner as in our probability calculations in Chapter 5, and this is given by

$$Pr(X = x) = \frac{\binom{4}{x}\binom{6}{8-x}}{\binom{10}{8}}. \tag{6.1}$$

Note, however, that, since the total number of employees working on Project I is only four, $x$ can be no larger than four. On the other hand, since eight employees are selected and there are only six employees working on Project II, the number of Project I employees selected can be no smaller than two. Hence the only values that $x$ can assume are 2, 3 and 4. It is also a simple matter to check that $Pr(X = 2) + Pr(X = 3) + Pr(X = 4) = 1$. Thus the probability structure of the random variable is completely described by the formula in Equation (6.1) together with the set of values $\mathcal{X}$ that the random variable $X$ can assume, which in this case is $\{2, 3, 4\}$.                                    ‖

**Example 6.2.** Suppose now that we have a set up similar to that of the previous example, but now the eight employees are selected *with replacement* (i.e., the selected employee is returned to the pool before the next selection), and the individuals who have already been selected are eligible for reselection. Once again let $X$ represent the number of Project I employees among the selected eight. This random variable is very different from the one considered in Example 6.1, and we now describe its probability structure.

Since the selections are with replacement, the number of Project I employees remains constant at four before each selection, and the number of Project II employees remains fixed at six; thus the corresponding probabilities of getting a Project I employee or a Project II employee also remain unchanged over the selections at $4/10 = 2/5$ and $6/10 = 3/5$, respectively. If a particular sequence of getting a total of $x$ Project I employees (and, therefore, $8 - x$ Project II employees) in the eight selections is specified, then such an event has the probability $(2/5)^x(3/5)^{8-x}$, which follows from the generalized multiplicative rule of counting. However, there are $\begin{pmatrix} 8 \\ x \end{pmatrix}$ distinct ways of getting a total of $x$ Project I employees in eight selections, so that we get

$$Pr(X = x) = \begin{pmatrix} 8 \\ x \end{pmatrix}\left(\frac{2}{5}\right)^x\left(\frac{3}{5}\right)^{8-x}. \tag{6.2}$$

Note that $x$ can only be one of the possible values in the set $\mathcal{X} = \{0, 1, \ldots, 8\}$, and the formula given in Equation (6.2) works for each such value $x$. Thus, together with the set of possible values given by $\mathcal{X}$, the above formula completely describes the probability structure of the random variable $X$. It also turns out (see the appendix) that

$$\sum_{x=0}^{8} Pr(X = x) = \sum_{x=0}^{8} \begin{pmatrix} 8 \\ x \end{pmatrix}\left(\frac{2}{5}\right)^x\left(\frac{3}{5}\right)^{8-x} = 1.$$

‖

*Remark* 6.1. In the above two examples we have used both the symbols $X$ and $x$. While these are standard in any discussion involving random variables, they might confuse the first time reader, and it does not hurt to emphasize the notation in this case. The symbol $X$ stands for the random variable itself, whereas the symbol $x$ represents any generic value in the list of values that the random variable can assume.  ⊕

*Remark* 6.2. The two examples given above are representative of two special classes of distributions, the hypergeometric and the binomial. Some special random variables, including these two, will be described in Section 6.2, where the derivation of the probability $Pr(X = x)$ for a binomial random variable will be discussed in more detail.  ⊕

*Remark* 6.3. Through Examples 6.1 and 6.2 we have also introduced the two standard forms of sampling, which are *sampling without replacement* and *sampling with replacement*. In sampling without replacement, the previously drawn elements of the population are not returned before the next unit is chosen and hence are not eligible for reselection. In sampling with replacement, the previously selected elements are returned to the population before selecting the next element, and it is possible for the same element to show up again and again.  ⊕

*Remark* 6.4. In Example 6.2, each selection is *independent* of any other selection in the sense that the outcome of any selection does not alter the probabilities of any other selection. While we will give a more formal definition of independence later on, the sense here is clear. Denoting each selection as a trial, the selection of the eight employees in Example 6.2 may be viewed as a sequence of eight independent trials.                                         ⊕

## 6.1    Discrete and Continuous Random Variables

A random variable, therefore, is characterized by the set of possible values it can assume together with the probability structure on this set. If the random variable can only assume finitely many distinct values $x_1, x_2, \ldots, x_k$, or at most countably many distinct values $x_1, x_2, \ldots$, the random variable is called a *discrete random variable*. Clearly, such a random variable concentrates the entire probability on a set of discrete points. The set of possible values that the random variable can assume is called its support, and this is denoted by the symbol $\mathcal{X}$ in the previous subsection. The random variables defined in Examples 6.1 and 6.2 are both discrete random variables.

For a discrete random variable $X$ supported on $\mathcal{X}$, we define a function $f$, referred to as the probability mass function (PMF) of $X$, as

$$f(x) = \begin{cases} Pr(X = x) & \text{for } x \in \mathcal{X}, \\ 0 & \text{otherwise.} \end{cases}$$

Often we simply define the PMF only on the support, it being implicitly understood that it is equal to zero everywhere else. This function $f$ together with the support $\mathcal{X}$ completely characterizes the random variable $X$. Note that the PMF $f(x)$ satisfies the following two conditions:

$$(i) \ f(x) \geq 0 \text{ for all } x, \text{ and } (ii) \ \sum_{x \in \mathcal{X}} f(x) = 1.$$

**Example 6.3.** The motivating example with which we began this chapter included tossing a fair coin three times and counting the number of heads (which is our random variable $X$). In this case the support of the random variable $X$ is $\{0, 1, 2, 3\}$, and the PMF is

| | |
|---|---|
| $f(0) = 1/8$ | $f(1) = 3/8$ |
| $f(2) = 3/8$ | $f(3) = 1/8$. |

**Example 6.4.** Consider a discrete random variable having the PMF given by

$$f(x) = \begin{cases} 1/2 & x = 0, \\ 1/4 & x = 1, \\ 1/4 & x = 2. \end{cases}$$

It is easily seen that the function $f$ is nonnegative and $f(0)+f(1)+f(2) = 1$, so that this is a valid probability mass function. We present this PMF graphically in Figure 6.1.

On the other hand, if the random variable had the same support, but if the function $f(x)$ has the form

$$f(x) = \begin{cases} 1/2 & x = 0, \\ 1/4 & x = 1, \\ 1/3 & x = 2, \end{cases} \quad \text{or} \quad f(x) = \begin{cases} 1/2 & x = 0, \\ 3/4 & x = 1, \\ -1/4 & x = 2, \end{cases}$$

then, in either case, the conditions of a PMF are violated. In the first case the masses do not add to 1. In the second case one of the masses is negative, although the sum total is equal to 1. ‖

**Example 6.5.** Suppose that two fair dice are rolled simultaneously. We denote by $X$ the larger of the two numbers that show up on this roll. We want to find the probability mass function of this random variable $X$.

Clearly, in this case the maximum of these two numbers must belong to the set $\mathcal{X} = \{1, 2, 3, 4, 5, 6\}$, so that $\mathcal{X}$ is the support of this random variable. There are $6 \times 6 = 36$ possibilities when two fair dice are rolled, and the table below gives the values of the random variable in question in the case of each combination of the outcomes of the two dice.

| (1,1); $X = 1$ | (1,2); $X = 2$ | (1,3); $X = 3$ | (1,4); $X = 4$ | (1,5); $X = 5$ | (1,6); $X = 6$ |
|---|---|---|---|---|---|
| (2,1); $X = 2$ | (2,2); $X = 2$ | (2,3); $X = 3$ | (2,4); $X = 4$ | (2,5); $X = 5$ | (2,6); $X = 6$ |
| (3,1); $X = 3$ | (3,2); $X = 3$ | (3,3); $X = 3$ | (3,4); $X = 4$ | (3,5); $X = 5$ | (3,6); $X = 6$ |
| (4,1); $X = 4$ | (4,2); $X = 4$ | (4,3); $X = 4$ | (4,4); $X = 4$ | (4,5); $X = 5$ | (4,6); $X = 6$ |
| (5,1); $X = 5$ | (5,2); $X = 5$ | (5,3); $X = 5$ | (5,4); $X = 5$ | (5,5); $X = 5$ | (5,6); $X = 6$ |
| (6,1); $X = 6$ | (6,2); $X = 6$ | (6,3); $X = 6$ | (6,4); $X = 6$ | (6,5); $X = 6$ | (6,6); $X = 6$ |

The simple events supporting the values $x = 1, 2, \ldots, 6$ are marked by expanding blocks from the upper left corner of the table. Direct counting of the number of cases yields the PMF given in the following table.

| | |
|---|---|
| $f(1) = 1/36$ | $f(4) = 7/36$ |
| $f(2) = 3/36$ | $f(5) = 9/36$ |
| $f(3) = 5/36$ | $f(6) = 11/36$ |

It is a simple matter to check that this is a valid PMF. ‖

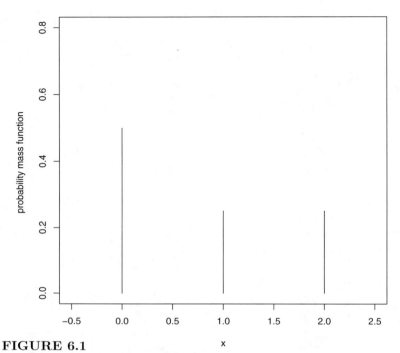

**FIGURE 6.1**
A discrete probability mass function.

We noted earlier that a discrete random variable is completely character-ized by its support and its probability mass function. The probability mass function $f(x)$ is equal to the probability $Pr(X = x)$ for every value $x$ in the support of the random variable $X$. Sometimes the PMF $f(x)$ can be expressed as an algebraic function of $x$ and does not have to be separately written down for each value in the support. There are some popular and well known classes of distributions for which this is possible. In the following we describe some of them.

## 6.2   Some Special Discrete Distributions

### 6.2.1   Binomial Distribution

Consider the following set up.

1. Perform $n$ independent binary trials in the sense of Remark 6.4.

2. Each trial can result in one of two possible outcomes (call them *success* and *failure*).

3. The trials are identical; that is, the probability of success, say $p$, remains constant over the trials.

4. At the end of $n$ trials, let $X$ be the number of successes observed among these $n$ trials.

Sequences of independent trials that satisfy the above properties are called *Bernoulli trials*. Our interest is in the random variable $X$ defined above. This random variable is said to have a *binomial distribution* with parameters $n$ and $p$, symbolically expressed as $X \sim binomial(n, p)$. (Similar symbolic representation will be used for other special random variables as well.) The binomial distribution also arises when one randomly samples with replacement from a group of objects which contains items of only two distinct types, as in Example 6.2.

The development of the PMF of the binomial distribution, already partially described in Example 6.2, is taken up in the appendix. The PMF of a binomial random variable having parameters $n$ and $p$ is given as

$$f(x) = \binom{n}{x} p^x q^{n-x}, \quad x = 0, 1, \ldots, n. \tag{6.3}$$

It is easy to see that each term in Equation (6.3) is nonnegative; in the appendix it is demonstrated that they sum to 1.

The binomial distribution with parameters $n = 1$ and $p$ is called a *Bernoulli(p)* distribution. Each Bernoulli trial with success probability $p$ generates a *Bernoulli(p)* random variable. If $X \sim Bernoulli(p)$, then the corresponding probability mass function is given by

$$f(x) = p^x q^{1-x}, \quad x = 0, 1. \tag{6.4}$$

**Example 6.6.** Suppose that a multiple choice test has six questions, each of them having five choices. A student taking the test does not know the answer to any of the questions and guesses the answer randomly on each of them independently of the others. Suppose that we want to find the probability that the student answers exactly two questions correctly. Notice that this satisfies all the requirements of Bernoulli trials, and the number of questions correctly answered by the student has a binomial distribution with parameters $n = 6$ and $p = 1/5$. Thus the probability that the student answers exactly two questions correctly is

$$\binom{6}{2} \left(\frac{1}{5}\right)^2 \left(\frac{4}{5}\right)^4 = 0.24576.$$

The one line R code which generates this probability is the following:

```
> dbinom(2,6,1/5)
[1] 0.24576
```

If the probability of correctly answering *at most two questions* was required (rather than exactly two questions), we would need to find the cumulative probabilities over the values 0, 1 and 2, and this can be obtained in R by replacing the dbinom() command with pbinom(). Thus the required probability is $\sum_{x=0}^{2} f(x) = 0.90112$, where $f$ represents the PMF of a binomial distribution with parameters $n = 6$ and $p = 0.2$. Summing the individual terms using the dbinom() command gives the same answer.

```
> pbinom(2,6,0.2)
[1] 0.90112
> dbinom(0,6,0.2)+dbinom(1,6,0.2)+dbinom(2,6,0.2)
[1] 0.90112
```

‖

**Example 6.7.** From past experience, the manager of a showroom knows that the probability is 0.25 that he will make a sale of a particular piece of furniture every time a customer walks in, independently of any other sale. What is the probability that, if 20 customers visit the showroom on a particular day, the manager will manage to sell the item on 8 out of the 20 occasions?

Assuming that this experiment satisfies all the conditions of a Bernoulli trial, the number of sales has a *binomial*(20, 0.25) distribution. Thus the required probability is

$$\binom{20}{8} (0.25)^8 (0.75)^{1} 2 = 0.06088669.$$

The corresponding R code is

```
> dbinom(8,20,0.25)
[1] 0.06088669
```

On the other hand, if the problem asked for the probability of eight or fewer sales (rather than exactly eight sales), the required probability and the corresponding R code are as given below.

$$\sum_{x=0}^{8} \binom{20}{x} (0.25)^x (0.75)^{20-x} = 0.9590748.$$

```
> pbinom(8,20,0.25)
[1] 0.9590748
```

‖

**Example 6.8.** Suppose that 1.5% of the customers receiving a loan from a bank default when paying it back. What is the probability that no more than 3 customers among the next 100 customers receiving a loan will default when paying it back?

The number of defaulting customers in this group will have a binomial distribution with parameters $n = 100$ and $p = 0.015$, assuming all the binomial conditions are satisfied. Then the required probability is

$$\sum_{x=0}^{3} \binom{100}{x} (0.015)^x (0.985)^{20-x} = 0.9357841.$$

```
> pbinom(3,100,0.015)
[1] 0.9357841
```

‖

## 6.2.2 Hypergeometric Distribution

The *hypergeometric distribution* is another important distribution in statistics. The random variable $X$ (the number of red balls) in Example 6.1 has a hypergeometric distribution (with parameters 4, 6 and 8).

In general, if there are $N$ objects, of which $N_1$ are of type I and $N_2$ are of type II with $N = N_1 + N_2$, and $n(< N)$ objects are chosen from this set randomly *without replacement*, then $X$, the number of objects of type I among the $n$ objects chosen, is said to have a hypergeometric distribution with parameters $N_1$, $N_2$ and $n$. The set of values $x$ in the support of this random variable is defined by the restrictions $x \leq n$, $x \leq N_1$, and $n - x \leq N_2$.

The PMF $f(x)$ of the hypergeometric random variable with parameters $(N_1, N_2, n)$ is given by

$$f(x) = Pr(X = x) = \frac{\binom{N_1}{x}\binom{N_2}{n-x}}{\binom{N}{n}}$$

where the support of this random variable is determined from the above conditions; symbolically, we write $X \sim hypergeometric(N_1, N_2, n)$.

**Example 6.9.** Suppose that a pond has 100 fish. Twenty of these fish are caught, tagged and released in the pond again. Later on, 30 fish are freshly caught from the pond (again from the original pool of 100 fish). Suppose we want to find the probability that out of these 30, 10 fish are already tagged. (All catches are at random and without replacement.) Note that the number $X$ of already tagged fish in the second sample of 30 has a hypergeometric distribution with parameters $N_1 = 20$, $N_2 = 80$ and $n = 30$. Thus the probability of having 10 tagged fish in a sample of 30 equals

$$\frac{\binom{20}{10}\binom{80}{20}}{\binom{100}{30}} = 0.02223762. \tag{6.5}$$

The relevant R code is

```
> dhyper(10,20,80,30)
[1] 0.02223762
```

The PMF of the random variable $X$ is

$$f(x) = \frac{\binom{20}{x}\binom{80}{30-x}}{\binom{100}{30}}.$$

Under the necessary conditions, the support reduces to the set $\mathcal{X} = \{0, 1, \ldots, 20\}$. Also, the above PMF reduces to the expression in (6.5) when $x = 10$.       ‖

### 6.2.3    Poisson Distribution

The *Poisson distribution* is another very common discrete distribution. The other distributions that we have looked at so far (binomial and hypergeometric) have very simple set ups from which they have been generated. The set up for the Poisson distribution is a bit more complex. Instead of getting into that at this stage, we will simply define it through its PMF and give appropriate applications. Advanced books might deal with the Poisson distribution through the *Poisson Process* set up, which is beyond the scope of this book.

A random variable $X$ is said to the have the Poisson distribution if the PMF of the random variable is given by

$$f(x) = \frac{e^{-\lambda}\lambda^x}{x!}, \quad x = 0, 1, 2, \ldots, \tag{6.6}$$

where $\lambda > 0$ is the parameter of the distribution. It is easy to see that this PMF is nonnegative. That the probabilities add to 1 is verified in the appendix.

The Poisson distribution is a very useful distribution in statistics. One the reasons for this is that the Poisson distribution provides a good approximation to the binomial probabilities. Consider the binomial probability for $x$ corresponding to parameters $n$ and $p$; when $n$ becomes very large and $p$ is very small (so that $np$ is a moderately small number), this probability is well approximated by the corresponding Poisson probability of $x$ with $\lambda = np$. Thus, for such values of $n$ and $p$,

$$\binom{n}{x} p^x q^{n-x} \approx \frac{e^{-\lambda}\lambda^x}{x!},$$

where $\lambda = np$. The proof of this result is not difficult. It simply consists of writing down the binomial probabilities, writing out the factorials, and taking the limit as $n \to \infty$ and $np \to \lambda$, and is available in most standard textbooks of statistics.

Often it happens that we are faced with the problem of finding a binomial probability for which the combination term involves factorials of very large numbers. Such terms may be very difficult, sometimes practically impossible, to compute accurately even with high-speed computers. In such cases the probabilities may be determined approximately using the Poisson approximation to the binomial probability.

**Example 6.10.** An individual buys a lottery ticket 1000 times, and each time his probability of winning is 0.01 independently of any other time. We want to find the probability that he will win exactly eight times.

The number of times this individual will win has a binomial distribution with parameters $n = 1000$ and $p = 0.01$. The actual binomial probability that he wins exactly eight times, which equals

$$\binom{1000}{8}(0.01)^8(0.99)^{992} = 0.1128,$$

is well approximated by the Poisson probability that he will win exactly eight times with $\lambda = np = 1000 \times 0.01 = 10$. Under this approximation, this probability is given by

$$e^{-10}10^8/8! = 0.1125,$$

which is a reasonable approximation of the true value. ∥

**Example 6.8.** (Continued) In this example we had found, using exact binomial calculations, the probability the number of people who default in repaying their loans is no more than three to be 0.9357841. Using the Poisson approximation with parameter $\lambda = 100 \times 0.015$, the required probability of three or fewer defaulters turns out to be 0.9343575, which is quite close to the actual binomial probability.

```
> ppois(3,1.5)
[1] 0.9343575
```

∥

The Poisson distribution is very important in statistics for modeling *count data* or *frequency of frequencies*, which involves counting the number of events in a fixed interval of time (or even counting the number of events in a fixed spatial domain). The observed data in a large-scale household survey will record, among other things, the number of households having size 1, the number of households having size 2, etc. Some standard examples of naturally occurring random variables which follow the Poisson probability law are

1. The number of accidents in a given stretch of highway on a given day;

2. The number of misprints on a page of a book or a newspaper;

3. The number of customers entering a store during a given time period;

4. The number of letters received by an individual on a given day;

5. The number of phone calls received by a call center during an hour;

6. The number of radioactive particles discharged by some radioactive material during a given period of time.

In most of the above cases, the use of the Poisson law is justified by the *Poisson approximation to the binomial* property.

**Example 6.11.** Suppose that the number of accidents occurring on a highway on each day is a Poisson random variable with parameter $\lambda = 3$. We want to find the probability that three or more accidents will occur today.

Letting $X$ be the number of accidents that will occur today,

$$
\begin{aligned}
Pr(X \geq 3) &= 1 - Pr(X \leq 2) \\
&= 1 - [Pr(X = 0) + Pr(X = 1) + Pr(X = 2)] \\
&= 1 - [e^{-3} + 3e^{-3} + \frac{3^2 e^{-3}}{2!}] \\
&= 1 - 0.4232 = 0.5768.
\end{aligned}
$$

```
> 1-ppois(2,3)
[1] 0.5768099
```

## 6.3  Distribution Functions

We observed earlier that the probability structure of a discrete random variable is completely characterized by the probability mass function of the random variable together with its support. In practice, it may also be characterized by the cumulative distribution function (or simply distribution function)

$F(x)$ of the random variable. For each $x$ in the support of $X$, this function is defined as

$$F(x) = Pr[X \leq x].$$

It is easily seen that this function may be constructed from the probability mass function $f(x)$. If the possible values of $X$ are listed in an increasing order as $x_1 < x_2 < \ldots$, then we have the relation

$$F(x) = \sum_{x_i \leq x} f(x_i).$$

Also, $F(x) = 0$ for any $x < x_1$. If the largest value in the support of $X$ is finite, then $F(x) = 1$ for any $x$ equal to or greater than it. Also, for any $i > 1$, we have

$$f(x_i) = F(x_i) - F(x_{i-1}),$$

while for $i = 1$ we have $f(x_1) = F(x_1)$. Thus one can construct the cumulative distribution function (CDF) from the PMF by aggregating from the left, while one can also recover the PMF from the CDF by taking successive differences.

**Example 6.12.** Continuing with Example 6.4, consider the PMF considered therein and a random variable having such a PMF. By cumulating the successive probabilities, we see that the CDF corresponding to this random variable is given as

$$
\begin{aligned}
F(x) &= 0 \text{ for } x < 0, \\
F(x) &= 1/2 \text{ for } 0 \leq x < 1, \\
F(x) &= 3/4 \text{ for } 1 \leq x < 2, \\
F(x) &= 1 \text{ for } x \geq 2.
\end{aligned}
$$

A graph of this CDF is presented in Figure 6.2. ∥

Note also that the CDF given in Figure 6.2 has the nature of a step function, which is the general characteristic of the CDF of any discrete random variable. The graph has jumps at the support points and stays constant between the support points.

There are some other properties which every CDF must satisfy (including the CDFs of continuous distributions to be defined in the next chapter). These properties are important, although to some extent technical. We discuss these properties in the appendix, with particular reference to Example 6.12. Among other things, this discussion clarifies the meaning of the empty dots and the filled dots in Figure 6.2.

## 6.4 Bivariate and Multivariate Distributions

So far we have been studying the distribution of one random variable at a time. However, sometimes we are interested in the simultaneous behavior of

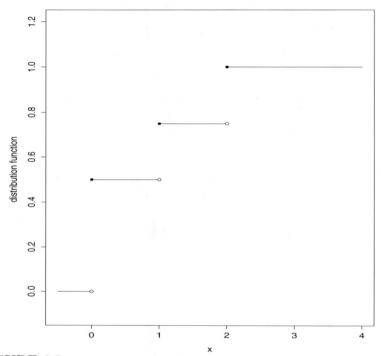

**FIGURE 6.2**
A discrete cumulative distribution function.

two or more random variables. In this case we will be talking about the *joint* or *multivariate* (*bivariate* in the case of two) distribution of these variables.

We begin with the joint distribution of two discrete random variables. While discussing the joint distribution of two discrete random variables, we will no longer be talking about $Pr(X = x)$ or $Pr(Y = y)$ in isolation, but about the joint probability

$$Pr(X = x, Y = y). \tag{6.7}$$

The expression in Equation (6.7) is the shortened notation for the probability of the intersection $Pr((X = x) \cap (Y = y))$, i.e., the probability that the events $(X = x)$ and $(Y = y)$ occur together.

The function $f(x, y)$ is called the bivariate PMF (probability mass function) of the two discrete random variables $X$ and $Y$ if

$$f(x, y) = Pr(X = x, Y = y).$$

In this case the support of the joint distribution of $X$ and $Y$ will be a subset of the real plane (rather than the real line). The joint distribution of two random variables is discrete when the corresponding two dimensional support is either a finite or a countable set.

As in the one dimensional case, the bivariate PMF satisfies the conditions:

(i) $f(x, y) \geq 0$ for all $x, y$.

(ii) $\sum_x \sum_y f(x, y) = 1$, where the sum is over all the points in the bivariate support of $(X, Y)$.

**Example 6.13.** Two fair coins are tossed. Let $X = 1$ if we get a head on the first coin; $X$ is zero otherwise. Similarly $Y = 1$ if we get a head on the second coin, and is zero otherwise. We are interested in the bivariate probability structure of $(X, Y)$.

Here the support of $(X, Y)$ is the set $\{(0,0), (0,1), (1,0), (1,1)\}$. The joint PMF of $(X, Y)$ is given in the following probability table. The cells in the body of the table give the joint probabilities. The values in the support of $X$ are enumerated in the first row, while those in the support of $Y$ are enumerated in the first column. As the coins are fair, the probability of each outcome is the same. Thus $Pr(X = 0, Y = 0) = 1/4$ and so on.

|         | $x = 0$ | $x = 1$ |
|---------|---------|---------|
| $y = 0$ | $1/4$   | $1/4$   |
| $y = 1$ | $1/4$   | $1/4$   |

$\parallel$

**Example 6.14.** Now consider the joint distribution of $X$ and $Y$ where $X$ is as in Example 6.13, but now $Y$ is the number of heads in the two tosses. Now the support of the joint distribution is

$$\{(0,0), (0,1), (1,1), (1,2)\},$$

and the joint PMF is given by

|         | $x = 0$ | $x = 1$ |
|---------|---------|---------|
| $y = 0$ | $1/4$   | $0$     |
| $y = 1$ | $1/4$   | $1/4$   |
| $y = 2$ | $0$     | $1/4$   |

The additional cells $(x = 0, y = 2)$ and $(x = 1, y = 0)$ that show up for making the probability table rectangular are actually outside the bivariate support of $(X, Y)$ and have zero probability. $\parallel$

**Example 6.15.** Three balls are randomly selected from an urn containing 3 red, 4 white and 5 black balls. Let $X$ and $Y$ denote, respectively, the number of red and white balls chosen. The joint PMF of $X$ and $Y$ are as given below.

|        | $x = 0$  | $x = 1$  | $x = 2$  | $x = 3$ |
|--------|----------|----------|----------|---------|
| $y = 0$ | 10/220  | 30/220   | 15/220   | 1/220   |
| $y = 1$ | 40/220  | 60/220   | 12/220   | 0       |
| $y = 2$ | 30/220  | 18/220   | 0        | 0       |
| $y = 3$ | 4/220   | 0        | 0        | 0       |

The actual calculations of the probabilities of any cell easily follows from the rules of probability described in Chapter 5. For example $Pr(X = 1, Y = 1)$ essentially asks for the probability that there are one red and one white ball in the chosen selection (and therefore the third remaining ball is black). This probability is equal to

$$\frac{\binom{3}{1}\binom{4}{1}\binom{5}{1}}{\binom{12}{3}} = \frac{3 \times 4 \times 5}{\frac{12 \times 11 \times 10}{1 \times 2 \times 3}} = \frac{15}{55}.$$

The probability of any particular subset of the support is obtained by adding over the probabilities (the values of the PMF) of the points of the support lying in that region. Thus in this example $Pr(X \geq 2, Y \leq 2)$ equals

$$Pr(X \geq 2, Y \leq 2) = \frac{15}{220} + \frac{1}{220} + \frac{12}{220} = \frac{7}{55}.$$

$\parallel$

## 6.4.1   Marginal Distributions

Suppose we are given the joint (bivariate) distribution of $X$ and $Y$. Often we may want to know whether we can recover the univariate distribution of $X$ from this joint distribution. This can be easily done by observing that the probability $Pr(X = x)$ can be obtained from the joint distribution by cumulating $Pr(X = x, Y = y)$ over all the values $y$ for which $Pr(X = x, Y = y)$ has a positive probability. In Example 6.13, the probability $Pr(X = 0)$ may be obtained as

$$Pr(X = 0) = Pr(X = 0, Y = 0) + Pr(X = 0, Y = 1) = 1/4 + 1/4 = 1/2.$$

Similarly, $Pr(X = 1)$ is also equal to 1/2.

Since we use the symbol $f$ to represent the bivariate (joint) PMF in this case, the univariate PMFs of $X$ and $Y$ will be represented by $f_X$ and $f_Y$ respectively. In this case, therefore, $f_X(0) = 1/2$ and $f_X(1) = 1/2$. Notice that these probabilities are given in the bottom row (column margins) of the probability table below, and are given in bold fonts to indicate that they represent marginal probabilities.

|         | $x = 0$ | $x = 1$ |         |
|---------|---------|---------|---------|
| $y = 0$ | 1/4     | 1/4     | **1/2** |
| $y = 1$ | 1/4     | 1/4     | **1/2** |
|         | **1/2** | **1/2** | 1       |

Similarly the probabilities for the univariate distribution of $Y$ are given, in bold fonts, in the last column (row margins). As the univariate probabilities appear in the margins (row or column), the univariate distributions of $X$ and $Y$, as obtained from their joint distributions, are also called their *marginal* distributions. In this case, if $f_X(x)$ and $f_Y(y)$ represent the two marginal distributions, we have,

$$f_X(x) = 1/2, \ x = 0, 1,$$

and

$$f_Y(y) = 1/2, \ y = 0, 1.$$

Similarly the marginal distributions in Example 6.14 are given, in bold fonts, in the margins of the following probability table:

|         | $x = 0$ | $x = 1$ |         |
|---------|---------|---------|---------|
| $y = 0$ | 1/4     | 0       | **1/4** |
| $y = 1$ | 1/4     | 1/4     | **1/2** |
| $y = 2$ | 0       | 1/4     | **1/4** |
|         | **1/2** | **1/2** | 1       |

In this case, the support of $X$ equals $\{0, 1\}$, and

$$f_X(0) = f_X(1) = 1/2.$$

The support of $Y$ equals $\{0, 1, 2\}$ and

$$f_Y(0) = 1/4, f_Y(1) = 1/2, f_Y(2) = 1/4.$$

**Example 6.16.** The marginal distributions of $X$ and $Y$ in Example 6.15 are given, in bold fonts, in the margins of the following probability table:

|         | $x = 0$ | $x = 1$ | $x = 2$ | $x = 3$ | Total    |
|---------|---------|---------|---------|---------|----------|
| $y = 0$ | 10/220  | 30/220  | 15/220  | 1/220   | **56/220** |
| $y = 1$ | 40/220  | 60/220  | 12/220  | 0       | **112/220** |
| $y = 2$ | 30/220  | 18/220  | 0       | 0       | **48/220** |
| $y = 3$ | 4/220   | 0       | 0       | 0       | **4/220** |
| Total   | **84/220** | **108/220** | **27/220** | **1/220** | 1 |

In this case

$$f_X(x) = \begin{cases} 84/220 & \text{for } x = 0, \\ 108/220 & \text{for } x = 1, \\ 27/220 & \text{for } x = 2, \\ 1/220 & \text{for } x = 3, \end{cases}$$

which are given in the column margins. Similarly the marginal PMF $f_Y(y)$ of $Y$ is given in the row margins.                                                    ∥

### 6.4.2   Independence

Two random variables $X$ and $Y$ are *independent* if the joint PMF can be factored into the marginal PMFs for all values $(x, y)$ in their support, i.e., if

$$f(x,y) = f_X(x)f_Y(y), \text{ for all } x, y, \qquad (6.8)$$

where $f(x,y)$ is the joint PMF of $(X, Y)$, and $f_X$ and $f_Y$ are the marginals of $X$ and $Y$ respectively. It is obvious that a necessary condition for the independence of two discrete random variables is that the bivariate support of $(X, Y)$ is a rectangular lattice with sides parallel to the rectangular axes. Also, when the joint probability distribution is given in the form of a table as in the illustrations of Section 6.4.1, independence would require that the probability of every cell in the table should equal the product of the corresponding row and column margins.

**Example 6.17.** In Example 6.16, the variables $X$ and $Y$ are clearly dependent. Although the support is written in terms of a rectangular lattice in the table, the actual support is the upper triangular region, and all the lower triangular probabilities are zero. In the following we present another example of a two-dimensional table of probabilities where each cell probability given in the body of the table is the product of the corresponding row and column margins. This guarantees that the joint probabilities are the product of the two marginals. Once again, the numbers in the row and column margins of the table are given in bold fonts. Hence the random variables $X$ and $Y$, whose joint distribution is described in the following table, are independent.

|        | $x = 0$ | $x = 1$ | $x = 2$ | $x = 3$ | Total |
|--------|---------|---------|---------|---------|-------|
| $y = 0$ | 0.02 | 0.02 | 0.03 | 0.03 | **0.10** |
| $y = 1$ | 0.04 | 0.04 | 0.06 | 0.06 | **0.20** |
| $y = 2$ | 0.06 | 0.06 | 0.09 | 0.09 | **0.30** |
| $y = 3$ | 0.08 | 0.08 | 0.12 | 0.12 | **0.40** |
| Total | **0.20** | **0.20** | **0.30** | **0.30** | 1.00 |

### 6.4.3    Conditional Distributions

Just as $Pr(A|B)$ represents the probability of a conditional event, we can also find the conditional distribution of one random variable given a specific value of another. Thus, given the joint PMF $f(x, y)$, the conditional PMF of $X$ given a fixed value $y$ of $Y$ is defined as $f_{X|Y=y}(x) = Pr(X = x|Y = y)$, where the conditional support (support of the conditional distribution) includes all the values of $x$ of $X$ for which $Pr(X = x|Y = y)$ is positive. It is easy to see that the function $f_{X|Y=y}(\cdot)$ is a valid probability mass function which adds to 1 over the conditional support of $X$ given $Y = y$. Formally,

$$f_{X|Y=y}(x) = \frac{f(x, y)}{f_Y(y)}, \tag{6.9}$$

so that the conditional PMF is obtained as the ratio of the joint and the marginal PMFs.

When the random variables $X$ and $Y$ are independent, Equation (6.9) implies that

$$f_{X|Y=y}(x) = f_X(x) \tag{6.10}$$

for each $x$ which is in the marginal support of $X$ (which is the same as the conditional support of $X$ given $Y = y$ in this case) and each $y$. Under independence, therefore, marginal and conditional PMFs are the same. This implication works the other way also, and Equation (6.10) can be taken as the alternative equivalent definition of independence. In fact, the condition

$$f_{Y|X=x}(y) = f_Y(y) \tag{6.11}$$

for all $x, y$, provides a third equivalent definition of independence.

**Example 6.18.** As a follow up of Example 6.16, consider the conditional distribution of $X$ given $Y = 1$. Using the rule in Equation (6.9), we see that the support of the conditional distribution is $\mathcal{X} = \{0, 1, 2\}$, with the corresponding probabilities $f_{X|Y=1}(0) = 40/112$, $f_{X|Y=1}(1) = 60/112$ and $f_{X|Y=1}(2) = 12/112$. The probabilities of this conditional distribution add to 1 over the support points, but this conditional distribution is clearly distinct from the marginal distribution of $X$ given in the column margins of the table in Example 6.16 (the last row of the table). This is a consequence of the dependence of the random variables $X$ and $Y$.

On the other hand, calculation of the conditional PMF $f_{X|Y=1}(\cdot)$ of $X$ given $Y = 1$ using the rule in Equation (6.9) in the case of Example 6.17 produces the same PMF as that of the marginal of $X$ in the column margins of the table in Example 6.17. This is as one would expect, as the variables $X$ and $Y$ are independent in this case.                                          ‖

---

## 6.5   Expectation

Consider a random binary experiment where the outcome $X$ can take the values 0 or 1 each with probability $1/2$. We may want to know the *average value* of this observed $X$. As we are observing two values, each with probability $1/2$, the natural answer is that this average is $0 \times 1/2 + 1 \times 1/2 = 1/2$, although we are never going to observe the value $1/2$ itself.

However, if the above experiment were not equiprobable, i.e., if the probabilities of 0 and 1 were different, it would certainly not be appropriate to say that $1/2$ is the average value that will be observed. This average should now be weighted more toward the more probable value. In practice, if the probabilities of 0 and 1 are $(1-p)$ and $p$, respectively, our average value will be $0 \times (1-p) + 1 \times p = p$.

In the above spirit, the average of a discrete random variable, which we will call the *expectation*, is obtained as a weighted average of the values in the support of the random variable with the weights being equal to the probabilities of these points. Formally, therefore, the expectation $E(X)$ of a random variable $X$ with PMF $f(x)$ is defined as

$$E(X) = \sum_{x \in \mathcal{X}} x f(x). \tag{6.12}$$

We often represent the expectation of a random variable by the symbol $\mu$. We also use the terms expectation (or the expected value) and mean interchangeably, both representing the same quantity defined in Equation 6.12.

The expectation or the mean is the most commonly used measure that quantifies the center of the distribution. It has the interpretation of the center of gravity, or the balancing point of the random variable. And while technically it requires a formal proof, it should be immediately obvious to the reader that if $X$ is a constant equal to $a$ (in other words, $X$ takes the value $a$ with probability 1), then the expectation of $X$ is also $a$.

Often we are interested in the expectations of some function of the random variable $X$, rather than that of $X$ itself. Just as the expectation of $X$ represents a weighted mean of the values in the support of $X$ with the weights as given by the probabilities of the PMF $f(x)$, the expectation of a function of $X$ represents the weighted means of the functions of the values in the support of $X$. Formally, if $h(x)$ represents any function of $X$, then

$$E(h(X)) = \sum_x h(x)f(x).$$

An obvious consequence of the last definition is the following theorem. While the proof is simple, we have placed it in the Appendix.

**Theorem 6.1.** *Let $X$ be a discrete random variable with PMF $f(x)$. Given any real values $a$ and $b$, we have*

$$E(aX + b) = aE(X) + b.$$

Thus in taking the expectation of a linear function of a random variable, one can run the expectation operator through the linear part of the function. However this would not be true for nonlinear functions. For example

$$E(X^2) \neq [E(X)]^2,$$

and hence

$$E(a + X^2) = a + E(X^2) \neq a + [E(X)]^2.$$

Together with the center of the distribution, it is important to also characterize the dispersion of the distribution. This is generally done with the variance measure, which finds the expectation of the squared deviation against the mean. Thus, the variance of $X$ equals

$$Var(X) = E[X - E(X)]^2 = E[X - \mu]^2. \tag{6.13}$$

We will reserve the symbol $\sigma^2$ for the variance of the distribution. The *standard deviation* $\sigma = [Var(X)]^{1/2}$ is also useful for measuring variability.

Some quick algebra shows that

$$
\begin{aligned}
\sigma^2 &= E[X - \mu]^2 = \sum_x (x - \mu)^2 f(x) \\
&= \sum_x x^2 f(x) - 2\mu \sum_x x f(x) + \mu^2 \sum_x f(x) \\
&= E(X^2) - 2\mu^2 + \mu^2 = E(X^2) - \mu^2. \tag{6.14}
\end{aligned}
$$

It is often easier to calculate the variance (or the standard deviation) using this particular form, rather than using the definition of variance directly as presented in Equation (6.13).

Under linear transformations, the variance satisfies the following relation, which indicates that the variance is not affected by the change of base, but does depend on the change of scale.

**Theorem 6.2.** *Let $X$ be a discrete random variable with PMF $f(x)$. Given any real values $a$ and $b$, we have*

$$Var(aX + b) = a^2 Var(X).$$

If $X$ is a constant equal to $a$, the variance of $X$ is equal to zero.

### 6.5.1 Moments

In discussions involving functions of random variables, the moments of a distribution are of particular interest. For a positive integer $r$, the $r$-th raw moment of a random variable is, simply,

$$\mu'_r = E(X^r) = \sum_x x^r f(x).$$

The first raw moment of the random variable (corresponding to $r = 1$) is the expectation $E(X)$ which describes the center of the distribution. Higher moments, particularly the second, third and fourth moments $\mu'_2, \mu'_3$ and $\mu'_4$, are useful in describing different characteristics of the distribution, like dispersion, skewness and kurtosis.

As opposed to the raw moments,

$$\mu_r = E(X - E(X))^r = E(X - \mu)^r$$

represents the $r$-th central moment of the distribution, where $\mu = E(X)$ is the expectation of the random variable. The second central moment (corresponding to $r = 2$) is the variance of the distribution.

To do the groundwork for the covariance and the correlation measures, we also briefly define the expectation in the bivariate case. If $f(x, y)$ is the bivariate PMF of $(X, Y)$, we define

$$E(XY) = \sum xy f(x, y), \tag{6.15}$$

where the sum is over all the points $(x, y)$ in the bivariate support of $(X, Y)$. When $X$ and $Y$ are independent, we get

$$E(XY) = E(X)E(Y). \tag{6.16}$$

More generally, the expectation of any function $h(x, y)$ is obtained as

$$E(h(X, Y)) = \sum h(x, y) f(x, y), \tag{6.17}$$

where the sum is again over the bivariate support.

If $h(x, y) = g_1(x)g_2(y)$, where $g_1(\cdot)$ and $g_2(\cdot)$ are functions of $x$ and $y$ alone, and if the random variables $X$ and $Y$ are independent,

$$E[h(X, Y)] = E[g_1(X)g_2(Y)] = E[g_1(X)]E[g_2(Y)]. \tag{6.18}$$

This result is a direct generalization of Equation (6.16).

## 6.5.2 Covariance and Correlation

Now we consider the joint distribution of two random variables $X$ and $Y$. Let $\mu_X$ and $\mu_Y$ be the means of $X$ and $Y$. Similarly, let $\sigma_X$ and $\sigma_Y$ be the indicated standard deviations. Consider the quantity

$$Cov(X,Y) = E[(X - \mu_X)(Y - \mu_Y)]. \qquad (6.19)$$

This measure, referred to as the covariance of $X$ and $Y$, is a measure of linear relationship between the two variables. If an increase in $X$ leads to, on the average, a corresponding increase in $Y$, the correlation measure is positive. This may be intuitively understood from the fact that the quantity $(X - \mu_X)(Y - \mu_Y)$ is positive when both $X$ and $Y$ are simultaneously large (compared to their respective means) or simultaneously small in the same sense; on the other hand, when an increase in $X$ leads, on the average, to a drop in $Y$, the quantity $(X - \mu_X)(Y - \mu_Y)$ is expected to be negative most of the time, so that the covariance measure is also expected to be negative.

The covariance measure may be viewed as a bivariate moment of order 2. Note that the covariance of $X$ with itself is the variance of $X$, which is always positive by definition. Some simple algebra establishes that

$$Cov(X,Y) = E[(X - \mu_X)(Y - \mu_Y)] = E(XY) - \mu_X\mu_Y.$$

From Equation (6.16) it then follows that the covariance of two random variables $X$ and $Y$ is always zero whenever they are independent.

The covariance is an important measure in statistical distribution theory and data analysis. However, as a measure of linear relationship it has one drawback. While the sign of the covariance measure indicates the direction of the relationship of $X$ and $Y$, there is no absolute interpretation of the magnitude of the covariance with respect to the strength of the relationship of the two random variables. The same numerical value of the covariance measure may indicate different strengths of relationship in different cases, as this strength is the function of the variabilities of the two distributions also. In fact, both the magnitude and the sign of the covariance may be affected when either or both the variables are linearly transformed.

**Theorem 6.3.** *For two random variables $X$ and $Y$ and real constants $a, b, c$ and $d$, the following relations hold.*

(i) *$Cov(X,a) = 0$. Thus the covariance of a random variable with any constant is always zero.*

(ii) *$Cov(aX+b, cY+d) = acCov(X,Y)$. Thus both the magnitude and the sign of the covariance depend on the scale change of the variables (through the constant ac), although the base change is unimportant.*

A simple remedy which allows the measure to be independent (at least in terms of its magnitude) of the factors $a$ and $c$ is to scale the covariance measure by the standard deviations of the two variables. This generates the correlation coefficient defined by

$$\rho_{XY} = Corr(X,Y) = \frac{Cov(X,Y)}{\sqrt{Var(X)}\sqrt{Var(Y)}} = \frac{Cov(X,Y)}{\sigma_X \sigma_Y}. \qquad (6.20)$$

The measure $\rho_{XY}$, or simply $\rho$, is a very useful measure in that it is constrained to lie in the bounded interval $[-1, 1]$. We get $\rho = 1$ only when the variables have a perfect, positive, linear relationship; similarly $\rho = -1$ corresponds to a perfect, negative, linear relationship. Since the upper bound of the magnitude is now known, the strength of the relationship is also well understood from the magnitude of the correlation coefficient. Thus a value like $\rho = 0.8$ implies strong positive correlation, while, say, $\rho = -0.3$ implies somewhat weak negative correlation.

The correlation measure is independent of a change in the base of the variables. It is also independent of a change in scale in terms of the magnitude. The relevant result, which may be easily proved, is

$$Corr(aX + b, cY + d) = \begin{cases} Corr(X,Y) & \text{if } ac > 0, \\ -Corr(X,Y) & \text{if } ac < 0. \end{cases}$$

We do not consider the case $ac = 0$, as this implies that at least one of $a$ or $c$ is zero, and the correlation of any random variable with a constant must always be identically zero. Thus the transformation of $(X, Y)$ to $(aX + b, cY + d)$ does not change the magnitude of the correlation; it also keeps the sign unchanged whenever $ac$ is positive.

A value $\rho = 0$ indicates that there is no linear relationship between the variables. When initially exposed to the concept of the correlation measure, students are sometimes confused about the meaning of having a zero correlation. Is this the same as claiming that the variables are independent? Actually, having a zero correlation is a weaker property than having independence. As we have already demonstrated, two independent random variables must have covariance equal to zero (and hence correlation equal to zero as well). However, a value of zero for the correlation coefficient does not necessarily indicate independence of the variables. This is because correlation only quantifies the linear relationship, and, even when the correlation is zero, some nonlinear dependence between the random variables may exist, making them dependent.

**Example 6.19.** Consider the joint distribution of $X$ and $Y$, where the distribution is presented in the following probability table. The positive probabilities are highlighted in the body of the table to indicate the circular nature of the relationship.

|  | $x = -1$ | $x = 0$ | $x = 1$ | Total |
|---|---|---|---|---|
| $y = -1$ | 0 | **1/4** | 0 | 1/4 |
| $y = 0$ | **1/4** | 0 | **1/4** | 1/2 |
| $y = 1$ | 0 | **1/4** | 0 | 1/4 |
| Total | 1/4 | 1/2 | 1/4 | 1 |

The marginal distributions given in the row and column margins (the last column and the last row of the above table) show that the marginal expectations satisfy $E(X) = E(Y) = 0$. Also

$$
\begin{aligned}
E(XY) = &\ (-1) \times (-1) \times 0 + (-1) \times 0 \times 1/4 + (-1) \times 1 \times 0 \\
&+0 \times (-1) \times 1/4 + 0 \times 0 \times 0 + 0 \times 1 \times 1/4 \\
&+1 \times (-1) \times 0 + 1 \times 0 \times 1/4 + 1 \times 1 \times 0 = 0.
\end{aligned}
$$

Thus $Cov(X, Y) = E(XY) - E(X)E(Y) = 0$, and hence $Corr(X, Y)$ is zero as well. Yet the variables are clearly dependent, as the true support is not a rectangle and the condition given in Equation (6.8) is clearly violated. We have therefore demonstrated that, for this pair of dependent variables, the correlation coefficient is zero. As we have already indicated, the relationship between $X$ and $Y$ is of a circular nature, but no linear relationship exists between them. ‖

*Remark* 6.5. The terms mean, variance, covariance and correlation have all been introduced in Chapter 3 as descriptive measures for a set of numerical data. The mean, variance, correlation and covariance that are considered here (and will be considered later in Chapter 7) are the theoretical analogs of the quantities introduced in Chapter 3. Thus, while $\mu = E(X)$ is the theoretical mean (average) of the random variable $X$, the quantity $\frac{1}{n} \sum_{i=1}^{n} X_i = \bar{X}$ is the sample mean (average) where the sample of size $n$ is drawn from the distribution of $X$. Similarly, the quantity

$$
s^2 = \frac{1}{n-1} \sum_{i=1}^{n} (X_i - \bar{X})^2
$$

is the sample variance as obtained from the data and corresponds to the theoretical variance $\sigma^2 = E(X - \mu)^2$. The sample and the theoretical correlation coefficients are distinguished by the symbols $r$ and $\rho$.

Although the sample variance and sample covariance are averages of $n$ quantities, in practice we use the divisor $n-1$ rather than $n$ in their definitions. This is because the sample quantities are closer to the theoretical quantities with a divisor of $(n-1)$ compared to a divisor of $n$ in the sense of having a smaller *bias*, a concept which is discussed in Section 7.8.3. ⊕

### 6.5.3   Moment Generating Functions

The moment generating function $M_X(\cdot)$ of a random variable $X$ is defined, as a function of a real argument $t$, by the relation

$$M_X(t) = E(e^{tX}),$$

provided it exists in an interval of $t$ around zero (note that it must be 1 at $t = 0$). The successive raw moments may be recovered from the moment generating function (MGF) through the process of differentiation, a property which lends the function its name. This process is described in the appendix.

**Example 6.20.** (Binomial) Suppose that $X \sim binomial(n,p)$. In the appendix it is shown that the moment generating function of $X$ is given by

$$M_X(t) = (pe^t + q)^n. \tag{6.21}$$

How one can use the above moment generating function to recover the mean and variance of the binomial random variable is also shown in the appendix. For the $binomial(n,p)$ variable, these measures turn out to be

$$E(X) = np, \quad Var(X) = npq.$$

See Example 6.23 in the appendix.                                              ‖

   Apart from being useful in finding the moments of a distribution, moment generating functions have another major utility. They have a one to one relation with the distribution. If a particular distribution leads to a specific MGF, then it is the only distribution that can generate that MGF. Thus a distribution is completely characterized by its MGF. So instead of defining a $binomial(n,p)$ distribution by its PMF, given in Equation (6.3), we can also define it by its MGF, given in Equation (6.21). As a result the technique based on MGFs is a very powerful tool for identifying distributions of random variables, particularly because the MGF may be easier to derive than the PMF in many cases.

**Example 6.21.** Suppose $X \sim binomial(n_1,p)$ and $Y \sim binomial(n_2,p)$, where $X$ and $Y$ are independent. What can we say about the distribution of $X + Y$? We will provide an intuitive answer to this question in this example. Suppose a coin lands heads with a probability $p$ when tossed. Person $A$ tosses the coin $n_1$ times. We note $X$, the number of heads observed in these tosses. Subsequently, person $B$ tosses the same coin $n_2$ times. Let $Y$ be the number of heads among these $n_2$ tosses. We know that $X \sim binomial(n_1,p)$ and $Y \sim binomial(n_2,p)$. In this case $X + Y$ is simply the number of heads observed when the coin is independently tossed $n_1 + n_2$ times. Intuitively, it is clear that, the distribution will be a binomial with parameters $(n_1 + n_2)$ and $p$. However, this can be mathematically established as well, as we will shortly do.                                                                      ‖

**Theorem 6.4.** *Let $M_X(t)$ and $M_Y(t)$ be the moment generating functions of $X$ and $Y$. Also suppose that $X$ and $Y$ are independent. Then*

$$M_{X+Y}(t) = M_X(t)M_Y(t),$$

*i.e., the moment generating function of the sum $X + Y$ is the product of the individual moment generating functions.*

This theorem is a direct corollary of the result given in Equation (6.18). Armed with the result of the above theorem, we continue with Example 6.21.

**Example 6.21.** (Continued) Let $X$ and $Y$ be independent binomial random variables with parameters $(n_1, p)$ and $(n_2, p)$. Consider the MGF of the sum $X + Y$. From the previous theorem it follows that

$$
\begin{aligned}
M_{X+Y}(t) &= M_X(t)M_Y(t) \\
&= (q + pe^t)^{n_1}(q + pe^t)^{n_2} \\
&= (q + pe^t)^{n_1+n_2}.
\end{aligned}
$$

But the right hand side of the above equation represents the moment generating function of a binomial random variable with parameters $(n_1 + n_2, p)$. By the one to one relationship between the MGF and the distribution it then follows that $X + Y \sim binomial(n_1 + n_2, p)$. ‖

More generally, the result proved in Theorem 6.4 applies to $n$ independently distributed random variables $X_1, X_2, \ldots, X_n$. An obvious extension of the proof of this theorem yields

$$M_{X_1+X_2+\cdots X_n}(t) = M_{X_1}(t)M_{X_2}(t)\cdots M_{X_n}(t). \tag{6.22}$$

As an immediate application, we see that, if $X_i \sim binomial(n_i, p), i = 1, 2 \ldots, k$, where all the $k$ random variables are independent, then

$$\sum_{i=1}^{k} X_i \sim binomial\left(\sum_{i=1}^{k} n_i, p\right).$$

We end this section with some results about the Poisson distribution (without proof).

**Theorem 6.5.** *Suppose that the random variable $X$ has the Poisson distribution with parameter $\lambda$. Then the following results are available.*

*(i) $E(X) = \lambda$ and $Var(X) = \lambda$.*

*(ii) $M_X(t) = e^{\lambda(e^t-1)}$.*

Just as in the binomial case, distributions of sums of independent Poisson random variables can also be obtained using the moment generating function technique. See Section 7.14.6 for some results involving sums of independent random variables for the binomial, Poisson and normal cases.

## 6.6    Appendix

### 6.6.1    Binomial PMF

Here we will develop the probability mass function of a random variable $X$ having a binomial distribution, described earlier in Section 6.2.1. For illustration we consider the following question. Suppose a fair coin is tossed five times. What is the probability that exactly three heads are observed among the five tosses? If we had specified the positions where we want the heads to occur, this would be a very easy task. For example, if we wanted the probability of the event where the outcomes are in the sequence $HHHTT$, it would be

$$\left(\frac{1}{2}\right)^3 \left(\frac{1}{2}\right)^2 = \left(\frac{1}{2}\right)^5.$$

However, this would also be the probability of any other sequence which has three heads and two tails, such as $HTHTH$ or $TTHHH$. Thus it is important that we systematically count how many such sequences there are which lead to exactly three heads in five tosses. This number is the same as the number of ways for choosing three objects (here representing the three tosses resulting in heads) from a group of five objects (representing the totality of five tosses). From what we have learned in our earlier discussion of probability, this is simply $\binom{5}{3}$, so that the probability of having exactly three heads in five tosses is

$$\binom{5}{3}\left(\frac{1}{2}\right)^5.$$

If the problem were more generally phrased as "What is the probability that exactly $x$ heads are observed among the five tosses?" the answer will be

$$\binom{5}{x}\left(\frac{1}{2}\right)^5$$

for any integer $x$ between 0 and 5. Note that, if exactly $x$ heads are observed, that means exactly $5 - x$ tails are observed.

This coin tossing experiment satisfies all the conditions of the binomial set up. We have $n$ (here, five) independent Bernoulli trials (coin tosses), and $X$ represents the number of successes (heads) observed among all the trials. In the above example, $X \sim binomial(5, \frac{1}{2})$. However, if the coin being tossed had a probability of head (probability of success) equal to $p$ (possibly different from $1/2$), then the probability of having three heads in five tosses would be

$$\binom{5}{3}p^3 q^2,$$

where $q = 1 - p$. More generally, if the experiment consists of a series of $n$

Bernoulli trials, and $X$ represents the number of successes observed, the probability $Pr(X = x)$ is given by

$$\binom{n}{x} p^x q^{n-x}.$$

Thus the PMF of a binomial random variable having parameters $n$ and $p$ is given to be

$$f(x) = \binom{n}{x} p^x q^{n-x}, \quad x = 0, 1, \ldots, n. \tag{6.23}$$

It is easy to see that each term in Equation (6.23) is nonnegative; this PMF also satisfies the second necessary condition, since

$$\sum_{x=0}^{n} f(x) = \sum_{x=0}^{n} \binom{n}{x} p^x q^{n-x} = (p+q)^n = 1.$$

The last result makes use of the binomial expansion of the series $(p+q)^n$. For a finite integer $n$, when the quantity $(p+q)^n$ is expanded in a series, we get

$$(p+q)^n = \sum_{x=0}^{n} \binom{n}{x} p^x q^{n-x}.$$

## 6.6.2 Geometric Distribution

Consider again a sequence of independent Bernoulli trials. Each trial can result in only one of two possible outcomes, denoted success and failure. However, in this case the number of trials are not fixed (unlike the binomial case). We continue performing trials until the first success is observed. Let $X$ be the number of trials needed to get the first success. What is the probability that $X = x$?

Let $p$ be the probability of success at any trial (and the probability of a failure, therefore, is $q = 1-p$). If $X = x$, then trial number $x$ must have resulted in a success, and all the previous $x - 1$ trials must have resulted in failures. Thus $Pr(X = x) = q^{x-1}p$. The possible values of $x$ are 1, 2, .... This random variable $X$, the number of trials required for the first success, is said to have a *geometric distribution* with parameter $p$. It has the probability mass function

$$f(x) = q^{x-1}p, \tag{6.24}$$

with support $\mathcal{X} = \{1, 2, \ldots\}$. If a random variable $X$ has the geometric distribution with parameter $p$, we represent it as $X \sim geometric(p)$.

**Example 6.22.** Suppose that the probability of having a child of either sex at any random birth is 0.5. A family plans to continue having children until the first son (daughter) is born. If $X$ be the number of births necessary for this, what is the distribution of $X$? What is $Pr(X = 3)$?

The sequence of childbirths represent independent Bernoulli trials. The number of trials needed to get the first son (daughter) has a geometric distribution with parameter 0.5. Thus $Pr(X = 3) = (0.5)^2 \times 0.5 = 0.125$. ‖

*Remark* 6.6. The binomial and the geometric distributions are generated through the same set up, a series of Bernoulli trials. However, for the binomial distribution, the number of trials is fixed (at $n$) and one counts the number of successes, which is the random variable. For the geometric case, on the other hand, it is the number of successes that is fixed (at one) and the number of trials required is now the random quantity.

*Remark* 6.7. When we developed the geometric random variable $X$, we defined it as the number of trials required to get the first success. Consider, instead, the random variable $Y$, which is the number of failures preceding the first success. Clearly, $X = Y + 1$. Some authors would also refer to this random variable $Y$ as having a geometric distribution with parameter $p$; thus, when describing a geometric random variable, it helps to be clear about which definition is used. The probability mass function of the random variable $Y$ is

$$f(y) = q^y p, \ y = 0, 1, \ldots.$$

Note that, unlike the support of $X$, the support of $Y$ starts from zero; this is as it should be, given the relation $X = Y + 1$.

### 6.6.3 Negative Binomial Distribution

We now complicate the set up of the geometric random variable a little further. Consider a sequence of independent Bernoulli trials (with success probability $p$), where we continue the experiment until we observe $r$ ($\geq 1$) successes.

Let $X$ be the number of trials needed for observing $r$ successes. This random variable is said to have a *negative binomial distribution* with parameters $r$ and $p$; the geometric random variable is a special case of this for $r = 1$. To find $Pr(X = x)$ for this random variable, note that $X = x$ means that the $r$-th success has been observed at trial number $x$. Therefore $(r - 1)$ successes and hence $(x - r)$ failures must have been observed in the first $(x - 1)$ trials. Thus

$$
\begin{aligned}
Pr(X = x) &= \binom{x - 1}{r - 1} p^{r-1} q^{x-r} p \\
&= \binom{x - 1}{r - 1} p^r q^{x-r}.
\end{aligned}
$$

Thus a random variable $X$ having the negative binomial distribution with parameters $r$ and $p$ has the PMF

$$f(x) = \binom{x - 1}{r - 1} p^r q^{x-r}. \tag{6.25}$$

The support of this random variable is $x = r, r + 1, \ldots$. This reduces to the PMF in Equation (6.24) when $r = 1$.

### 6.6.4 Poisson Distribution

We have already studied the Poisson distribution in some detail; here we will check that the Poisson probabilities add to one. Note that

$$
\begin{aligned}
\sum_{x=0}^{\infty} f(x) &= e^{-\lambda} + \lambda e^{-\lambda} + \frac{\lambda^2 e^{-\lambda}}{2!} + \cdots \\
&= e^{-\lambda}\left(1 + \lambda + \frac{\lambda^2}{2!} + \cdots\right) \\
&= e^{-\lambda} e^{\lambda} \\
&= 1.
\end{aligned}
$$

Here we have used the expansion $e^{\lambda} = 1 + \lambda + \frac{\lambda^2}{2!} + \cdots$ of the exponential series.

### 6.6.5 Distribution Functions

We elaborate on the right continuity property of the CDF through a continuation of Example 6.12.

**Example 6.12.** (Continued) We have already stated that the CDF of a random variable is right continuous. Here we clearly explain this concept. This right continuity implies

$$
\lim_{h \downarrow 0} F(x + h) = F(x).
$$

For example, in Figure 6.2 we have $\lim_{h \downarrow 0} F(1+h) = F(1) = 0.75$. This relation will not hold under the limit $h \uparrow 0$ at the jump points; for example, $\lim_{h \uparrow 0} F(1+h) = 0.5 \neq F(1) = 0.75$. In particular, the value of the function $F(\cdot)$ at $x = 0$ is 0.5, at $x = 1$ is 0.75, and at $x = 2$ is 1. The filled dots (and the empty dots) in the figure indicate what the values of the function $F(x)$ are (and are not) at the jump points. For example, $F(0) = 0.5$ (and not 0), and so on. ‖

### 6.6.6 Proof of Theorem 6.1

*Proof.* By definition,

$$
\begin{aligned}
E(aX + b) &= \sum_{x}(ax + b)f(x) \\
&= a\sum_{x} x f(x) + b\sum_{x} f(x) \\
&= aE(X) + b.
\end{aligned}
$$

□

### 6.6.7 Moment Generating Functions

The moment generating function (MGF), defined in Section 6.5.3 is so named because the successive raw moments may be recovered from this function through the process of differentiation. For example,

$$\frac{d}{dt} M_X(t) = \frac{d}{dt} \sum_x e^{tx} f(x) = \sum_x x e^{tx} f(x),$$

which is equal to $E(X) = \mu$ when evaluated at $t = 0$. Similarly, the second derivative of $M_X(t)$ gives the second moment $E(X^2)$ when evaluated at $t = 0$, the third moment $E(X^3)$ is recovered from the third derivative, etc.

**Example 6.23.** (Binómial) Suppose that $X \sim binomial(n,p)$. We want to find the moment generating function, and hence the moments of a binomially distributed random variable. Direct evaluation gives us

$$
\begin{aligned}
M_X(t) &= \sum_x e^{tx} \binom{n}{x} p^x q^{n-x} \\
&= \sum_x \binom{n}{x} (pe^t)^x q^{n-x} \\
&= (pe^t + q)^n.
\end{aligned}
\tag{6.26}
$$

Taking a derivative of the above form, and evaluating at $t = 0$, we get $E(X) = np$. Another derivative reveals the second raw moment to be $E(X^2) = np + n^2 p^2 - np^2$, so that the variance of the binomially distributed random variable becomes

$$Var(X) = E(X^2) - \{E(X)\}^2 = np - np^2 = np(1 - p) = npq.$$

$\parallel$

### 6.6.8   Table of Parameters

In Table 6.1, the means, variances and the moment generating functions of some of the common discrete distributions are provided. For the first three

**TABLE 6.1**
Table of means, variances and moment generating functions

| Distribution | Mean | Variance | Moment Generating Function |
|---|---|---|---|
| $Bernoulli(p)$ | $p$ | $pq$ | $pe^t + q$ |
| $binomial(n,p)$ | $np$ | $npq$ | $(pe^t + q)^n$ |
| $Poisson(\lambda)$ | $\lambda$ | $\lambda$ | $e^{\lambda(e^t - 1)}$ |
| $geometric(p)$ | $1/p$ | $\dfrac{q}{p^2}$ | $\dfrac{pe^t}{1 - qe^t},\ t < -\ln q$ |
| $hypergeometric$ | $\dfrac{nN_1}{N}$ | $\dfrac{nN_1 N_2}{N^2}\dfrac{N - n}{N - 1}$ | — |

distributions in the table, the moment generating function is defined for all $t$. For the hypergeometric distribution, the parameter set is $(N_1, N_2, n)$, with $N = N_1 + N_2$. The moment generating function of the hypergeometric distribution has a complicated expression and is not useful to us at this stage.

In the case of the geometric distribution, the version that is used has the support $\{1, 2, 3, \ldots\}$.

## 6.7 Suggested Further Reading

Several of the texts referenced in Chapter 5 are useful in connection with the present section as well. In particular, Ross (2010), DeGroot and Schervish (2012) and Walpole et al. (2010) are good resources. Akritas (2014) provides a good treatment supported by R illustrations. Other basic texts of our preference include include Pal and Sarkar (2009), Anderson et al. (2014), Levin and Rubin (2011), Black (2012) and Mann (2013). An extensive list of univariate and multivariate discrete distributions and their properties are provided in Johnson et al. (2005) and Johnson et al. (1997).

# 7

## Continuous Random Variables

In the previous chapter we observed that the number of support points for a discrete random variable is either finite or at most countably infinite. Masses assigned to these points add up to one. In the case of a *continuous random variable*, however, the support contains at least one entire interval on the real line, so that the number of points in the support of such a random variable is necessarily uncountable. Unfortunately, the mathematics does not work when one starts assigning positive probabilities to each individual point in the support in this case, since the total mass (probability) can then no longer be bounded by one. Thus, under continuous models, single points cannot carry any mass. So if $X$ is a continuously distributed random variable, $Pr[X = k]$ is *always* equal to zero for any fixed value $k$ in the support of the random variable. By the same rule, the probability statements for open and the corresponding closed intervals are always equal for a continuous random variable; for example, $Pr[X < k] = Pr[X \leq k]$.

## 7.1 PDF and CDF

In dealing with continuous random variables, our real interest will be in probabilities of intervals on the real line of the type

$$Pr[a \leq X \leq b].$$

This is unlike the discrete case where our interest is in the probabilities of specific values in the support. The characterizing function $f(x)$ will now be a smooth function such that the total area enclosed between the curve $f(x)$ and the $X$-axis will be equal to 1. The probability $Pr[a \leq X \leq b]$ will have the interpretation of being the area bounded by the curve $f(x)$ and the $X$-axis between the vertical lines drawn at $a$ and $b$ (see Figure 7.1). Readers equipped with the background of calculus will know that this is also the value of the definite integral obtained by integrating the function $f(x)$ between $a$ and $b$. When $a$ and $b$ coincide, it simply results in the cross section through the point $a$, and that does not enclose a two-dimensional region, which intuitively explains why $Pr[X = a] = 0$. For continuous random variables, the concept of probability must therefore be understood in terms of two-dimensional areas, and not in terms of masses of single points.

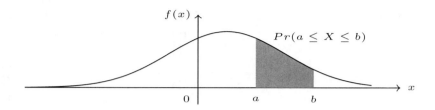

**FIGURE 7.1**
A probability density function.

The characterizing function $f(x)$ in the case of a continuous random variable is called its probability density function (PDF), which must satisfy the following conditions.

(i) $f(x) \geq 0$ for all $x \in \mathbb{R}$.

(ii) $\int_{-\infty}^{\infty} f(x)dx = 1$. If the support of the corresponding random variable is known to be $[\alpha, \beta]$, then $\int_{\alpha}^{\beta} f(x)dx = 1$. In non-technical language this means that the area bounded by the curve $f(x)$ and the $X$-axis over the region $[\alpha, \beta]$ (or whatever may be the endpoints of the region of support of $X$) must be unity.

There is one important distinction between the PMF and the PDF. In the discrete case, the PMF $f(x)$ represents an actual probability and must be bounded above by 1. In the case of continuous random variables, $f(x)$ does not represent a probability. Although the total area under the curve is bounded by 1, there is nothing to guarantee that the function $f(x)$ itself must be everywhere bounded by 1 in this case. We will frequently come across perfectly legitimate PDFs where $f(x)$ exceeds 1 for a part of its support (or even the entire support).

As in the case of the discrete random variable, we can construct the cumulative distribution function $F(x)$ of a continuous random variable $X$ as

$$F(x) = Pr[X \leq x] = \int_{-\infty}^{x} f(t)dt \qquad (7.1)$$

where $f(\cdot)$ is the PDF of the random variable $X$. Unlike the discrete case, the cumulative distribution function (CDF) of a continuous random variable $X$ is a smooth, increasing function. As a representative CDF in the continuous case, the CDF of the $N(0, 1)$ distribution, to be described in the subsequent sections, is displayed in Figure 7.2. It may be observed that this CDF is continuously increasing, and not just increasing in discrete jumps. At any

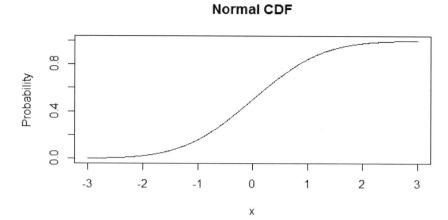

**FIGURE 7.2**
The cumulative distribution function of a continuous random variable.

point $x$, the value of the CDF $F(x)$ simply represents the area under the PDF curve to the left of the point $x$ on the $X$-axis.

As in the case of discrete random variables, the CDF completely characterizes the distribution. One can construct the CDF from the PDF, and also recover the latter from the former. We will not particularly emphasize this, as this will require techniques from calculus. However, enough additional material and examples are given in the appendix for the interested reader.

## 7.2 Special Continuous Distributions

A list of some standard and common continuous random variables is given below.

1. Uniform: A random variable $X$ is said to have a uniform distribution over the interval $(a, b)$ (or $[a, b]$, as the inclusion or exclusion of the terminal points makes no difference), represented as $X \sim uniform(a, b)$, if the PDF of the random variable has the form

$$f(x) = \begin{cases} \dfrac{1}{b-a} & a \leq x \leq b, \\ 0 & \text{otherwise.} \end{cases} \tag{7.2}$$

The boundaries $a$ and $b$ are necessarily finite. Uniform random variables have the same PDF over the entire support of the random variable. Of special interest is the $uniform(0,1)$ random variable, which is the basic random variable used in the generation of random numbers.

In the following we will use the style $f(x) = 1/(b - a), a \le x \le b$ for describing the PDF of the above random variable. It will be understood that the PDF is zero everywhere else.

2. Exponential: The random variable $X$ is said to have an exponential distribution with a (positive) parameter $\theta$ if its PDF is given by

$$f(x) = \frac{1}{\theta} e^{-x/\theta}, x \ge 0. \tag{7.3}$$

The exponential distribution is used in many contexts, including problems of reliability and survival analysis. It models the interarrival times of certain stochastic processes. One of its very important properties is the *memoryless property*, which states

$$Pr[X \ge x] = Pr[X \ge x + y | X \ge y].$$

Thus the conditional probability that $X$ is greater than an additional $x$ units, given that it is already greater than $y$ units, is equal to the unconditional probability of being greater than $x$ units. The exponential distribution has a mode at 0 (zero). We will use the symbol $expo(\theta)$ to represent the exponential distribution with parameter $\theta$.

3. Normal: A random variable having a normal distribution with parameters $\mu$ and $\sigma^2$ has the density function

$$f(x) = \frac{1}{\sigma\sqrt{2\pi}} e^{-\frac{(x-\mu)^2}{2\sigma^2}}, \quad -\infty < x < \infty. \tag{7.4}$$

This random variable is of vital importance to us, and we will discuss this in much greater detail later on in this chapter and the rest of the book. If $X$ has the density given in Equation (7.4), we write $X \sim N(\mu, \sigma^2)$. Note that the form of the distribution is completely known when the parameters $\mu$ and $\sigma^2$ are known.

The $\chi^2$, $t$ and $F$ distributions will also be of interest to us, and we will develop them later as sampling distributions arising from the normal distribution.

## 7.3   Expectation

We have already described the concept of the expectation of a random variable in connection with discrete random variables in the previous chapter.

The concept naturally extends, with similar interpretations, to its continuous analogs. Let $X$ be a random variable with PDF $f(x)$. Then the expectation of $X$ is defined by

$$E(X) = \int x f(x) dx,$$

where the integral is over the support of $X$. This is the obvious extension of the expectation concept to continuous variables where summations are now replaced by integrals. The variance of the random variable $X$ is

$$Var(X) = E(X - E(X))^2 = \int (x - E(X))^2 f(x) dx.$$

One can similarly define the moments as in the discrete case. We will usually denote the mean parameter by the symbol $\mu$ and the variance by the symbol $\sigma^2$.

We now state some useful results involving expectations. Expectations of linear combinations of random variables transform exactly in the same way as in the discrete case. The proof is basically similar to the discrete case, with integrals replacing summations.

**Theorem 7.1.** *(Means and variances of linearly transformed variables) Let $\mu$ and $\sigma^2$ represent the mean and the variance of the random variable $X$. For specified real values $a$ and $b$, let $Y = aX + b$ be a linear transformation of $X$. Then the mean and the variance of $Y$ are given by*

$$E(Y) = aE(X) + b = a\mu + b \text{ and } Var(Y) = a^2 Var(X) = a^2 \sigma^2.$$

*Note that the variance of $Y$ is not affected by the location shift $a$, but does depend on the scale shift $b$.*

Since the scale shift is squared, the result is always a positive number. In particular, if $Y = -X$, then $Y$ has mean $E(Y) = -\mu$, but its variance is still $\sigma^2$!

Once again we see that the expectation operator runs through the linear part of the transformation. This result, in fact, extends to the case of more than one random variable. For example, if we have $k$ random variables $X_1, X_2, \ldots, X_k$, then

$$E(a_1 X_1 + a_2 X_2 + \ldots a_k X_k + b) = a_1 E(X_1) + a_2 E(X_2) + \cdots + a_k E(X_k) + b. \quad (7.5)$$

## 7.3.1 Transformations of Random Variables

We have defined a random variable as a function from the sample space of a statistical experiment to the real line. Sometimes it may be of interest to describe the properties of functions of random variables when the random variables themselves have already been well characterized. Thus, given a random variable $X$, we may be interested in the properties of the transformed quantity $Y = g(X)$, which is itself a random variable, where $g(\cdot)$ is a suitable function. When the random variable is discrete, finding the distribution of the transformed random variable is usually a simple matter. Some technical results regarding the distributions of functions of continuous random variables are presented in the appendix. Here we present the result which is most relevant in our context.

**Theorem 7.2.** *Suppose that $X$ follows a normal distribution with parameters $\mu$ and $\sigma^2$, and consider the transformation $Y = aX + b$. Then*

$$Y \sim N(a\mu + b, a^2\sigma^2).$$

**Corollary 7.3.** *Suppose $X \sim N(\mu, \sigma^2)$, with $\sigma > 0$. Consider the transformation $Z = (X - \mu)/\sigma$. An application of Theorem 7.2 shows that $Z$ has a normal distribution with parameters 0 (mean) and 1 (variance).*

### 7.3.2 Moment Generating Function

The moment generating function (MGF) of a continuous random variable $X$ is defined as

$$M_X(t) = E(e^{tX}) = \int e^{tx} f(x)dx$$

where $f(x)$ is the PDF of $X$, provided it exists for all $t$ in an interval around $t = 0$. Like the discrete case, the MGF is useful for finding the moments of a given continuously distributed random variable but also has several other uses. The MGF, provided it exists, completely characterizes the distribution. For example, if $X \sim N(\mu, \sigma^2)$, then it has the moment generating function

$$M_X(t) = e^{\mu t + \frac{1}{2}\sigma^2 t^2} \tag{7.6}$$

(which exists for all $t$). On the other hand, if a random variable has a moment generating function given by Equation (7.6), it must have a $N(\mu, \sigma^2)$ distribution.

## 7.4 Normal Distribution

The normal distribution is by far the most important distribution in probability and statistics. It is also called the Gaussian distribution; the name honors the nineteenth century German mathematician Carl Friedrich Gauss, who used it, among other things, for the analysis of errors of measurement made in astronomical observations. The normal distribution is found to be suitable for many natural processes.

We will denote by $Z$ the random variable having the probability density function

$$\phi(z) = \frac{1}{\sqrt{2\pi}} e^{-z^2/2}, \quad -\infty < z < \infty. \tag{7.7}$$

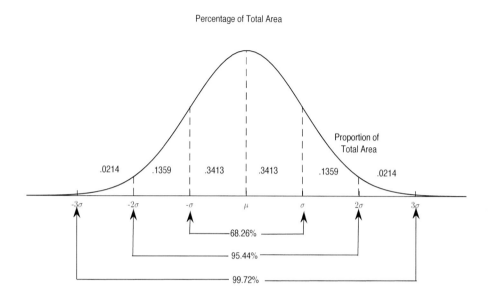

**FIGURE 7.3**
Normal probabilities.

Comparing with the density in Equation (7.4) it is immediately seen that this is the PDF of a normal density with $\mu = 0$ and $\sigma^2 = 1$. This represents a symmetric unimodal density with a peak at zero, the point of symmetry. It may be proved that the function defined by Equation (7.7) is actually a valid probability density function; we will omit the proof, which is of no particular importance to us. The support of the $Z$ distribution (or, indeed, any normal distribution) is the entire real line, although the probability that $Z$ lies outside the interval $[-3, 3]$ is, approximately, only 0.0028. The random variable defined by Equation (7.7) is said to have a standard normal distribution and it is so special to us that we will reserve the symbol $Z$ specifically for this random variable. Similarly, while $f$ will represent the PDF of a generic random variable, we will reserve the symbol $\phi$ specifically for the PDF of the random variable $Z$.

For any normal distribution, the probabilities bounded within one, two and three standard deviations from the mean are approximately 68.26%, 95.44% and 99.72%. Mathematically, for any normal distribution with mean $\mu$ and

**Normal Density Curve**

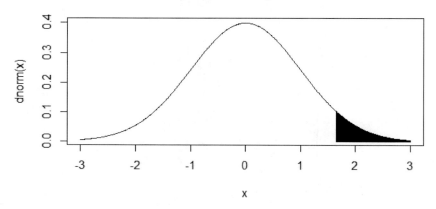

**FIGURE 7.4**
The normal density curve.

standard deviation $\sigma$, we have

$$Pr(\mu - \sigma \leq X \leq \mu + \sigma) = 0.6826, \quad Pr(\mu - 2\sigma \leq X \leq \mu + 2\sigma) = 0.9544$$

and

$$Pr(\mu - 3\sigma \leq X \leq \mu + 3\sigma) = 0.9972.$$

This is the theoretical version of the empirical rule mentioned in Section 3.4.2 and is clearly illustrated in Figure 7.3.

The form of the density of a normal distribution, which has a greater probability concentration around its central part, is often described as a bell-shaped curve. Figure 7.4 displays the unimodal bell-shaped PDF curve $\phi$ corresponding to the standard normal random variable.

Note that a segment of the lower right hand part of the area under the normal curve in Figure 7.4 is represented in a dark shade. This has been done to introduce the concept of a quantile for a random variable, which we will do through the example of a normal distribution. The $q$-th quantile of the normal distribution is the value on the $X$-axis such that $100q\%$ of the probability of the corresponding normal distribution is below (to the left of) this value, and $100(1 - q)\%$ is above it. They are the same as percentiles, but expressed in fractions rather than percentages. What is shown in Figure 7.4 is the quantile corresponding to $q = 0.95$. The shaded area represents 5% of the total area under the curve. The $q$-th quantile for $q = 0.95$ in the case of the standard normal is 1.644854 (often approximated by 1.645), as verified by the R code below. Thus the cut-off point for the shaded area in the figure is the point 1.644854 on the $X$-axis.

```
> qnorm(0.95)
[1] 1.644854
```

The argument of the `qnorm()` command takes the parameters $\mu = 0$ and $\sigma = 1$ by default, and adding these values to the arguments would have made no difference to the output of the previous command. However, a different set of parameters produces a different quantile.

```
> qnorm(0.95,0,1)
[1] 1.644854
> qnorm(0.95,2,4)
[1] 8.579415
```

The quantiles for other distributions can also be similarly determined by identifying the points on the $X$-axis which bound the required fraction of the total area under the PDF curve to their left.

The cumulative distribution function of the standard normal distribution is denoted exclusively by $\Phi$, just as the PDF is denoted by $\phi$. Because of the symmetry of the PDF, we have $\phi(z) = \phi(-z)$, and $Pr(Z \geq z) = 1 - \Phi(z)$ is equal to $Pr(Z \leq -z) = \Phi(-z)$.

The general form of the PDF of a random variable $X$ having the normal distribution (with parameters $\mu$ and $\sigma^2$) is given in Equation (7.4). Some simple algebraic calculations show that the parameters $\mu$ and $\sigma^2$ are actually the mean and the variance of the distribution. This also indicates that the variable $Z$ has a normal distribution with mean 0 and variance 1, i.e., $Z \sim N(0,1)$.

The normal distribution is often found suitable for modeling and analyzing data that arise in real life problems in many different fields, which makes it the most widely used distribution in practice. One of the most important reasons for the popularity of the normal distribution is the large domain of applicability opened up by the *Central Limit Theorem*, which we will describe and discuss in the subsequent sections of this chapter. However, the normal distribution possesses many other useful properties. One such property, already stated in Theorem 7.2, is that it is closed under linear transformations.

It also follows from Corollary 7.2 that, if $X \sim N(\mu, \sigma^2)$, then $Z = (X - \mu)/\sigma$ will follow a normal distribution with parameters 0 and 1. Similarly, if $Z$ follows the standard normal distribution, then $X = \mu + \sigma Z$ has the $N(\mu, \sigma^2)$ distribution. In the traditional calculation of normal probabilities using normal distribution tables, this fact is heavily used, as it allows the computation of the probabilities relating to any normally distributed random variable with mean $\mu$ and variance $\sigma^2$ using the $Z$ distribution tables alone. Based on this transformation, one gets

$$Pr(X \leq a) = Pr\left(\frac{X - \mu}{\sigma} \leq \frac{a - \mu}{\sigma}\right) = Pr\left(Z \leq \frac{a - \mu}{\sigma}\right) = \Phi\left(\frac{a - \mu}{\sigma}\right).$$

The last quantity can be easily read from the tables of the $Z$ distribution. We will, of course, directly obtain the quantity $Pr(X \le a)$ from R, but will provide the reduction to the standard form under the $Z$ transformation wherever possible.

**Example 7.1.** The resistance $X$ of certain resistors in an electrical circuit has a normal distribution with a mean $\mu = 10$ ohms and a standard deviation $\sigma = 0.3$ ohms.

(a) Find the probability that the resistance of a randomly selected resistor is less than 9.4.

Although the question asks for the probability of "less than 9.4," we will usually compute the "$\le$" probability, which makes no difference in the case of the normal distribution. Note that

$$
\begin{aligned}
Pr(X \le 9.4) &= Pr\left(\frac{X - 10}{0.3} \le \frac{9.4 - 10}{0.3}\right) \\
&= Pr(Z \le -2) \\
&= 1 - Pr(Z \le 2) \\
&= 0.0228.
\end{aligned}
$$

From R, we get the verification of the above result, both as a direct calculation of the normal probability from the $N(10, 0.3^2)$ distribution, and also as the transformed probability from the $Z$ distribution.

```
> pnorm(9.4,10,0.3)
[1] 0.02275013
> pnorm(-2)
[1] 0.02275013
```

(b) Suppose 12 different resistors are taken from the class of resistors having the above normal distribution. What is the distribution of the number of resistors having a resistance less than 9.4 ohms (among the above 12)?

We note that all the conditions of a binomial distribution are satisfied here. We have a series of 12 independent trials, each having probability of success equal to 0.0228 (probability of having a resistance less than 9.4) and the random variable is the number of successes. Thus the number of resistors $X$ having resistance less than 9.4 has a binomial distribution with parameters $n = 12$ and $p = 0.0228$.

(c) What is the probability that exactly two of those resistors will have resistances below 9.4 ohms?

Given that the random variable has the binomial distribution as above, the required probability is

$$
\binom{12}{2}(0.0228)^2(0.9772)^{10}.
$$

A routine computation shows that the above value is 0.02724. ‖

Both the normal and the binomial distributions have been used to find the required probabilities in the different parts of the above problem.

## 7.5  Continuous Bivariate Distributions

If the random variables $X$ and $Y$ have a joint continuous distribution, there will be a function $f(x, y)$, called the joint PDF (probability distribution function) of $X$ and $Y$ such that the probability that $(X, Y)$ lies in any subset of the real plane is obtained by integrating the joint probability density function over that subset.

The joint PDF $f(x, y)$ will satisfy the following conditions:

(i)  $f(x, y) \geq 0$, for all $x, y$.

(ii)  $\int_x \int_y f(x, y) dy dx = 1$. This means that the total three-dimensional volume enclosed between the two-dimensional $(X, Y)$ plane and the two-dimensional $f(x, y)$ curve is unity.

The actual range of integration in part (ii) above is the joint (bivariate) support of $(X, Y)$. In this case the probabilities may be interpreted as the volumes enclosed between the $(X, Y)$ plane and the two-dimensional $f(x, y)$ curve over the regions of interest. Although the PDF integrates to 1 over the entire support, we do not, as in the univariate case, require that the joint PDF be restricted to less than 1 everywhere on the bivariate support. We will, in fact, frequently come across distributions where the joint density will exceed 1 over parts of its support. Example 7.2, later, gives one such illustration where the PDF is greater than 1 over the entire support.

### 7.5.1  Marginal and Conditional Distributions

Given the joint PDF of $X$ and $Y$, the marginal PDF of either random variable can be obtained by integrating the joint PDF over the other variable. This is an exact analog of the discrete case with the summation replaced by integration. Thus, if $f(x, y)$ is the joint PDF, the marginal PDFs $f_X(x)$ and $f_Y(y)$ of $X$ and $Y$ are obtained as

$$f_X(x) = \int_y f(x, y) dy \quad \text{and} \quad f_Y(y) = \int_x f(x, y) dx.$$

Similarly, the conditional distributions may be obtained by dividing the joint PDFs by the marginal PDFs. Under the exact same notation as in the discrete case, the conditional PDFs may be obtained as

$$f_{X|Y=y}(x|y) = \frac{f(x,y)}{f_Y(y)} \quad \text{and} \quad f_{Y|X=x}(y|x) = \frac{f(x,y)}{f_X(x)}.$$

**Example 7.2.** Let the joint probability density function of $X$ and $Y$ be given by

$$f(x,y) = \begin{cases} 2 & \text{if } 0 \leq y \leq x \leq 1, \\ 0 & \text{otherwise.} \end{cases}$$

We want to verify whether this is a valid joint PDF, and find $Pr(X < 2Y)$.

The PDF is clearly nonnegative everywhere. It is, in fact, uniform over the lower triangular half of the unit square. We will push the actual integral calculations verifying part (ii) of the definition to the appendix. However, intuitively it is not difficult to see why the total volume under the PDF curve is unity. Note that the base of the region (i.e., the support) has an area of $1/2$. On the other hand, the joint density over this entire area is 2. The total volume of this triangular block (with base equal to $1/2$ and height equal to 2) is $1/2 \times 2 = 1$.

In the same spirit, we can calculate the probability $Pr(X < 2Y)$ without doing any integral calculations. The region where the given condition is satisfied is the shaded lower half of the support in Figure 7.5. However, the PDF is a constant over the entire support, so that the given probability is simply the ratio of the fraction of the area covered by the shaded area in relation to the entire support. As this ratio is $1/2$, so is the given probability. The same probability is derived using calculus in the appendix. ‖

**Example 7.3.** Let the joint probability density function of $X$ and $Y$ be given by

$$f(x,y) = \begin{cases} 8xy & \text{if } 0 \leq y \leq x \leq 1, \\ 0 & \text{otherwise.} \end{cases}$$

We want to find the marginal and conditional PDFs of $X$ and $Y$.

The marginal PDF of $X$ is given by $f_X(x) = 4x^3$, $0 \leq x \leq 1$. Similarly, the marginal PDF of $Y$ equals $f_Y(y) = 4y(1 - y^2)$, $0 \leq y \leq 1$. The calculations are given in the appendix.

The conditional PDFs are $f_{X|Y=y}(x) = 2x/(1 - y^2), y \leq x \leq 1$, and $f_{Y|X=x}(y) = 2y/x^2, 0 \leq y \leq x$. These may be easily obtained by dividing the joint PDF by the marginal PDFs. ‖

## 7.6    Independence

Two continuous random variables $X$ and $Y$ are said to be independent if the joint PDF can be factored into the marginal PDFs for all values $(x,y)$ in its

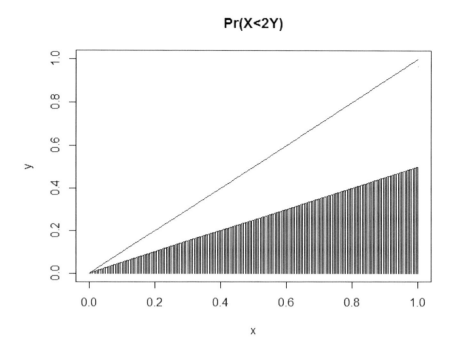

**FIGURE 7.5**
Geometrical representation of $Pr(X < 2Y)$.

support, i.e., if

$$f(x, y) = f_X(x) f_Y(y) \text{ for all } x, y.$$

A simple way to check whether two random variables $X$ and $Y$ having a continuous bivariate distribution are independent is to verify the following conditions.

1. The joint PDF is the product of two factors where the first factor $g(x)$ is a function of $x$ alone, while the second factor $h(y)$ is a function of $y$ alone.

2. The (bivariate) support of $(X, Y)$ is a two-dimensional rectangle in $\mathbb{R}^2$ with sides parallel to the rectangular coordinate axes. Thus the univariate support of $X$ does not depend on the specific value $y$ of $Y$.

If $f(x, y)$ satisfies both the above conditions, then $X$ and $Y$ are independent. In this case $g(x)$ and $h(y)$ are the marginal PDFs of $X$ and $Y$, respectively, except for multiplicative constants independent of $x$ and $y$.

**Example 7.4.** Suppose

$$f(x,y) = \begin{cases} 6xy^2 & \text{if } 0 \le x \le 1, 0 \le y \le 1, \\ 0 & \text{otherwise.} \end{cases}$$

We want to determine whether the two random variables $X$ and $Y$ are independent. It is easily seen that the bivariate support of $(X, Y)$ is the unit square, which has sides parallel to the rectangular coordinate axes. On the other hand, the joint PDF $6xy^2$ can obviously be factored into components (such as $2x$ and $3y$) where each component is a function of only one of the variables. Thus $X$ and $Y$ are independent in this particular instance. This approach does not require the determination of any marginal or conditional densities.

It is also verified in the appendix that the marginal PDF $f_X(x)$ of $X$ is $2x, 0 \le x \le 1$, and the marginal PDF $f_Y(y)$ of $Y$ is $3y^2, 0 \le y \le 1$. Of course one could factor $6xy^2$ in many other ways, such as $6x$ and $y^2$, but this factorization differs from the $3x$ and $2y^2$ factorization only in the multiplicative constants. It may be easily verified that in this example the conditional PDFs are the same as marginals, as they should be if they are independent.

On the other hand, Examples 7.2 and 7.3 deal with pairs of dependent random variables. In both cases the supports are non-rectangular.  ‖

## 7.7 Bivariate Normal Distribution

Two random variables $X$ and $Y$ are said to have a (joint) bivariate normal distribution if the joint PDF is given by

$$f(x,y) = \frac{1}{2\pi\sigma_x\sigma_y\sqrt{1-\rho^2}} e^{-\frac{1}{2(1-\rho^2)}q(x,y)}, \tag{7.8}$$

where

$$q(x,y) = \left[ \left(\frac{x-\mu_x}{\sigma_x}\right)^2 - 2\rho\left(\frac{x-\mu_x}{\sigma_x}\right)\left(\frac{y-\mu_y}{\sigma_y}\right) + \left(\frac{y-\mu_y}{\sigma_y}\right)^2 \right],$$

where $-\infty < x, y < \infty$.

The bivariate normal distribution has five parameters, $\mu_x, \mu_y, \sigma_x^2, \sigma_y^2$ and $\rho$. The form of the bivariate normal density looks complicated, but this is the only time we will see it. Besides, the form becomes substantially simpler when $\rho = 0$. In this case, we have

$$f(x,y) = \frac{1}{\sqrt{2\pi}\sigma_x} e^{-\frac{1}{2\sigma_x^2}(x-\mu_x)^2} \times \frac{1}{\sqrt{2\pi}\sigma_y} e^{-\frac{1}{2\sigma_y^2}(y-\mu_y)^2}, \tag{7.9}$$

so that the joint density factors into the product of two univariate densities, corresponding to the $N(\mu_x, \sigma_x^2)$ and $N(\mu_y, \sigma_y^2)$ distributions, respectively. These individual univariate densities actually turn out to be the marginal PDFs of $X$ and $Y$, respectively. This result is true in general, not just for the case $\rho = 0$.

From the above discussion, it also follows that the quantities $\mu_x$ and $\sigma_x^2$ are the mean and the variance of the marginal distribution of $X$, while $\mu_y, \sigma_y^2$ represent the same characteristics of the marginal distribution of $Y$. The quantity $\rho$ is actually the correlation coefficient of $X$ and $Y$. Equation (7.9) shows that, when $\rho = 0$, the joint PDF of $X$ and $Y$ factors into the product of the marginal PDFs of $X$ and $Y$, indicating independence. Thus, although zero correlation is in general a weaker property than independence, in the bivariate normal case, zero correlation and independence are equivalent concepts.

The conditional distributions can then be obtained rather easily by dividing the joint PDF by the marginal PDFs. For ready reference, we provide the marginal and the conditional PDFs in the table below.

| | | |
|---|---|---|
| marginal $X$ | : | $X \sim N(\mu_x, \sigma_x^2)$, |
| marginal $Y$ | : | $Y \sim N(\mu_y, \sigma_y^2)$, |
| Conditional ($X$ given $Y$) | : | $X\|Y = y \sim N(\mu_x + \rho\frac{\sigma_x}{\sigma_y}(y - \mu_y), \sigma_x^2(1 - \rho^2))$, |
| Conditional ($Y$ given $X$) | : | $Y\|X = x \sim N(\mu_y + \rho\frac{\sigma_y}{\sigma_x}(x - \mu_x), \sigma_y^2(1 - \rho^2))$. |

Note that the conditional distribution depends on the conditioning value of the other variable only in the expression of the mean. Thus the distribution of $X$ given $Y = y$ has a mean depending on $y$, but the variance is independent of $y$. Also, the mean of the conditional distribution of $X$ given $Y = y$ is linear in $y$. This structure (mean linear in $y$, variance independent of $y$) will be exploited later in fitting the least squares regression line, which we will study later under linear regression in Chapter 9.

Note that, if $\rho$ is zero, the conditional distribution of $X$ given $Y = y$ is $N(\mu_x, \sigma_x^2)$, i.e., the conditional distribution is identical to the unconditional (marginal) distribution. The same is true in the case of the conditional distribution of $Y$. Thus, in this case, the random variables are independent when $\rho = 0$, a phenomenon that we have already observed in Equation (7.9).

**Example 7.5.** Let $X$ denote yearly income and $Y$ denote yearly spending in taking holidays in a certain monetary unit for a certain upwardly mobile urban population. Assume that $X$ and $Y$ have a bivariate normal distribution with parameters $\mu_x = 180$, $\mu_y = 84$, $\sigma_x^2 = 100$, $\sigma_y^2 = 64$ and $\rho = 0.6$.

(a) Find the conditional distribution of $Y$ given $X = 190$.

Given $X = 190$, the conditional distribution of $Y$ is normal with mean

$$84 + 0.6 \times \frac{\sqrt{64}}{\sqrt{100}}(190 - 180) = 84 + 4.8 = 88.8$$

and variance

$$(1 - 0.6^2) \times 64 = 0.64 \times 64 = 6.4^2.$$

(b) Find $Pr(86.4 \le Y \le 95.36 | X = 190)$.

$$
\begin{aligned}
Pr(86.4 \le Y \le 95.36 | X = 190) &= Pr\left(\frac{86.4 - 88.8}{6.4} \le Z \le \frac{95.36 - 88.8}{6.4}\right) \\
&= Pr(-0.375 \le Z \le 1.025) \\
&= 0.4935.
\end{aligned}
$$

‖

**Example 7.6.** In a retail shop, let $X$ be the daily footfall and $Y$ be the daily transaction amount in a certain monetary unit. Assume that $X$ and $Y$ have a bivariate normal distribution with $\mu_x = 415, \mu_y = 347, \sigma_x^2 = 611, \sigma_y^2 = 400$, and $\rho = 0.65$.

(a) Find $Pr(309.2 \le Y \le 360.6)$.

$$
\begin{aligned}
Pr(309.2 \le Y \le 360.6) &= Pr\left(\frac{309.2 - 347}{\sqrt{400}} \le \frac{Y - 347}{\sqrt{400}} \le \frac{360.6 - 347}{\sqrt{400}}\right) \\
&= Pr(-1.89 \le Z \le 0.68) \\
&= 0.7224.
\end{aligned}
$$

(b) Find $E(Y|x)$.

$$
\begin{aligned}
E(Y|x) &= 347 + 0.65 \frac{\sqrt{400}}{\sqrt{611}}(x - 415) \\
&= 128.75 + 0.5259x.
\end{aligned}
$$

The conditional expectation $E(Y|x)$ is a function of $x$. For specific values of $x$, one will plug in that value in the above expression to get $E(Y|x)$.

(c) Find $Var(Y|x)$.

$$
\begin{aligned}
Var(Y|x) &= 400(1 - 0.65^2) \\
&= 231
\end{aligned}
$$

Note that the conditional variance $Var(Y|x)$ is independent of the value of $x$.

(d) Find $Pr(309.2 \le Y \le 380.6 | X = 385.1)$.

Here $E(Y|x) = 331.2741$ as obtained from (b) with $x = 385.1$. Thus

$$
\begin{aligned}
&Pr(309.2 \le Y \le 360.6 | X = 385.1) \\
&= Pr\left(\frac{309.2 - 331.2741}{\sqrt{231}} \le Z \le \frac{360.6 - 331.2741}{\sqrt{231}}\right) \\
&= Pr(-0.8906 \le Z \le 1.1849) = 0.9.
\end{aligned}
$$

Thus this conditional probability is higher than the unconditional probability in part (a).

‖

## 7.8   Sampling Distributions

The main ingredient of any statistical data analysis is a suitably chosen sample from the population of interest. Once such a sample is available to us, our aim would be to create appropriate functions of the observations in the sample to describe different characteristics of the population, or check for the validity of different hypotheses involving the population quantities. This often requires the derivation of the distribution of these functions of the sample observations that are used in these inference problems.

**Definition 7.1.** The most important case for us will be the one where the sample observations $X_1, X_2, \ldots, X_n$ chosen from the population of interest have a common distribution represented by the density $f(x)$, and the observations are mutually independent. We will call such a sample an *i.i.d. random sample* from the parent population. Because of the independent nature, the overall joint density of the data $(X_1, X_2, \ldots, X_n)$ will simply be the product of the individual (marginal) densities. Thus, if $f_O(x_1, x_2, \ldots, x_n)$ represents the joint probability density function of the data, then we have the representation

$$f_O(x_1, x_2, \ldots, x_n) = f(x_1)f(x_2)\cdots f(x_n).$$

Many authors would use the terms *i.i.d. sample* and *random sample* interchangeably in this connection. However, the literature is not entirely consistent in this respect and in order to avoid confusion, we will attach both the qualifications (*i.i.d.* and *random*) when describing such a sample.    ◈

Given an i.i.d. random sample $X_1, X_2, \ldots, X_n$, a function

$$T = T(X_1, X_2, \ldots, X_n)$$

of the data is called a *statistic*. A statistic must not depend on the value of the possibly unknown parameter, and the experimenter should be able to determine the exact numerical value of the statistic when the sample is available. We will primarily emphasize the situation where the random variables $X_1, X_2, \ldots, X_n$ and the statistic $T$ are one dimensional. We will describe and deal with higher-dimensional variables and statistics as and when required. Since the statistic is based on the $n$ random observations of the sample, its value will change when a fresh sample is chosen. The statistic $T$ is thus itself a random variable, and its probability distribution will be called its *sampling distribution*.

Sampling distributions can arise when sampling from any particular distribution. The most common statistics are usually the sample total (the sum of all the observations in the sample), and the sample mean (the simple average of all the observations in the sample). If $X_1, X_2, \ldots, X_n$ form an i.i.d. random sample from a *Bernoulli*$(p)$ distribution, the sum $\sum_{i=1}^{n} X_i$ can be shown to have the binomial distribution with parameters $n$ and $p$. If $X_1, X_2, \ldots, X_n$ form an i.i.d. random sample from a *Poisson*$(\lambda)$ distribution, the sum $\sum_{i=1}^{n} X_i$ has a *Poisson*$(n\lambda)$ distribution. However, the most interesting and useful sampling distributions are those that arise in connection with sampling from normal distributions, and that is what we will be concerned with in the rest of this chapter. In particular, our primary focus will be on the $\chi^2$, $t$ and $F$ distributions.

For the rest of the chapter we will, unless otherwise mentioned, assume that the sample at our disposal is an i.i.d. random sample drawn from a normal distribution.

### 7.8.1   Linear Combinations of Independent Variables

Linear combinations of independent random variables are used in statistical data analysis all the time and it is of great importance to have a handle on the distribution of such quantities. In the appendix we prove a general result about the moment generating function of such linear combinations (see Theorem 7.15). The following theorem is a consequence of that.

**Theorem 7.4.** *Suppose $X_1, X_2, \ldots, X_n$ are independent and $X_i \sim N(\mu_i, \sigma_i^2)$. Then the moment generating function of the linear combination $Y = \sum_{i=1}^{n} a_i X_i + b$ is given by*

$$M_Y(t) = e^{At + B\frac{t^2}{2}},$$

*where*

$$A = \sum_{i=1}^{n} a_i \mu_i + b, \quad B = \sum_{i=1}^{n} a_i^2 \sigma_i^2.$$

As moment generating functions completely characterize the distribution, comparing the above result with the moment generating function of a univariate normal random variable as given in Equation 7.6, we get the following corollary.

**Corollary 7.5.** *Suppose $X_1, X_2, \ldots, X_n$ are independent and $X_i \sim N(\mu_i, \sigma_i^2)$. Then*

$$\sum_{i=1}^{n} a_i X_i + b \sim N\left(\sum_{i=1}^{n} a_i \mu_i + b, \sum_{i=1}^{n} a_i^2 \sigma_i^2\right).$$

A further simplification occurs in case of an i.i.d. random sample. The following result is of paramount importance to us.

**Corollary 7.6.** *Suppose $X_1, \ldots, X_n$ represent an i.i.d. random sample from the $N(\mu, \sigma^2)$ distribution. Then*

$$\sum_{i=1}^{n} X_i \sim N(n\mu, n\sigma^2).$$

*Alternatively,*

$$\frac{1}{n} \sum_{i=1}^{n} X_i = \bar{X} \sim N\left(\mu, \frac{\sigma^2}{n}\right).$$

## 7.8.2 Distribution of the Sample Mean

Suppose we have an i.i.d. random sample from a distribution with mean $\mu$ and variance $\sigma^2$. Let us consider the sample mean $\bar{X} = \frac{1}{n} \sum_{i=1}^{n} X_i$. It is an easy matter to check that

$$E(\bar{X}) = E\left(\frac{1}{n} \sum_{i=1}^{n} X_i\right) = \frac{1}{n} \sum_{i=1}^{n} E(X_i) = \frac{n\mu}{n} = \mu \qquad (7.10)$$

and

$$Var(\bar{X}) = Var\left(\frac{1}{n} \sum_{i=1}^{n} X_i\right) = \frac{1}{n^2} \sum_{i=1}^{n} Var(X_i) = \frac{n\sigma^2}{n^2} = \frac{\sigma^2}{n}. \qquad (7.11)$$

Thus the sampling distribution of the sample mean retains the same mean as the original distribution, but the variance of the distribution is reduced by a factor of $n$. Thus, relative to $X$, $\bar{X}$ will be much more concentrated around $\mu$. This is illustrated in Figure 7.6, where the top panel gives a histogram of 1500 observations generated from $N(0,1)$, whereas the bottom panel gives the histograms of the sample means of 1500 different samples from $N(0,1)$, each based on 36 observations.

The mean and the variance given in Equations (7.10) and (7.11) are true in general and have nothing to do with the normal distribution in particular. However, if the original variables represent an i.i.d. random sample from the $N(\mu, \sigma^2)$ distribution, the sample mean $\bar{X}$ has the $N(\mu, \sigma^2/n)$ distribution, as we have already seen in Corollary 7.6.

When we have an i.i.d. random sample $X_1, X_2, \ldots, X_n$ from the normal distribution, an additional important result follows. Given its impact on the subsequent material, we put it in the form of a theorem. Let

$$s^2 = \frac{1}{n-1} \sum_{i=1}^{n} (X_i - \bar{X})^2 \qquad (7.12)$$

be the sample variance.

**Theorem 7.7.** *Suppose $X_1, X_2, \ldots, X_n$ form an i.i.d. random sample from a normal distribution. Then the sample mean $\bar{X}$ and sample variance $s^2$ are independent.*

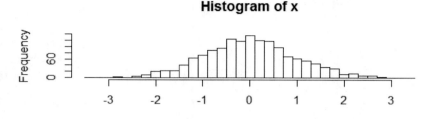

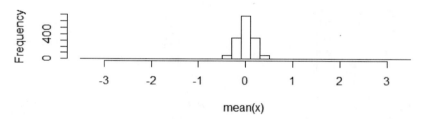

**FIGURE 7.6**
The histogram of the random variable and the sample mean.

### 7.8.3    Bias and Standard Error

A statistic is used for describing some characteristic of the population. However, a statistic is a random variable, and sometimes it will match the target characteristic closely and sometimes not. So it is necessary to have an overall idea about the precision of the statistic as an estimator.

If the population parameter of interest is $\theta$, and if the statistic in question is $T$, we say that the statistic $T$ is *unbiased* for $\theta$, if $E(T) = \theta$. A statistic which does not have this property is *biased* as an estimator of $\theta$ with $bias(T) = E(T) - \theta$. The property of unbiasedness indicates that the estimator is doing the right thing on the average.

However, the property of unbiasedness is not sufficient by itself to describe the closeness of the estimator and the statistic. In general, to quantify how closely the estimator is clustered around the target parameter, we use the mean square error ($MSE$) criterion

$$MSE(T) = E(T - \theta)^2.$$

The smaller the $MSE$, the better the estimator. It may be shown that

$$MSE(T) = (bias(T))^2 + Var(T).$$

Thus it is important to control the variance of the estimator together with the bias. For a statistic $T$ which is unbiased for $\theta$, the mean square error is equal to the variance.

Most of the statistics that we will consider will either be unbiased as estimators of their respective target parameters, or their bias will go down to zero with increasing sample size. Often the variance of the statistic will turn out to be our primary concern. The variances, however, will themselves depend on unknown parameters most of the time. For example, variance of the sample mean $\bar{X}$ is $\sigma^2/n$, as observed in the previous subsection, depending on a possibly unknown $\sigma$. In such a case, we estimate them with their usual estimators. The estimated standard deviation of a statistic is called its *standard error*. For example, the standard error of $\bar{X}$ is $s/\sqrt{n}$, where $s^2$ is the sample variance defined in Equation (7.12). The standard error is often reported together with an estimator as an indicator of its precision.

We use the divisor $(n-1)$ in Equation (7.12) to keep it unbiased. Using a denominator of $n$ forces it to be a biased estimator of $\sigma^2$, although the bias becomes small as $n$ increases.

## 7.9 Central Limit Theorem

The position of prime importance enjoyed by the normal distribution in statistics is due, in a large part, to the central limit theorem (CLT).

Suppose that $X_1, X_2, \ldots, X_n$ form an i.i.d. random sample from any distribution (not necessarily normal) with mean $\mu$ and variance $\sigma^2$. Let $\bar{X}$ represent the sample mean. We know that $E(\bar{X}) = \mu$ and $Var(\bar{X}) = \sigma^2/n$, regardless of the original distribution. As an immediate corollary it may be seen that $(\bar{X} - \mu)/(\sigma/\sqrt{n}) = \sqrt{n}(\bar{X} - \mu)/\sigma$ has mean 0 and variance 1. Yet, although the mean and variance are known, without knowledge of the distribution of $X$, it is not possible to make any general statements about the distribution of $\sqrt{n}(\bar{X} - \mu)/\sigma$ for small values of the sample size $n$.

A major breakthrough is available, however, when the sample size is large. Under some minimum assumptions on the structure of the distribution,

$$\frac{(\bar{X} - \mu)}{\sigma/\sqrt{n}} = \frac{\sqrt{n}(\bar{X} - \mu)}{\sigma}$$

has an asymptotic (approximately true for large $n$) standard normal or a $Z$ distribution. While the sample size necessary for this approximation to be reasonable depends on many factors, generally we use a cut off of $n = 30$ as a rule of thumb; thus we will approximate the distribution of $\sqrt{n}(\bar{X} - \mu)/\sigma$ with a standard normal distribution for $n \geq 30$ regardless of the nature of the original distribution. In some cases even smaller sample sizes give good approximations, such as in cases where the underlying distribution is symmetric and has a single peak at around the center (i.e., is unimodal).

*Remark* 7.1. There are many versions of the central limit theorem, and the regularity conditions vary according to the version used. For the most standard case we will simply require that the variance $\sigma^2$ of $X$ is finite and positive. ✦

**Example 7.7.** Suppose $\bar{X}$ is the sample mean for a random sample of size 36 based on an exponential distribution with parameter $\theta = 3$. Approximate $Pr(2.5 \leq \bar{X} \leq 4)$.

In this case the mean of the distribution is $\theta = 3$, and the variance is $\theta^2 = 9$ (see Table 7.1 in the appendix). As the sample size is larger that 30, the distribution of $\sqrt{36}(\bar{X} - 3)/\sqrt{9}$ can be reasonably approximated by a $N(0,1)$ distribution. Thus

$$
\begin{aligned}
Pr(2.5 \leq \bar{X} \leq 4) &= Pr\left( \frac{2.5 - 3}{\sqrt{9/36}} \leq \frac{\bar{X} - 3}{\sqrt{9/36}} \leq \frac{4 - 3}{\sqrt{9/36}} \right) \\
&= Pr(-1 \leq Z \leq 2) \\
&= 0.8186.
\end{aligned}
$$

∥

**Example 7.8.** Let $X$ be the birth weight, in gms, of a baby born in a developing Asian country. Assume that $X$ has mean 2860 and variance $660^2$. If $\bar{X}$ represents the sample mean based on an i.i.d. random sample of 100 babies born in this country, find $Pr(2730.64 \leq \bar{X} \leq 2989.36)$.

As the sample size is very large, we can safely apply the central limit theorem. We thus have

$$
\begin{aligned}
&Pr(2730.64 \leq \bar{X} \leq 2989.36) \\
&= Pr\left( \frac{2730.64 - 2860}{660/\sqrt{100}} \leq \frac{\bar{X} - 2860}{660/\sqrt{100}} \leq \frac{2989.36 - 2860}{660/\sqrt{100}} \right) \\
&= Pr(-1.96 < Z < 1.96) \\
&= 0.95.
\end{aligned}
$$

∥

### 7.9.1 Central Limit Theorem with Continuity Correction

Often the central limit theorem is used to determine probabilities in connection with discrete distributions. Sometimes this is slightly anomalous, since a discrete distribution takes discrete values only, but the approximating normal distribution can assume any value on the real line. In such cases a continuity correction improves the result. The most common approximations are with respect to binomial and random variables!

Let $X_1, X_2, \ldots, X_n$ represent an i.i.d. random sample from the *Bernoulli(p)* distribution. We know that, in this case, the sum $X = \sum_{i=1}^{n} X_i$ has a binomial distribution with parameters $n$ and $p$. A direct application of the CLT shows that $\sqrt{n}(X/n - p)/\sqrt{p(1-p)}$ has an asymptotic $N(0,1)$ distribution. Since

$$\frac{\sqrt{n}(X/n - p)}{\sqrt{p(1-p)}} = \frac{X - np}{\sqrt{np(1-p)}}, \tag{7.13}$$

we can directly use a $N(0,1)$ approximation for the expression on the right hand side of Equation (7.13).

Similarly, if $X_1, X_2, \ldots, X_n$ represent an i.i.d. random sample from the *Poisson($\lambda$)* distribution, we will use the standard normal distribution to approximate the distributions of

$$\frac{\sqrt{n}(\bar{X} - \lambda)}{\sqrt{\lambda}} = \frac{(\sum_{i=1}^{n} X_i - n\lambda)}{\sqrt{n\lambda}}. \tag{7.14}$$

Note that $\sum_{i=1}^{n} X_i$ itself has a *Poisson($n\lambda$)* distribution. Thus, if $X$ follows a *Poisson($\lambda$)* distribution,

$$\frac{X - \lambda}{\sqrt{\lambda}} \tag{7.15}$$

has an approximate $N(0,1)$ distribution for large $\lambda$.

**Example 7.9.** Let the distribution of $X$ be *binomial*$(15, 1/2)$. Approximate the probability

$$Pr(5 < X \leq 7)$$

using continuity correction.

Notice that within the probability expression we have the inequality $X \leq 7$ on the right hand side. In considering the transformation from continuous to discrete variables, all values up to 7.5 will be rounded down to 7. Similarly the left side shows $5 < X$, and by a similar argument we should be counting the values 5.5 or higher in this case. Thus in calculating $Pr(5 < X \leq 7)$ by the normal approximation we should really be calculating $Pr(5.5 \leq X \leq 7.5)$.

Since $E(X) = np = 15 \times \frac{1}{2} = 7.5$, and $Var(X) = 15 \times \frac{1}{2} \times (1 - \frac{1}{2}) = 3.75$. Thus

$$
\begin{aligned}
Pr(5 < X \leq 7) &= Pr(5.5 \leq X \leq 7.5) \\
&= Pr\left(\frac{5.5 - 7.5}{\sqrt{3.75}} \leq \frac{X - 7.5}{\sqrt{3.75}} \leq \frac{7.5 - 7.5}{\sqrt{3.75}}\right) \\
&= Pr(-1.032796 \leq Z \leq 0) \\
&= 0.3491503.
\end{aligned}
$$

The actual probability of 0.3491211 as obtained directly from the binomial distribution is very close to the normal approximation. The calculated probability using the normal approximation without the continuity correction is 0.2996, which is clearly a substantially poorer approximation. ∥

**Example 7.10.** The number $X$ of flaws in a certain tape of length 1 yard follows a Poisson distribution with mean 0.3. We examine $n = 100$ such tapes and count the total number $Y$ of flaws.

(a) Assuming independence, what is the distribution of $Y$?

The distribution of the sum of independent Poisson random variables is itself a Poisson with mean equal to the sum of the means. Thus the total number of flaws in 100 tapes will have a Poisson distribution with mean $0.3 \times 100 = 30$.

(b) Approximate $Pr(Y \le 25)$.

Using the continuity correction and the normal approximation in Equation (7.15), we get

$$
\begin{aligned}
Pr(Y \le 25) &= Pr(Y \le 25.5) \\
&= Pr\left(\frac{Y - 30}{\sqrt{30}} \le \frac{25.5 - 30}{\sqrt{30}}\right) \\
&= Pr(Z \le -0.8216) \\
&= 0.2056.
\end{aligned}
$$

‖

## 7.10    Sampling Distributions Arising from the Normal

### 7.10.1    $\chi^2$ Distribution

The $\chi^2$ distribution is the first of the three basic distributions arising from samples drawn from the normal distribution. It is indexed by its *degrees of freedom*, which plays the role of the parameter of the distribution and can take positive integral values. The probability density function of a random variable $X$ having a $\chi^2$ distribution with degrees of freedom $n$ is given by

$$
f(x) = \frac{1}{2^{n/2}\Gamma(n/2)} x^{n/2-1} e^{-x/2}, \quad x \ge 0.
$$

This is a special case of the *gamma*$(\alpha, \theta)$ distribution with $\alpha = n/2$ and $\theta = 2$; however, we will never use the actual form of the PDF of the $\chi^2$ distribution.

And invoking the moment generating function of the gamma distribution or otherwise, it is easy to prove that, when $X$ follows a $\chi^2(n)$ distribution, we have

$$
E(X) = n \quad \text{and} \quad Var(X) = 2n.
$$

The following theorem is a very useful result that is extremely helpful in determining the sampling distributions of many of the basic statistics resulting from samples drawn from normal distributions.

**Theorem 7.8.** *If $Z \sim N(0,1)$, then $Z^2 \sim \chi^2(1)$.*

The following result is another one which we will make frequent use of.

**Theorem 7.9.** *If $X_i \sim \chi^2(n_i)$, $i = 1, 2, \ldots, k$, and the $X_i$s are independent, then $\sum_{i=1}^{k} X_i \sim \chi^2\left(\sum_{i=1}^{k} n_i\right)$. This is known as the reproductive property of the chi-square.*

If $X_i \sim N(\mu_i, \sigma_i^2)$, $i = 1, 2, \ldots, n$, then $Z_i = (X_i - \mu_i)/\sigma_i$ has a $Z$ distribution for each $i = 1, 2, \ldots, n$. If, in addition, the $n$ variables are also independent, then the above two theorems lead to the following corollary.

**Corollary 7.10.** *Suppose $X_i \sim N(\mu_i, \sigma_i^2)$, $i = 1, 2, \ldots, n$, and the $n$ variables are independent. Then*

$$\sum_{i=1}^{n} \left(\frac{X_i - \mu_i}{\sigma_i}\right)^2 \sim \chi^2(n).$$

*In particular, if the $X_i$s are i.i.d. with mean $\mu$ and variance $\sigma^2$, then*

$$\frac{1}{\sigma^2} \sum_{i=1}^{n} (X_i - \mu)^2 \sim \chi^2(n).$$

A final result of interest arises when one has to deal with the quantity $\sum_{i=1}^{n} (X_i - \mu)^2 / \sigma^2$ described in Corollary 7.10, but the quantity $\mu$ is unknown. The obvious remedy is to use $\bar{X}$, the sample mean, as a substitute for $\mu$. The distribution of the resulting quantity $\sum_{i=1}^{n} (X_i - \bar{X})^2 / \sigma^2$ is given in the following theorem.

**Theorem 7.11.** *Suppose $X_1, X_2, \ldots, X_n$ are i.i.d. $N(\mu, \sigma^2)$. Then*

$$\frac{\sum_{i=1}^{n} (X_i - \bar{X})^2}{\sigma^2} \sim \chi^2(n-1).$$

*As the sample variance $s^2$ equals $\dfrac{1}{n-1} \sum_{i=1}^{n} (X_i - \bar{X})^2$, the above result may also be expressed as*

$$\frac{(n-1)s^2}{\sigma^2} \sim \chi^2(n-1).$$

The chi-square distribution is very heavily used in statistics, including parametric tests of hypotheses, confidence intervals and goodness-of-fit tests, as we will see in subsequent chapters.

## 7.10.2 Student's $t$ Distribution

Suppose we have two independent random variables $Z$ and $X$, such that $Z \sim N(0,1)$ and $X \sim \chi^2(n)$. Then the ratio

$$Y = \frac{Z}{\sqrt{X/n}} \tag{7.16}$$

is said to have a $t$ distribution with degrees of freedom $n$.

To understand the importance of this distribution, let us consider the following. When $X_1, X_2, \ldots, X_n$ form a random sample from the $N(\mu, \sigma^2)$ distribution, $\sqrt{n}(\bar{X} - \mu)/\sigma$ has the $N(0,1)$ distribution. This transformation is used in statistical problems all the time. However, on certain occasions it will so happen that although the value of the mean $\mu$ is known (at least tentatively), the value of the standard deviation $\sigma$ is unknown. In such cases the usual procedure is to replace the unknown value of $\sigma$ by the sample standard deviation $s$. That however will no longer preserve the distribution. Note that

$$\sqrt{n}(\bar{X} - \mu)/s \quad = \quad \frac{\sqrt{n}(\bar{X} - \mu)/\sigma}{\sqrt{\frac{(n-1)s^2}{\sigma^2}/(n-1)}}.$$

But the expression on the right hand side of the above equation is in the form of a ratio where the numerator has a $N(0,1)$ distribution, while the term under the square root in the denominator is a chi-square random variable divided by its degrees of freedom $(n-1)$. The random variables in the numerator and denominator are independent. By definition this has a $t$ distribution with $(n-1)$ degrees of freedom.

Like the standard normal distribution, the $t$ distributions have bell shaped density curves symmetric around the origin. The curves are, however, heavier tailed than $\phi(\cdot)$, the $Z$ density curve, although the $t$-curves approach $\phi(\cdot)$ when the degrees of freedom approaches infinity. For all practical purposes, we approximate the quantiles of the $t$ distribution with those of the $Z$ when the degrees of freedom exceeds 30. See Figure 7.7.

Like the $Z$, we will make frequent use of the quantiles of the $t$ distribution in our statistical analysis; we will use R to obtain the quantiles or the other probabilities related to $t$ distributions.

## 7.10.3 The $F$ Distribution

suppose that $X_i$, $i = 1, 2$, have $\chi^2$ distributions with degrees of freedom $n_i$, $i = 1, 2$ respectively. Then the ratio

$$F = \frac{X_1/n_1}{X_2/n_2}$$

is said to have an $F$ distribution with parameters $n_1$ and $n_2$, symbolically represented as $F(n_1, n_2)$. Like the chi-square and the $t$ distributions, the $F$

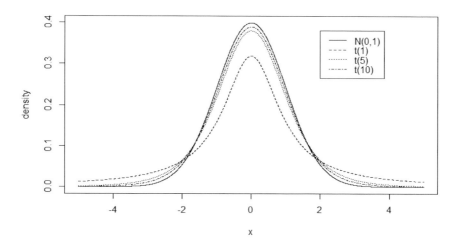

**FIGURE 7.7**
The $t$ density curves.

distribution is frequently used in calculating probabilities relating to the normal distribution. In particular, when we deal with a set up where the corrected sum of squares for a model has to be computed under several different sets of constraints, often these sum of squares are independent and have appropriate chi-square distributions. One can then use the $F$ distribution to compare these sum of squares and test hypothesis involving different model parameters in the process.

Being a ratio of chi-square random variables, a random variable having an $F$ distribution is necessarily nonnegative. Also note that if $X \sim t(n)$, then $X^2 \sim F(1, n)$.

## 7.11 Sampling from Two Independent Normals

(a) Suppose $X \sim N(\mu_1, \sigma_1^2)$ and $Y \sim N(\mu_2, \sigma_2^2)$. Also suppose that $X$ and $Y$ are independent. Then $X - Y \sim N(\mu_1 - \mu_2, \sigma_1^2 + \sigma_2^2)$.

(b) Suppose $X_1, X_2, \ldots, X_{n_1}$ form an i.i.d. random sample from a $N(\mu_1, \sigma_1^2)$ distribution, and $Y_1, Y_2, \ldots, Y_{n_2}$ form an i.i.d. random sample from a $N(\mu_2, \sigma_2^2)$ distribution. Suppose that the $X$ and $Y$ sample are independent. Then $\bar{X} - \bar{Y} \sim N(\mu_1 - \mu_2, \frac{\sigma_1^2}{n_1} + \frac{\sigma_2^2}{n_2})$. If $n_1$ and $n_2$ are large ($> 30$), this result is still approximately true even when the original distributions are not normal.

**Example 7.11.** Let $X$ denote the daily garment sales in a retail shop on weekends and $Y$ denotes the daily sales on a weekday. Suppose

$$X \sim N(184.09, 39.37) \quad \text{and} \quad Y \sim N(171.93, 50.88).$$

Assuming that the distributions of $X$ and $Y$ are independent, find $Pr(X \geq Y)$.

The distribution of $X - Y$ is normal with mean $184.09 - 171.93 = 12.16$ and variance $39.37 + 50.88 = 90.25$. Then

$$
\begin{aligned}
Pr(X \geq Y) &= Pr(X - Y \geq 0) \\
&= Pr\left( \frac{X - Y - 12.16}{\sqrt{90.25}} \geq \frac{-12.16}{\sqrt{90.25}} \right) \\
&= Pr(Z \geq -1.28) \\
&= 0.8997.
\end{aligned}
$$

$\parallel$

**Example 7.12.** The distribution of $X$ is $N(820, 2500)$ and that of $Y$ is $N(800, 3000)$. Let $\bar{X}$ be the sample mean based on a sample of size 20 from the population of $X$, and $\bar{Y}$ be the sample mean based on a sample of size 30 from the population of $Y$. Find $Pr(\bar{X} \geq \bar{Y})$.

Note that $\bar{X} - \bar{Y}$ follows a normal distribution with mean $820 - 800 = 20$ and variance $2500/20 + 3000/30 = 225$.

$$
\begin{aligned}
Pr(\bar{X} > \bar{Y}) &= Pr(\bar{X} - \bar{Y} > 0) \\
&= Pr\left( \frac{\bar{X} - \bar{Y} - 20}{\sqrt{225}} > \frac{-20}{\sqrt{225}} \right) \\
&= Pr(Z > -1.333) \\
&= 0.9087.
\end{aligned}
$$

$\parallel$

## 7.12 Normal Q-Q Plots

We often use Q-Q plots to have a quick check on whether a given sample comes from a specific theoretical distribution. Here we will run exploratory checks for the normal distribution, and for this purpose we will consider normal Q-Q plots. In the normal Q-Q plot, the sample quantiles are plotted against

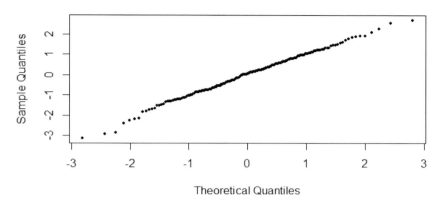

**FIGURE 7.8**
Normal Q-Q plot: Parent distribution is $N(0, 1)$.

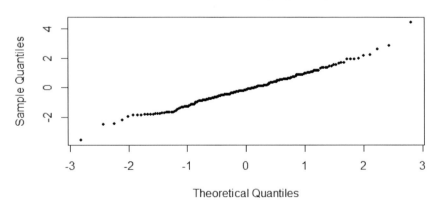

**FIGURE 7.9**
Normal Q-Q plot: Parent distribution is $t(20)$.

the theoretical quantiles of the normal distribution. If the sample itself is generated from a normal distribution, the Q-Q plot against the theoretical quantiles will be approximately linear. If the true data generating distribution is non-normal, departures from linearity will be expected. As the distribution moves farther and farther away from the normal (by becoming more skewed, more heavy tailed, etc.), the departures are expected to be more pronounced.

Here we present four normal Q-Q plots corresponding to samples drawn from four different distributions. These are (i) the standard normal distribution, (ii) the $t$ distribution with 20 degrees of freedom, (iii) the $t$ distribution with 2 degrees of freedom, and (iv) the exponential distribution with unit mean (i.e., parameter $\theta = 1$). The sample size in each case is 200.

For the distribution in case (i), the Q-Q plot is naturally expected to be linear. This is what is observed in Figure 7.8. The reasonable empirical inference from this exploratory check is that the parent distribution is normal.

In case (ii), the parent distribution, although not exactly normal, is fairly close to it. We have observed and discussed the fact the $t$ distribution approaches the standard normal as the degrees of freedom increases indefinitely. This is reflected in Figure 7.9, where only a slight – if any – departure from normality is observed.

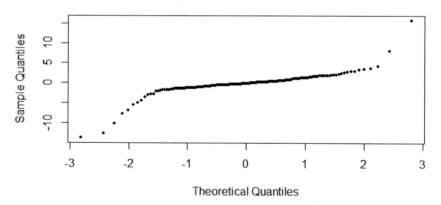

**FIGURE 7.10**
Normal Q-Q plot: Parent distribution is $t(2)$.

However, clear departures from normality are observed in Figures 7.10 and 7.11, although the nature of the departure is quite different. In Figure 7.10 the graph is linear in the central part but drops off sharply in the left and shoots up steeply in the right. This is along expected lines, as the $t(2)$ distribution is symmetric (around 0) and unimodal like the standard normal, but unlike the latter has very heavy tails.

**Normal Q-Q Plot: expo(1) Sample**

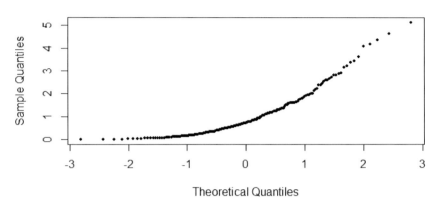

**FIGURE 7.11**
Normal Q-Q plot: Parent distribution is *expo*(1).

In Figure 7.11, on the other hand, we see a curved relationship, appropriate for a skewed distribution. The exponential distribution is positively skewed, and the curve is of a convex nature. If the parent distribution was negatively skewed, the nature of the curve in the normal Q-Q plot would be concave.

## 7.13 Summary

Sampling distributions arising from a normal distribution is of paramount importance to us. To re-emphasize the results, we provide a short and focused summary of some of the results discussed in this chapter.

Suppose that we have an i.i.d. random sample $X_1, X_2, \ldots, X_n$ from a $N(\mu, \sigma^2)$ distribution. The sample mean $\bar{X}$ and the sample variance $s^2$ are defined as

$$\bar{X} = \frac{1}{n} \sum_{i=1}^{n} X_i, \quad \text{and} \quad s^2 = \frac{1}{n-1} \sum_{i=1}^{n} (X_i - \bar{X})^2.$$

Then the following results hold.

(a) The sample mean $\bar{X}$ has a $N(\mu, \sigma^2/n)$ distribution.

(b) The sample mean $\bar{X}$ and sample variance $s^2$ are independent.

(c) $\sum\limits_{i=1}^{n} \left( \dfrac{X_i - \mu}{\sigma} \right)^2$ has a $\chi^2(n)$ distribution.

(d) $\sum\limits_{i=1}^{n} \left( \dfrac{X_i - \bar{X}}{\sigma} \right)^2 = \dfrac{(n-1)s^2}{\sigma^2}$ has a $\chi^2(n-1)$ distribution. Estimation of the unknown parameter $\mu$ by $\bar{X}$ leads to the loss of one degree of freedom.

(e) The sample mean and the sample variance are unbiased for their respective target parameters. Thus

$$E(\bar{X}) = \mu \quad \text{and} \quad E(s^2) = \sigma^2.$$

For the second result note that (Table 7.1) the mean of $\chi^2(n)$ random variable is $n$. Thus using the result from the previous item,

$$(n-1) = E\left[ \frac{(n-1)s^2}{\sigma^2} \right] = \frac{(n-1)E(s^2)}{\sigma^2}$$

which implies $E(s^2) = \sigma^2$.

(f) When both parameters $\mu$ and $\sigma$ are unknown, the standard error of $\bar{X}$ is $s/\sqrt{n}$.

## 7.14    Appendix

### 7.14.1    PDF and CDF

In the case of continuous random variables, readers equipped with knowledge of calculus can construct the CDF from the PDF, and can also recover the latter from the former. For a continuous distribution, while we obtain the CDF $F(x)$ as in (7.1), the PDF is recovered from the CDF through the derivative relation

$$f(x) = \frac{d}{dx} F(x).$$

**Example 7.13.** Consider a continuous random variable having a PDF of the form

$$f(x) = e^{-x}, x \geq 0.$$

This is the exponential random variable with unit mean (parameter $\theta = 1$). Direct integration of the PDF gives

$$F(x) = Pr[X \leq x] = \int_0^x e^{-t} dt = 1 - e^{-x}, x \geq 0.$$

On the other hand, by differentiating the above CDF, we recover the PDF as

$$f(x) = \frac{d}{dx}F(x) = \frac{d}{dx}(1 - e^{-x}) = e^{-x}, x \geq 0.$$

$\parallel$

## 7.14.2   Other Continuous Distributions

In Section 7.2 we list some distributions which we believe would be the most important ones for the analytics professional. Here we provide some additional distributions which are frequently used in statistical data analysis and could be of some peripheral interest in analytics.

1. Gamma: The gamma distribution, as a function of two positive parameters $\alpha$ and $\theta$, has the pdf

$$f(x) = \begin{cases} \dfrac{1}{\Gamma(\alpha)\theta^\alpha}x^{\alpha-1}e^{-x/\theta}, & x \geq 0, \\ 0 & \text{otherwise.} \end{cases} \qquad (7.17)$$

   The symbol $\Gamma(\cdot)$ represents the gamma function. The parameter $\alpha$ is called the shape parameter, while $\theta$ is the scale parameter. Like the exponential distribution, the gamma distribution is right skewed. However, as $\alpha$ becomes larger, the mode moves away from the origin and the distribution becomes more symmetric. When $\alpha$ equals 1, the gamma distribution is the exponential distribution with parameter $\theta$. When $\alpha$ is an integer larger than 1, the gamma distribution represents the distribution of the sum of $\alpha$ independent exponential random variables, each with parameter $\theta$. The gamma distribution is used to model, among other things, financial data, rainfall and other weather data, and certain waiting times. It is also used as a conjugate prior in Bayesian analysis. The chi-square distribution, one of the important sampling distributions arising from the normal, is a special case of the gamma distribution. Symbolically, we will write $gamma(\alpha, \theta)$ for the gamma distribution with the indicated parameters.

2. Beta: A random variable $X$ is said to have a beta distribution with two positive parameters $p$ and $q$ if it has the density

$$f(x) = \begin{cases} \dfrac{1}{B(p,q)}x^{p-1}(1 - x)^{q-1}, & 0 \leq x \leq 1, \\ 0 & \text{otherwise.} \end{cases} \qquad (7.18)$$

   The symbol $B(\cdot, \cdot)$ represents the beta function. Among other things, the beta distribution, denoted as $beta(p, q)$, is used as a conjugate prior in the Bayesian estimation of the binomial success probability.

**TABLE 7.1**
Table of means, variances and moment generating functions

| Distribution | Mean | Variance | Moment Generating Function |
|---|---|---|---|
| $uniform(a,b)$ | $\dfrac{a+b}{2}$ | $\dfrac{(b-a)^2}{12}$ | $\dfrac{e^{tb}-e^{ta}}{t(b-a)},\ t \neq 0$ |
| $expo(\theta)$ | $\theta$ | $\theta^2$ | $(1-\theta t)^{-1},\ t < 1/\theta$ |
| $N(\mu,\sigma^2)$ | $\mu$ | $\sigma^2$ | $e^{\mu t + \frac{1}{2}\sigma^2 t^2}$ |
| $\chi^2(n)$ | $n$ | $2n$ | $(1-2t)^{-n/2},\ t < 1/2$ |
| $gamma(\alpha,\theta)$ | $\alpha\theta$ | $\alpha\theta^2$ | $(1-\theta t)^{-\alpha},\ t < 1/\theta$ |
| $beta(p,q)$ | $\dfrac{p}{p+q}$ | $\dfrac{pq}{(p+q)^2(p+q+1)}$ | — |

Table 7.1 provides the means, variances and moment generating functions of all the important continuous distributions considered in this book.

The moment generating function of the beta distribution has been left blank, as it is a complicated expression which is of little use to us. The moment generating function of the normal distribution is defined for all real $t$.

### 7.14.3   Transformations of Random Variables

Given a continuous random variable $X$, we may be interested in the properties of the transformed random variable $Y = g(X)$, where $g$ is a suitable function.

**Theorem 7.12.** *Suppose the random variable $X$ has a continuous distribution with PDF $f(x)$ over a given support, and let $Y = g(X)$ be a transformation of the random variable $X$. If $g(\cdot)$ is continuous and monotone, so that the inverse function $X = g^{-1}(Y)$ exists, then the PDF of the transformed variable $Y$ is given by*

$$h(y) = f(g^{-1}(y)) \left| \frac{d}{dy} g^{-1}(y) \right|,$$

*over the support (of $Y$) to which the original support (of $X$) is translated through the transformation $Y = g(X)$, provided the derivative $\dfrac{d}{dy} g^{-1}(y)$ is continuous over the support of $Y$.*

The above result is a very useful one and is frequently used to find distributions of transformations of random variables. Some useful corollaries of the above theorem are given below. However, the result is not applicable when the transformation is not monotone. For example, the distribution of the $Y = Z^2$, when $Z$ has a $N(0,1)$ distribution, cannot be derived from the above theorem.

**Corollary 7.13.** *Suppose that $F(\cdot)$ is the CDF of $X$, and let $F(\cdot)$ be continuous and strictly increasing. Consider the transformation $Y = F(X)$. Then the random variable $Y$ is distributed as $uniform(0,1)$. (Note that $F(\cdot)$ is continuous and increasing, hence monotone).*

**Example 7.14.** Suppose $X \sim expo(1)$; in Example 7.13, the CDF of $X$ has been observed to be $F(x) = 1 - e^{-x}$, $x \geq 0$. Given the transformation $Y = 1 - e^{-X}$, a direct application of Theorem 7.12 demonstrates that $Y$ has the uniform distribution on $[0,1]$. ‖

**Corollary 7.14.** *Suppose that $F(\cdot)$ is the CDF of $X$, and let $F(\cdot)$ be continuous and strictly increasing. Let $U \sim uniform(0,1)$. Then $F^{-1}(U)$ is distributed with CDF $F(\cdot)$, where $F^{-1}$ is the inverse function of the distribution function $F$.*

**Example 7.15.** Suppose $X \sim expo(1)$ so that the CDF of $X$ is $F(x) = 1 - e^{-x}$, $x \geq 0$. Letting $Y = F(X)$, the inverse transformation is given as $X = F^{-1}(Y) = -\log(1 - Y)$. Then an application of Corollary 7.14 shows that $-\log(1 - U)$ has an $expo(1)$ distribution, where $U \sim uniform(0,1)$. ‖

Sometimes linear functions of random variables are of particular interest. In the case of some distributions like the normal, the distribution is closed under linear transformations; by this we mean that the transformed random variable also has a normal distribution (with appropriately transformed parameters).

### 7.14.4   Continuous Bivariate Distributions

In Section 7.5 we introduced continuous bivariate distributions with minimal emphasis on the technicalities and the process of integration. Here we fill the technical gaps in the examples of Section 7.5 to make the description complete.

**Example 7.2.** (Continued) Let the joint probability density function of $X$ and $Y$ be given by

$$f(x, y) = \begin{cases} 2 & \text{if } 0 \leq y \leq x \leq 1, \\ 0 & \text{otherwise.} \end{cases}$$

We want to verify whether this is a valid joint PDF, and find $Pr(X < 2Y)$.
We get

$$\int_0^1 \int_y^1 2dxdy = 2 \int_0^1 (1 - y)dy = 2\left[y - \frac{y^2}{2}\right]_0^1 = 1.$$

Thus $f(x, y)$ is a valid probability density function. Also

$$Pr(X < 2Y) = 2\int_0^1 \int_{x/2}^x dydx = \int_0^1 xdx = \left[\frac{x^2}{2}\right]_0^1 = \frac{1}{2},$$

which we observed earlier. ‖

**Example 7.3.** (Continued) Let the joint probability density function of $X$ and $Y$ be given by

$$f(x,y) = \begin{cases} 8xy & \text{if } 0 \le y \le x \le 1, \\ 0 & \text{otherwise.} \end{cases}$$

We want to find the marginal PDFs of $X$ and $Y$.

The marginal PDF of $X$ is given by

$$f_X(x) = \int_0^x 8xy\, dy = 8x \int_0^x y\, dy = 8x \left[ \frac{y^2}{2} \right]_0^x = 4x^3,\ 0 < x < 1.$$

Similarly, the marginal PDF of $Y$ is

$$f_Y(y) = \int_y^1 8xy\, dx = 8y \int_y^1 x\, dx = 8y \left[ \frac{x^2}{2} \right]_y^1 = 4y(1-y^2),\ 0 < y < 1.$$

$\parallel$

## 7.14.5 Linear Combinations of Independent Variables

The moment generating function technique is useful in determining the distributions of sums of independent random variables. A simple application of this technique gives the following result.

**Theorem 7.15.** *Suppose $X_1, \ldots, X_n$ are independent random variables. Let $M_{X_i}(t)$ represent the moment generating function of $X_i$. Then the moment generating function $M_Y(t)$ of*

$$Y = a_1 X_1 + a_2 X_2 + \cdots + a_n X_n + b$$

*is given by*

$$M_Y(t) = e^{bt} \prod_{i=1}^n M_{X_i}(a_i t).$$

This, however, does not indicate by itself what the distribution of the linear combination $a_1 X_1 + a_2 X_2 + \cdots + a_n X_n + b$ is. But Theorem 7.4 and Corollary 7.5, taken together, establish that for the specific case of the normal distribution.

## 7.14.6 Sums of Random Variables in Some Special Cases

We have seen that, in several cases, the moment generating function technique is useful in deriving the distributions of the sums of independent random variables. We consolidate these results below.

(a) Let $X_1$ and $X_2$ be independent random variables such that $X_1 \sim binomial(n_1, p)$ and $X_2 \sim binomial(n_2, p)$. Then

$$X_1 + X_2 \sim binomial(n_1 + n_2, p).$$

(b) Let $X_1$ and $X_2$ be independent random variables such that $X_1 \sim Poisson(\lambda_1)$ and $X_2 \sim Poisson(\lambda_2)$. Then

$$X_1 + X_2 \sim Poisson(\lambda_1 + \lambda_2).$$

(c) Let $X_1$ and $X_2$ be independent random variables such that $X_1 \sim N(\mu_1, \sigma_1^2)$ and $X_2 \sim N(\mu_2, \sigma_2^2)$. Then

$$X_1 + X_2 \sim N(\mu_1 + \mu_2, \sigma_1^2 + \sigma_2^2).$$

In all of the above cases, the results can be easily generalized to the sum of more than two random variables. However, from the examples listed above, it might appear that the distributions of sums of independent random variables always belong to the parent distribution. This is not a rule. For example, if $X_1$ and $X_2$ are independent $expo(\theta)$ random variables, $X_1 + X_2$ is distributed as $gamma(\alpha = 2, \theta)$ rather than as an exponential. More generally, if $X_1, X_2, \ldots, X_n$ represent an i.i.d. random sample from $expo(\theta)$, the distribution of the sum $\sum_{i=1}^n X_i$ has a gamma distribution with parameters $\alpha = n$ and $\theta$.

## 7.14.7   Central Limit Theorem

From the central limit theorem, we know that the distribution of $\sqrt{n}(\bar{X} - \mu)/\sigma$ has an approximate normal distribution. While we have used this fact to approximate probabilities of sample means, the exact nature of this approximation has been kept somewhat vague. Technically, the approximation of the variable $\sqrt{n}(\bar{X} - \mu)/\sigma$ with the standard normal variable $Z$ has the following sense. We have the convergence

$$Pr\left(\frac{\sqrt{n}(\bar{X} - \mu)}{\sigma} \leq t\right) \to \frac{1}{\sqrt{2\pi}} \int_{-\infty}^{t} e^{-x^2/2} dx = \Phi(t)$$

as $n \to \infty$, so that the CDF $F(t)$ of $\sqrt{n}(\bar{X} - \mu)/\sigma$ converges to $\Phi(t)$ at every point $t$.

## 7.15   Suggested Further Reading

As in Chapter 6, the books by Ross (2010) and DeGroot and Schervish (2012) rate highly in our opinion as additional reading resources for the material covered here. Other useful texts include Walpole et al. (2010) and Akritas (2014); the latter text also provides supporting R applications. Other basic texts dealing with distributions in our range of interest include Pal and Sarkar

(2009), Anderson et al. (2014), Levin and Rubin (2011), Black (2012) and Mann (2013).

Detailed descriptions of a large list of univariate and multivariate continuous distributions are provided in Johnson et al. (1994), Johnson et al. (1995) and Kotz et al. (2000).

# 8

## Statistical Inference

In terms of its content, *statistical inference* has two major branches, which are *estimation* and *hypothesis testing*. The topic of estimation can be further subdivided into *point estimation* and *interval estimation*. In this chapter we will briefly review some popular inference techniques in respect to some common parametric models and explain how these methods are useful to the analytics professional.

In the following discussion, we will restrict ourselves to the case of *parametric inference*. In this approach to inference, we assume that the data is generated by a particular parametric family, whose form may be known, but the specific parameters are unknown.

The main ingredient of our inference, as described in the previous chapters, is a suitably chosen sample from the parent distribution. This is the source of our information about the population. As this information is only based on a subset of the population (the sample), it is necessarily incomplete in relation to the entire population. To do a good job of inference, the following two points are therefore extremely vital:

(a) The sample should be carefully chosen, so that it is as representative of the population as possible.

(b) Given that the sample is only a subset of the population, we must make the best use of this limited information to find out as much as possible about the population characteristics.

Inference procedures dealing with population parameters are of paramount importance in analytics and business research. Managers of business projects and decision makers in business firms have to deal with a large number of variables on which data may be available. To formulate proper business models, the analyst needs to have a good idea about the mean (average) of these values, as well as their extremes. In this case the analyst needs to use a suitable estimation rule to describe the possible values of the parameters in question. On the other hand, when a new product is to be launched in the market, it is important to demonstrate that this product is superior to the existing product to make it commercially viable. In such a case our primary interest is comparative; we want to determine whether the new product is superior to the current one in respect to some characteristic, but we may not be particularly interested in the actual numerical values of these characteristics. This is a situation where a test of hypothesis is the appropriate tool to use.

## 8.1    Inference about a Single Mean

The first case that we want to look at is the problem of doing inference regarding the mean parameter of a single population where the variable of interest is continuous. In this section we will assume that the parent distribution is normal with a known variance. In this connection we will discuss the problems of estimation and testing of hypothesis in sufficient detail so that the concepts are properly established and the subsequent cases may be introduced with brief descriptions of the additional features. Unless otherwise mentioned, we will assume that the populations are infinitely large and the samples are drawn without replacement.

### 8.1.1    Point Estimation

Let $X_1, X_2, \ldots, X_n$ represent an i.i.d. random sample from a $N(\mu, \sigma^2)$ distribution where $\sigma^2$ is known. Let $\bar{X} = \frac{1}{n} \sum_{i=1}^{n} X_i$ be the sample mean. The statistic $\bar{X}$ is the most common and the most intuitive estimator of the population mean. We will use $\bar{X}$ as our *point estimator* of the unknown mean $\mu$. In particular, it may be shown that the sample mean $\bar{X}$ is the *maximum likelihood estimator* of $\mu$ and is unbiased for the target parameter. The concept of the maximum likelihood estimator is discussed in Section 8.8.1.

It is important to have an idea of how good the sample mean $\bar{X}$ is as an estimator of the population mean. We observed in Chapter 7 that the expectation of the sample mean $\bar{X}$ is the population mean $\mu$. The standard deviation of the sample mean is $\sigma/\sqrt{n}$, where $\sigma^2$ is the variance of the parent population and $n$ is the sample size; we may call this the standard error of the sample mean in the spirit of our discussion in Chapter 7, although this standard deviation includes no estimated parameters. The standard error gives us an idea of how closely the sample mean is clustered around the population mean. Clearly, the standard error becomes smaller as the sample size becomes larger, so that the sample mean gets better as an estimator of the population mean with increasing sample size. As the sample size increases indefinitely, the sample mean can be brought as close to the population mean as desired with a high probability. This is a desirable property that an estimator should have. The quantity $\sigma/\sqrt{n}$ is the key term in describing the error when estimating the population mean with the sample mean.

### 8.1.2    Interval Estimation

In the above we presented a point estimator for $\mu$. However, one would have to be extremely lucky to have the sample mean hit upon the population mean exactly! In general, there will be some discrepancy between the values of the estimator $\bar{X}$ and the true unknown value of $\mu$. Thus, although we use $\bar{X}$ as the estimator of $\mu$, we are practically certain that, for a fixed sample size, there would be some error in this estimation.

As a point estimator is highly unlikely to exactly represent the true value of the unknown parameter, often we also set up a *confidence interval* for the unknown true value of $\mu$. In this approach, we construct an interval, rather than use a single value, such that the true unknown mean lies in that interval with some prespecified high degree of confidence. This degree of confidence is often referred to as the *confidence level* or the *confidence coefficient*, and we will describe it more formally in our subsequent discussion. The width of the confidence interval increases with the confidence level; the normal practice is to choose a confidence level close to, but strictly smaller than 1.

In describing the computation of the confidence interval in this section, we will assume that the population variance $\sigma^2$ is known. As the parent distribution is normal,

$$Z = \frac{\bar{X} - \mu}{\sigma/\sqrt{n}} \qquad (8.1)$$

has a $N(0, 1)$ distribution. Even when the parent distribution is not normal, the quantity $Z$ defined above has an approximate $N(0, 1)$ distribution by the central limit theorem when the sample size is large. We will make use of the transformation in Equation (8.1) to construct a $100(1-\alpha)\%$ confidence interval for $\mu$ where $0 < \alpha < 1$. The quantity $(1 - \alpha)$, expressed as a percentage, is the confidence level of the confidence interval and represents the degree of our trust that the interval covers the true value of the mean $\mu$. For illustration, let us choose $\alpha = 0.05$. Indeed, $\alpha = 0.1, 0.05$ and $0.01$ are the most common levels at which we construct our confidence intervals. When $\alpha = 0.05$, we are looking for the 95% confidence interval.

For a $N(0, 1)$ variable $Z$, the quantiles corresponding to $q = 0.025$ and $0.975$ are $-1.959964$ and $1.959964$, respectively. (For simplicity they are often replaced by $-1.96$ and $1.96$, respectively.) By the definition of quantiles,

$$Pr[-1.959964 \le Z \le 1.959964] = 0.975 - 0.025 = 0.95. \qquad (8.2)$$

The above expression provides the probability of the central 95% of the $N(0, 1)$ curve (see Figure 8.1), and we will make use of this expression in constructing the 95% confidence interval for $\mu$. We represent the quantile of the $Z$ distribution corresponding to the fraction $q$ by $z_q$, so that $z_{0.975} = 1.959964$. Note also that $z_{0.025} = -z_{0.975}$. Combining the relation $-1.959964 \le Z \le 1.959964$ with Equation (8.1), we get

$$-z_{0.975} \le Z \le z_{0.975} \iff -z_{0.975} \le \frac{\bar{X} - \mu}{\sigma/\sqrt{n}} \le z_{0.975}.$$

Rearranging terms, this leads to the relation

$$\bar{X} - z_{0.975} \frac{\sigma}{\sqrt{n}} \le \mu \le \bar{X} + z_{0.975} \frac{\sigma}{\sqrt{n}}.$$

**Confidence Interval Calculation**

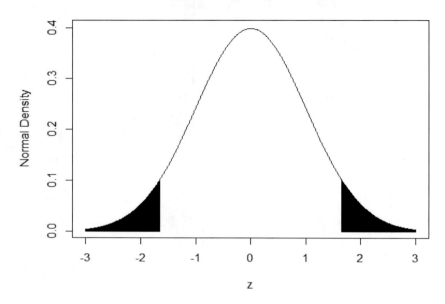

**FIGURE 8.1**

Confidence interval calculation for the normal mean.

Since the events in the last three sets of inequalities are equivalent, it follows that

$$P\left[\bar{X} - z_{0.975}\frac{\sigma}{\sqrt{n}} \le \mu \le \bar{X} + z_{0.975}\frac{\sigma}{\sqrt{n}}\right] = 0.95. \tag{8.3}$$

Thus the interval $(\bar{X} - z_{0.975}\frac{\sigma}{\sqrt{n}}, \bar{X} + z_{0.975}\frac{\sigma}{\sqrt{n}})$, often written as $\bar{X} \pm z_{0.975}\frac{\sigma}{\sqrt{n}}$, represents a region which has a 95% probability of capturing the true value of the mean $\mu$.

Note that the two endpoints of the above interval are random; they would change if one chooses a new i.i.d. random sample from the parent $N(\mu, \sigma^2)$ distribution. To graphically represent this, we have chosen 24 i.i.d. random samples, each of size 30, from the $N(0, 1)$ distribution. Assuming $\sigma = 1$ to be known, we have constructed the 95% confidence interval for the mean as described above, and displayed the spread of each interval. From Figure 8.2 it may be observed that 23 of the 24 confidence intervals capture the true value of the mean $\mu$ (here, zero), so that the empirical proportion of coverage is $23/24 = 0.9583$, reasonably close to the confidence level of 95%.

The use of terms like *95% confidence interval* or *interval with 95% con-*

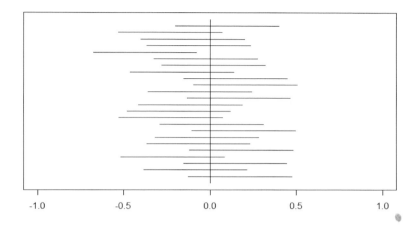

**FIGURE 8.2**
Coverage of the confidence interval.

*fidence level* must be clearly understood. From Equation (8.3) we know that there is a 95% probability that the interval $\bar{X} \pm z_{0.975}\frac{\sigma}{\sqrt{n}}$ will capture the unknown true value of the mean. In relation to Figure 8.2, it means that, if we were to increase the number of samples indefinitely (from the present 24), the proportion of samples which cover the true unknown value of the mean would get closer and closer to 0.95. However, when we are doing interval estimation in a practical situation, we would not be taking repeated samples but would present our inference based on a single sample only. Thus we would end up getting one particular interval from the totality of all possible confidence intervals. However, as we can see from Figure 8.2, once we have chosen the sample and constructed the interval, there is no random element left in the process and the interval either contains the mean or does not; thus the probability that the constructed interval contains the mean is either 0 or 1. Prior to the selection of the sample, it would be a legitimate statement to say that the probability is 95% that the interval to be constructed will contain the true mean. Subsequent to the selection of the sample and construction of the the actual numerical boundaries of the interval, we should only say that we are 95% confident of having covered the true unknown mean, but should not claim that the probability is 0.95 that the unknown mean $\mu$ is captured within the numerical endpoints of the constructed interval.

**Example 8.1.** We are interested in the quality of car tires manufactured by a certain tire manufacturing company in relation to a certain class of cars. In this case our sample will consist of an independent selection of tires produced by this company, where the experimenter will record the number of kilometers they run before they require replacement. (This will require following up on the tires over a period of time.) We will construct a 95% confidence interval for the mean $\mu$ of $X$, where $X$ represents the number of kilometers that a randomly chosen tire of this company will run before it needs replacement. Suppose it is known (from previous experience) that the standard deviation of the lifetime of the tires manufactured by this company is 2000 kilometers. Let the following be the realized values of an i.i.d. random sample of size 25 from the population of tires manufactured by this company.

51035 43059 49073 46139 50622 45325 52243 51777 49348 49515
51183 49035 45692 49578 46253 48370 50350 48988 49418 50758
47673 49825 47114 51965 49147

The above data yields $\bar{X} = 48,939.4$. Thus the 95% confidence interval for the mean lifetime of these tires is

$$\bar{X} \pm z_{0.975}\frac{\sigma}{\sqrt{n}} = 48,939.4 \pm 1.959964\frac{2000}{\sqrt{25}} = (48,155.41, 49,723.39).$$

Thus we are 95% confident that the interval (48,155.41, 49,723.39) contains the unknown true value of the mean lifetime of the tires, without claiming that there is a probability of 0.95 that the mean lies in this interval.  ‖

The $Z$ transformation in Equation (8.1) forms the basis of the confidence interval construction in this case. The quantity $Z$, defined as

$$Z = \frac{\bar{X} - \mu}{\sigma/\sqrt{n}},$$

is called the *pivotal element* in the calculation of the confidence interval. It is the quantity which paves the way for the construction of the final interval by rearranging the terms in the probability inequality it generates.

### 8.1.3    Hypothesis Testing

Apart from the theory of estimation, testing of hypothesis is the other fundamental paradigm of statistical inference. The key point where testing of hypothesis differs from estimation is that, in this case, there is a tentative idea about the unknown parameters under investigation. The experimenter wishes to check, on the basis of a suitably drawn sample from the population under study, whether the prevailing tentative idea is tenable or whether it should be rejected in favor of another system which better describes the random mechanism.

Let us go back to the example of the tire manufacturing company. Suppose, for example, that the company claims that their tires run, on the average, for

50,000 kilometers before they are worn out and need replacement. This information could serve as the tentative idea with which we begin our experiment. A natural procedure in this case would be to assume that the lifetime of the tires has a normal distribution with an unknown mean (which may be replaced by the tentative value to check out if that produces a good match with the observed data). Let us suppose that, from past experience, we know that the standard deviation of the distribution of the lifetime of these tires is 2000 kilometers. The job of the experimenter would be to decide, given the model assumptions and on the basis of a suitable sample, whether the claim of the tire manufacturer is accurate or not.

The *null hypothesis*, which quantifies the tentative belief that is to be tested, is generally expressed in terms of the parameter of interest. Given that the company has claimed that the mean number of kilometers that their tires would run is 50,000, we will begin with the assumption that the lifetime of the tires has a normal distribution with mean $\mu = 50,000$ (and a known standard deviation $\sigma = 2000$). The null hypothesis will then be stated as $H_0 : \mu = 50,000$. The testing procedure essentially checks whether the realized sample has the behavior which would be expected if the null hypothesis were true. If the behavior of the sample is too anomalous to what is expected under the null hypothesis, the experimenter *rejects* the null hypothesis.

If the null hypothesis is rejected (which implies that the experimenter no longer trusts the claim under the null hypothesis to be correct), there must be another hypothesis to move to. This other hypothesis is called the *alternative hypothesis*. The hypothesis testing procedure, therefore, pits the null hypothesis against the alternative hypothesis. What the form of the alternative hypothesis should be depends on the physical nature of the problem and what the experimenter wants to guard against. In this particular case, the consumer buying the tire should be concerned if the mean lifetime of the tire is *less* than what has been claimed. The consumer would not mind (in fact the consumer would be happier) if the mean lifetime is greater than what is claimed. Thus the alternative hypothesis, denoted by $H_A$, would be $H_A : \mu \leq 50,000$ for this problem. If the experimenter rejects the null, it would mean that she feels that there is enough reason to believe that the tire manufacturer's claim is overstated.

At the end of the testing process, the experimenter has to take a decision. Either she would reject the null hypothesis (in favor of the alternative hypothesis) or she would fail to reject the null. (It would not be a technically correct statement to say that one "accepts" the null.) The experimenter must create a suitable partition of the sample space. If the observed sample comes from the first part of the sample space, the null hypothesis is rejected; this part of the sample space is called the *rejection region* or the *critical region*. When the sample comes from the other part, the hypothesis cannot be rejected. The entire emphasis of the hypothesis testing procedure is on finding a suitable partition of the sample space so that the hypothesis testing procedure can be carried out efficiently.

Testing of hypothesis is a probabilistic process and is not entirely error free. One could end up rejecting $H_0$ when the null hypothesis is correct, which is called *type I error*. The quantity $Pr[\text{type I error}]$ is called the *size* of the test. A good test should have a small size. One could also fail to reject a false null hypothesis, which leads to *type II error*. The quantity $[1 - Pr(\text{type II error})]$ is called the power of the test. It is the probability of rejecting the null hypothesis when it is incorrect and quantifies the strength with which the test is able to recognize that the null is false. A good test should have a large power. The cross-classification of the errors with the true state of nature is described in the following table.

| True Natural State | Decision | |
|:---:|:---:|:---:|
| | Fail to Reject $H_0$ | Reject $H_0$ |
| $H_0$ True | Correct Decision | Type I Error |
| $H_0$ False | Type II Error | Correct Decision |

The irony is that the two different kinds of errors cannot be controlled simultaneously. The ideal situation would be the one where the probabilities for both types of error are zero; in the vast majority of practical problems this is impossible to achieve. If the rejection region shrinks, the test becomes more conservative and the probability of type I error goes down; at the same time, unfortunately, the probability of type II error goes up. These observations are reversed when the rejection region is enlarged. If the experimenter decides not to reject the null hypothesis regardless of the data, she would never commit type I error. However, the probability of type II error would be 1 in this case. If, on the other hand, the experimenter decides to reject the null hypothesis in any given situation, there would be no type II error, but the probability of type I error will be the maximum (equal to 1). Given these observations, the usual approach to hypothesis testing chooses tests which have a controlled small value of the probability of type I error, and subject to this the probability of type II error is minimized. An upper bound to the probability of type I error, called the *level* of the test, must be specified for conducting the test. In all the examples and illustrations we provide in this book, we will, unless otherwise mentioned, let the probability of type I error be equal to $\alpha$, the prespecified level of the test.

We explain the mechanics of a simple hypothesis testing problem by describing the different steps of the procedure through a continuation of the tire manufacturing example.

**Example 8.2.** Consider the data provided in Example 8.1. The company claims that their tires run for at least 50,000 kilometers before requiring replacement. We will perform a test of hypothesis in the following steps.

(a) Specify the model and the parameters of interest.

  · In our case the model is normal with unknown mean $\mu$ and known standard deviation $\sigma = 2000$.

(b) Specify the hypotheses.

Here our null hypothesis is $H_0 : \mu = 50,000$, to be tested against the alternative hypothesis $H_A : \mu < 50,000$.

(c) Specify the level of significance.

This is the maximum probability of type I error that the experimenter is ready to tolerate. Normally, this has to be specified for the test procedure to commence. In this case we will use the level of significance $\alpha = 0.05$.

(d) Select the test statistic.

The whole point of performing a test of hypothesis is to check whether the observed behavior of the sampled data matches the behavior one would expect if the null hypothesis were true. In this particular case we know that the sample mean $\bar{X}$ has a normal distribution with an unknown mean $\mu$ and known standard deviation $\sigma/\sqrt{n}$. Thus $\sqrt{n}(\bar{X} - \mu)/\sigma$ will have a $N(0,1)$ distribution. Of course, the last expression cannot be computed since $\mu$ is unknown, but if the null hypothesis is taken to be correct, then

$$Z = \sqrt{n}(\bar{X} - 50,000)/\sigma \qquad (8.4)$$

will have a $N(0,1)$ distribution, and this $Z$ will take values on the line according to the standard normal probability law. Once we have the sample, the quantity $Z$ in Equation (8.4) can be explicitly computed, since $n$, $\bar{X}$ and $\sigma$ are all known. If this observed value $Z$ is in a range which a $N(0,1)$ variable can reasonably assume, we will have no reason to suspect that the null is false.

If, however, the null hypothesis is incorrect and the true mean $\mu$ is smaller than 50,000, the transformed random variable in Equation (8.4) no longer has the $N(0,1)$ distribution. It will, in fact, have a normal distribution with a mean smaller than 0. In this case we will expect that the observed $Z$ value might drop below the usual range of the values of the $N(0,1)$ random variable. When such behavior is seen, the veracity of the null hypothesis is called into question.

The statistic given in Equation (8.4) is our *test statistic* in this case. We reject the null hypothesis when it is *too small*.

(e) Specify the rejection region.

As we explained in the previous item, we will reject the null hypothesis when the observed value of the statistic is too small. But how small is too small? Note that we are performing the test at level $\alpha = 0.05$, so that we will allow a probability of type I error equal to 0.05. Since lower values of observed $Z$ will provide evidence against the null (and for the alternative), we will set apart a region with probability equal to 0.05 on the lower tail of the standard normal distribution where we reject the null hypothesis.

Given that the quantile of $Z$ at 5% is $z_{.05} = -1.644854$, we must reject the null when our observed $Z$ is smaller than $-1.644854$; hence {observed $Z \leq -1.644854$} is our rejection region.

When the observed value of $Z$ falls in the rejection region, it may mean one of two things. Either the null hypothesis is still true, but an event of very small probability has occurred, leading to the small observed value of $Z$, or this may have happened because the null hypothesis is actually false. As the observed value of $Z$ keeps dropping, our faith in the null hypothesis becomes weaker. When $Z$ drops beyond the threshold of $-1.644854$, it crosses the limit that the experimenter is ready to tolerate, and the experimenter takes it as a signal that the null hypothesis is false.

(f) Provide a decision after performing the test.

In our case $\bar{X} = 48,939.4$, $n = 25$ and $\sigma = 2000$. The statistic in Equation (8.4) takes the value

$$Z = \frac{\sqrt{25}(48,939.4 - 50,000)}{2000} = -2.6515,$$

which is way below our threshold of $-1.644854$. Thus there is solid evidence that the null hypothesis is possibly untrue. The decision of the experimenter, at level $\alpha = 0.05$, will be to reject $H_0$. As far as the sample evidence is concerned, the claim of the tire manufacturing company appears to be false.

(g) Provide a p-value:

In the previous item, the experimenter will simply report if the test rejects the null or whether it fails to do so. This declaration, even when accompanied by the test statistic, will not provide the reader with a very clear idea about the strength of the decision. It will not clearly indicate, for example, if the rejection is a marginal one or an overwhelming one. In this respect, the p-value is a useful tool. The p-value is also called the *observed significance level*. It is the probability of actually observing something that is more extreme than the observed test statistic (in this case more extreme than $-2.6515$) if the null hypothesis were true. In this problem the p-value is $Pr[Z \leq -2.6515] = 0.0040$. This shows that the observed value is quite extreme, and if the null were true one would see a more extreme value only in 4 out of 1000 times. By itself, the p-value only gives the strength of the significance, and not a decision. One could, of course, give a decision by comparing the p-value with the level; we reject when the p-value is smaller than the level. However, the experimenter can also do the test without specifying a level and simply report the p-value. The reader can, on the basis of the reported p-value, decide which decision to take. This is, in fact, another advantage of the p-value. It lets the reader make her own decision based on the p-value. Thus although practically all standard levels of significance will reject the null hypothesis in this example, if the reader decides to be conservative and do this test at level $\alpha = 0.001$, the null hypothesis cannot be rejected.

∥

Tests of the above type can be one sided (greater than type, less than type) or two-sided (not equal to type) based on the nature of the alternative. The above test was done with a one-sided, less than type alternative. A one-sided, greater than type alternative will be similar to this with the rejection region shifted to the right tail. In this case large values of the test statistic $Z$ would be the more extreme values and would lead to rejection. For the two sided test in this connection the alternative would be of the type $H_A : \mu \neq 50,000$ (rather than being just less than or greater than 50,000). In this case one would reject the null for both very high and very low values of the test statistic $Z$ in Equation (8.4). Thus the 5% level of significance must be split up in both the tails in this case, so that the rejection region for the two sided test in the above example would be $\{Z > 1.959964$ or $Z < -1.959964\}$. Also, the p-value in this case will be double the usual one-sided p-value, because both very large and very small values provide evidence against the null.

In this section we have motivated the problem using the tire manufacturing example, where the null value was taken as $\mu = 50,000$. In many cases where such problems are theoretically examined, the null value of the mean is symbolically expressed as $\mu_0$, so that the hypotheses to be tested are expressed as $H_0 : \mu = \mu_0$ versus $H_A : \mu < \mu_0$ (or $\mu > \mu_0$, or $\mu \neq \mu_0$).

When the null is false (which we will never know for certain), we should reject $H_0$. The probability with which we do that is the power of the test. The power will depend on the actual true (the null being false) value of the parameter. The power will increase as the difference between the true mean and the null mean increases. When the true mean is $\mu = 48,000$, one is more likely to reject the null $H_0 : \mu = 50,000$ than when the true mean is $\mu = 49,000$. The sample size $n$ is another quantity which has a significant impact on the power of the test. The power (i.e., the probability of rejecting an incorrect null) increases with the sample size. While we will not get deeply inside the problem, the issue of determining the sample size to achieve a certain pre-specified power in a complicated hypothesis testing scenario is an important practical requirement and represents a fairly active research area.

## 8.2 Single–Population Mean with Unknown Variance

In the previous sections we described the problem of doing inference about the unknown population mean under the normal assumption where the population variance is known. This is not a very realistic scenario. If the population mean is unknown, the population variance will probably be unknown as well. In this section we reconsider the inference problems considered so far in this chapter without assuming that the population variance is known.

Suppose we have an i.i.d. random sample $X_1, X_2, \ldots, X_n$ from a normal $N(\mu, \sigma^2)$ distribution where both parameters are unknown. Our interest is still in doing inference about the unknown mean $\mu$. However, in this case we have to somehow account for this unknown value of $\sigma$. Such parameters, which do not form a part of our primary interest, but nevertheless have some impact on the inference, are called *nuisance parameters*. In the present context, $\sigma$ is a nuisance parameter. We will discuss the procedures to do point estimation, interval estimation and hypothesis testing about $\mu$ under this scenario.

The point estimator of $\mu$ is still the sample mean $\bar{X}$. Under the assumption of normality, it is not difficult to show that the sample mean has some optimal properties as an estimator of the population mean, regardless of the value of $\sigma$. For point estimation, therefore, $\bar{X}$ is still our preferred estimator. It still is the maximum likelihood estimator for $\mu$ and is also an unbiased estimator.

When finding a confidence interval for $\mu$, we now consider our pivotal element to be

$$t = \frac{\bar{X} - \mu}{s/\sqrt{n}}, \tag{8.5}$$

where $s$ is the sample standard deviation based on observed data. Note that this form has been obtained by replacing $\sigma$ with $s$ in Equation (8.1). In Section 7.10.2 we demonstrated that this quantity has a $t$ distribution with degrees of freedom $(n-1)$. Thus

$$Pr[-t_{1-\alpha/2}(n-1) \le t \le t_{1-\alpha/2}(n-1)]$$
$$= Pr[-t_{1-\alpha/2}(n-1) \le \frac{\bar{X} - \mu}{s/\sqrt{n}} \le t_{1-\alpha/2}(n-1)]$$
$$= 1 - \alpha,$$

where $t_q(n-1)$ represents the $q$-th quantile of the $t$ distribution with $(n-1)$ degrees of freedom. Because of the symmetric nature of the distribution, the quantiles satisfy the relation $t_q(n-1) = -t_{1-q}(n-1)$. Then, following the derivation in Section 8.1.2 exactly, the $100(1-\alpha)\%$ confidence interval for $\mu$ is found to be $\bar{X} \pm t_{1-\alpha/2}(n-1)\frac{s}{\sqrt{n}}$.

**Example 8.3.** Here we revisit Example 8.1. We are still interested, under the general assumptions of Example 8.1, in constructing a confidence interval for the mean lifetime of the tires produced by the company under scrutiny, but we are no longer ready to assume that the variance $\sigma^2$ is known (to be 2000, or any other prespecified value). In this case we replace $\sigma$ by the sample standard deviation $s$, the natural estimator of $\sigma$. For this data we have $\bar{X} = 48,939.4$ and $s = 2292.861$. As the sample size is 25, the pivotal element $t = \frac{\bar{X}-\mu}{s/\sqrt{n}}$ has a $t(24)$ distribution. The relevant quantile $t_{0.975}(24)$ equals 2.063899. Thus the required confidence interval is

$$\bar{X} \pm t_{0.975}(24)\frac{s}{\sqrt{n}} = 48,939.4 \pm 2.063899\frac{2292.861}{\sqrt{25}} = (47,992.95, 49,885.85).$$

Thus we are 95% confident that the unknown mean lies within the interval (47,992.95, 49,885.85). ‖

When we want to perform a test of hypothesis for the unknown mean $\mu$ of a normal distribution, adjusting for the unknown nuisance parameter $\sigma$, the test statistic corresponding to the null hypothesis is given by

$$t = \frac{\bar{X} - \mu_0}{s/\sqrt{n}}, \tag{8.6}$$

where $H_0 : \mu = \mu_0$ is the null hypothesis, and $\bar{X}$ and $s$ are the sample mean and the sample standard deviation based on an i.i.d. random sample $X_1, X_2, \ldots, X_n$, which is modeled by a normal $N(\mu, \sigma^2)$ distribution with unknown mean $\mu$ and unknown variance $\sigma$. As the quantity in Equation (8.6) has a $t$ distribution with $(n-1)$ degrees of freedom, the rejection region of this test can be easily determined. For example, if $\alpha$ is the level of significance and our alternative is $H_A : \mu < \mu_0$, we will reject the null hypothesis when the observed $t$ drops below the quantile $t_\alpha(n-1)$. For the greater than type alternative the rejection region is $\{t > t_{1-\alpha}(n-1)\}$. For the two-sided test the rejection region is $\{t > t_{(1-\alpha/2)}(n-1) \text{ or } t < -t_{(1-\alpha/2)}(n-1)\}$. Note that, by the symmetry of the $t$ distribution, $-t_{(1-\alpha/2)}(n-1) = t_{\alpha/2}(n-1)$.

**Example 8.4.** Suppose now we are interested, at level of significance $\alpha$, in testing the null hypothesis $H_0 : \mu = 50,000$ against the less than type alternative for the data and set up in Example 8.1, discarding the known $\sigma$ assumption. Our test statistic $t$ is as defined in Equation (8.6), and we will reject when the observed $t$ drops below $t_\alpha(n-1)$. Let $\alpha = 0.05$. The observed value of the test statistic in this case is

$$t = \frac{\bar{X} - \mu_0}{s/\sqrt{n}} = \frac{48,939.4 - 50,000}{2292.861/5} = -2.312831,$$

which is far less than the critical value $t_{.05}(24) = -1.710882$. Thus once again the null hypothesis is soundly rejected. The sampled information does not support the claim of the tire manufacturing company.

```
> t.test(tire.data,alternative="less",mu=50000)

One Sample t-test

data:  tire.data
t = -2.3128, df = 24, p-value = 0.01481
alternative hypothesis: true mean is less than 50000
```

The R output reconfirms our previous calculation, and also reports a p-value of 0.01481, sufficient to reject the null at level 0.05.     ‖

## 8.3    Two Sample $t$-Test: Independent Samples

So far we have been looking at inference problems involving the mean parameter of a single population. Until now we have been interested, directly of indirectly, in the actual numerical value of the parameter under study. In Examples 8.1–8.4 we have either tried to find numerical limits to the mean lifetime of the tires, or determine whether the mean lifetime is equal to 50,000 kilometers or less. However, we mentioned in the introduction to this chapter that, in statistical inference, sometimes our interest is comparative, and not necessarily in the actual numerical values of the characteristics under study. Suppose, for example, that the efficacy of a drug is measured by some quantitative standard. Our interest, when exploring the properties of a new drug, would be in checking whether the new one exceeds the existing one in terms of this standard, and not necessarily in the actual numerical efficacy attained by the new drug. Thus it is of vital importance to devise procedures which can statistically compare the behavior of two different samples.

Suppose that we have an i.i.d. random sample $X_1, X_2, \ldots, X_{n_1}$ from a normal $N(\mu_1, \sigma_1^2)$ population, and another i.i.d. random sample $Y_1, Y_2, \ldots, Y_{n_2}$ from a $N(\mu_2, \sigma_2^2)$ population. The samples are also independent among themselves. We want to undertake the following tasks.

(a) We want to find a point estimator for $\mu_1 - \mu_2$.

(b) We want to set up a confidence interval for $\mu_1 - \mu_2$.

(c) We want to test the hypothesis $H_0 : \mu_1 = \mu_2$, or $H_0 : \mu_1 - \mu_2 = 0$, against a suitable alternative.

All these tasks are to be performed without having explicit knowledge about the values of $\sigma_1^2$ and $\sigma_2^2$. For part (a), the obvious point estimator of the difference $\mu_1 - \mu_2$ is $\bar{X} - \bar{Y}$. However, we do not have a general, exact theory for constructing the confidence interval or doing the test of hypothesis required in parts (b) and (c). But the current literature does provide for this confidence interval and the corresponding test of hypothesis in one important special case. This case arises when the two populations may be assumed to have equal (although unknown) variances. Let this equal (but unknown) value be $\sigma^2$. It is then easy to check that the distribution of $\bar{X} - \bar{Y}$ is $N\left(\mu_1 - \mu_2, \sigma^2\left(\frac{1}{n_1} + \frac{1}{n_2}\right)\right)$ where $\bar{X}$ and $\bar{Y}$ are the indicated sample means. Thus, if the common variance $\sigma^2$ were known, we could use the statistic

$$Z = \frac{(\bar{X} - \bar{Y}) - (\mu_1 - \mu_2)}{\sqrt{\sigma^2\left(\frac{1}{n_1} + \frac{1}{n_2}\right)}} \tag{8.7}$$

as the pivotal element to set up confidence intervals for $(\mu_1 - \mu_2)$. However, the quantity $\sigma^2$ in the denominator is unknown, and we must replace it with some sample-based quantity. Since $\sigma^2$ is common to both populations, we must estimate it using the data from both populations. Noting that

$$\frac{(n_1 - 1)s_1^2}{\sigma^2} \sim \chi^2(n_1 - 1), \quad \frac{(n_2 - 1)s_2^2}{\sigma^2} \sim \chi^2(n_2 - 1),$$

where $s_1^2$ and $s_2^2$ are the two sample variances (see Section 7.10.1) and that these two estimators are from independent samples, we get

$$\frac{(n_1 - 1)s_1^2 + (n_2 - 1)s_2^2}{\sigma^2} \sim \chi^2(n_1 + n_2 - 2).$$

In the appendix we show that this implies that

$$s_p^2 = \frac{(n_1 - 1)s_1^2 + (n_2 - 1)s_2^2}{n_1 + n_2 - 2}$$

is an unbiased estimator of $\sigma^2$. The subscript $p$ stands for "pooled," since the estimator $s_p^2$ pools the information from both samples. We then replace $\sigma^2$ in Equation (8.7) with $s_p^2$; it can then be proved, exactly as was done in Section 7.10.2, that the quantity

$$t = \frac{(\bar{X} - \bar{Y}) - (\mu_1 - \mu_2)}{\sqrt{s_p^2\left(\frac{1}{n_1} + \frac{1}{n_2}\right)}} \tag{8.8}$$

has a $t(n_1 + n_2 - 2)$ distribution. Then, exactly as in Section 8.2, we can set up the $100(1 - \alpha)\%$ confidence interval for $(\mu_1 - \mu_2)$ as

$$(\bar{X} - \bar{Y}) \pm t_{1-\alpha/2}(n_1 + n_2 - 2)SE(\bar{X} - \bar{Y}),$$

where $t_{1-\alpha/2}(n_1 + n_2 - 2)$ is the quantile of the $t(n_1 + n_2 - 2)$ distribution corresponding to $q = 1 - \alpha/2$, and

$$SE(\bar{X} - \bar{Y}) = s_p\sqrt{\frac{1}{n_1} + \frac{1}{n_2}}$$

is the standard error of $(\bar{X} - \bar{Y})$.

**Example 8.5.** In Examples 8.1 and 8.2 we have done interval estimation and testing of hypothesis involving a single population mean. In the present example we compare two population means under a similar set up. We are interested in the comparison of the lifetime of the tires manufactured by two different companies; we refer to them as Company1 ($X$ population) and Company2 ($Y$ population) for notational convenience. In this example our purpose is to set up confidence intervals for the difference of the two population means; we will take up hypotheses involving the two means in Example 8.6. In order to set up the confidence interval under the theory described in this section, we assume that the variable (lifetime of tires) is normally distribution in both populations and the population variances of the lifetime of tires produced by the two companies are the same (although unknown). We have i.i.d. random samples of tires of size 25 from each of the companies, and their observed lifetimes (in kilometers) are as recorded below.

```
company1
50604 49945 47950 49812 52291 52591 47834 48274 51536 50072
49804 50326 50430 50105 49653 47230 51601 52321 48927 47048
49630 49314 48909 48809 49142
```

```
company2
48751 49557 47949 51924 47537 47956 48986 47712 46714 51758
49605 44630 51910 48318 51076 48339 45918 48092 50249 48628
51346 48386 50169 46509 47071
```

The summary statistics for the first company, including the sample mean and the sample standard deviation are

$$\bar{X} = 49,766.32, \quad s_1 = 1515.748.$$

Similarly the statistics for the second company are

$$\bar{Y} = 48,763.6, \quad s_2 = 1922.337.$$

Assuming that the population variances are equal, the pooled estimate of the standard deviation is obtained as

$$s_p = \sqrt{\frac{(n_1 - 1)s_1^2 + (n_2 - 1)s_2^2}{n_1 + n_2 - 2}} = \sqrt{\frac{24 \times 2,297,493 + 24 \times 3,695,380}{48}} = 1731.022.$$

Thus the standard error of $\bar{X} - \bar{Y}$ is

$$SE(\bar{X} - \bar{Y}) = s_p \sqrt{\frac{1}{25} + \frac{1}{25}} = 489.6069.$$

Thus the 95% confidence interval for the difference of the means $\mu_1 - \mu_2$ will be $(\bar{X} - \bar{Y}) \pm t_{0.975}(48) \times SE(\bar{X} - \bar{Y})$, which on simplification turns out to be

$$(49,766.32 - 48,763.6) \pm 2.010635 \times 489.6069 = (18.299, 1987.141).$$

The same information is given in the R output below, which gives the results for the confidence interval as well as the test of hypothesis.

Since the degrees of freedom of the $t$ statistic is large, we could have used the $Z$ quantiles also in this case instead of the actual $t$ quantiles.

```
> t.test(Company1,Company2,var.equal=TRUE)

Two Sample t-test

data:  Company1 and Company2
t = 2.048, df = 48, p-value = 0.04605
alternative hypothesis: true difference in means is not equal to 0
95 percent confidence interval:
  18.29936 1987.14064
sample estimates:
mean of x mean of y
 49766.32  48763.60
```

$\|$

Now we consider the problem of testing the null hypothesis $H_0 : \mu_1 = \mu_2$ under a suitable alternative. The statistic in Equation (8.8), which has a $t(n_1 + n_2 - 2)$ distribution, reduces to

$$t = \frac{(\bar{X} - \bar{Y})}{\sqrt{s_p^2 \left( \frac{1}{n_1} + \frac{1}{n_2} \right)}} \tag{8.9}$$

when the null hypothesis is correct; we will use this statistic for testing $H_0 : \mu_1 = \mu_2$. Depending on the alternative, we will use appropriate quantiles of the $t(n_1 + n_2 - 2)$ distribution to define the rejection region.

**Example 8.6.** Consider the problem of testing the null hypothesis $H_0 : \mu_1 = \mu_2$ against the not equal to alternative for the data and the model described in Example 8.5. As the hypothesis is two sided, we will reject the null when the observed $t$ exceeds $t_{1-\alpha/2}(n_1 + n_2 - 2)$ or drops below $-t_{1-\alpha/2}(n_1 + n_2 - 2)$. Let $\alpha = 0.05$ represent the level of significance. Thus $t_{0.025}(48)$ and $t_{0.975}(48)$ will represent the lower and upper cutoffs for rejection. Noting that $t_{0.025}(48) = -t_{0.975}(48)$, our rejection region includes all observed values of $t$ outside the interval

$$(t_{0.025}(48), t_{0.975}(48)) = (-2.010635, 2.010635).$$

As our observed value

$$t = \frac{(\bar{X} - \bar{Y})}{\sqrt{s_p^2 \left( \frac{1}{n_1} + \frac{1}{n_2} \right)}} = \frac{(49766.32 - 48763.6)}{489.6069} = 2.04801$$

falls in the rejection region, the null hypothesis is rejected; the p-value for this test is reported in the last set of R outputs as 0.04605. As far as the sample data is concerned, there is enough evidence to suggest that the population means are not the same.

We have also used the above data to construct side by side boxplots for the two distributions to get a visual idea of the comparison of the distributions (see Figure 8.3). Visually, the distributions appear to be too far separated for the means to be equal. Note, however, that the test statistic in the above example is pretty close to the threshold, and the decision to reject is somewhat of a tentative decision.                                                    ‖

**Side by Side Boxplots**

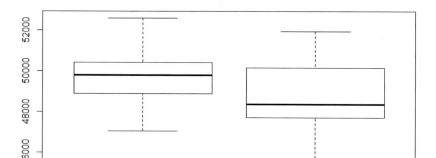

**FIGURE 8.3**
Side by side boxplots for the company data. Company 1 appears to have a higher mean.

*Remark* 8.1. We have done the above test under the assumption that the two variances are equal. Suppose that there is reason to believe that the two variances are not equal. In that case, the two sample $t$-test that we have just described is no longer the appropriate test. In fact in this case there does not exist any exact test. Welch's two sample $t$-test is an approximate test which allows one to test this hypothesis without the equality of variances assumption. We will not elaborate on this test, but the test can be performed in R by dropping the `var.equal=TRUE` subcommand. Welch's test has a complicated formula for calculating the degrees of freedom, and the standard error for the statistic $(\bar{X} - \bar{Y})$ is taken to be

$$\sqrt{\frac{s_1^2}{n_1} + \frac{s_2^2}{n_2}}.$$

Welch's two sample $t$-test, applied to the data considered in Example 8.6, gives the following result. The observed p-value is quite close to the equal variance case.

```
> t.test(Company1,Company2,alternative="two.sided")

Welch Two Sample t-test

data:  Company1 and Company2
t = 2.048, df = 45.523, p-value = 0.04635
alternative hypothesis: true difference in means is not equal to 0
95 percent confidence interval:
   16.91376 1988.52624
sample estimates:
mean of x mean of y
 49766.32  48763.60
```

## 8.4   Two Sample $t$-Test: Dependent (Paired) Samples

The situation considered in the previous section deals with two independent samples from two populations. Sometimes it so happens that the observations, although technically representing two distinct samples, are inherently dependent, so that they cannot be covered under the independent sample scenario. The most prominent example of this is the case involving matched pairs. Here the observations come as (dependent) pairs. For example, suppose we are taking samples from a group of individuals (such as workers in a particular factory) and the variable of interest represents a quantification of the work efficiency of the individual. If the (same group of) individuals are rated in terms of their work efficiency before and after a short-term training program, we have two samples which are dependent. Observations are paired because we have observations from the same individuals before and after the training program. In this case we may want to test whether the efficiency of the worker has increased after training; the appropriate procedure will be to test for the null hypothesis $H_0 : \mu_1 = \mu_2$ against $H_A : \mu_1 < \mu_2$ where $\mu_1$ and $\mu_2$ represent the mean efficiency before and after training.

The problem will remain the same if we want to test whether the efficiency of the workers remains the same before and after a motivational pep-talk or whether the blood pressure of a patient remains the same before and after the administration of a drug. However, not all paired sample examples are the before-and-after type. Darwin reported the results of an experiment where pairs of self-fertilized and cross-fertilized plants were germinated in the same pot. One could use the two resulting samples to test whether cross-fertilized plants have a superior growth rate in relation to self-fertilized plants. Paired data may be used to test for differences relating to observations from both eyes, both ears, both feet, etc., or observations coming from two identical twins.

As in the previous section, suppose that we have an i.i.d. random sample $X_1, X_2, \ldots, X_n$ from a $N(\mu_1, \sigma_1^2)$ population and another i.i.d. random sample $Y_1, Y_2, \ldots, Y_n$ from a $N(\mu_2, \sigma_2^2)$ population; however, the samples are no longer independent, and observations $(X_i, Y_i), i = 1, \ldots, n$ are paired. We want to test whether the means of the $X$ and $Y$ populations are different. The natural thing here is to take the differences $D_i = Y_i - X_i$. If the null hypothesis of the equality of means of the two populations (those of $X$ and $Y$) is true, then the $D$ observations represent an i.i.d. random sample from a normal distribution which has zero mean. Thus the problem is reduced to that of a single sample rather than two samples. One can test the hypothesis $H_0 : \mu_D = 0$, where $\mu_D$ is the population mean of the $D$ observations. Following the approach of Section 8.2, the relevant test statistic is seen to be

$$\frac{\bar{D}}{s_D/\sqrt{n}}, \tag{8.10}$$

where $\bar{D}$ and $s_D$ are the sample mean and sample standard deviations of the sample of $D$ observations; this statistic has a $t(n-1)$ distribution when the null hypothesis is true.

**Example 8.7.** A company which is pushing for the sales of a particular product wishes to determine whether an advertising campaign it is running has any real effect. Fifteen possible customers are randomly selected, and their purchase potentials are recorded before and after they see the advertisement (on a scale of 1 to 20, on the basis of their own assessment). The relevant R output is given in the table below. When the test given by the statistic in Equation (8.10) is performed for the null $H_0 : \mu_D = 0$ against the alternative $H_A : \mu_D > 0$, the null is easily rejected at level $\alpha = 0.05$; note that $t_{0.95}(14) = 1.76131$. As the observed statistic is $t = 3.5064$, there is enough evidence in the data to suggest that the advertising campaign has significantly enhanced the mean purchase potential.

| Serial | 1 | 2 | 3 | 4 | 5 | 6 | 7 | 8 | 9 | 10 | 11 | 12 | 13 | 14 | 15 |
|---|---|---|---|---|---|---|---|---|---|---|---|---|---|---|---|
| Before $(B)$ | 12 | 16 | 14 | 13 | 19 | 12 | 17 | 16 | 12 | 18 | 09 | 13 | 20 | 15 | 19 |
| After $(A)$ | 19 | 15 | 15 | 15 | 20 | 16 | 17 | 17 | 17 | 19 | 12 | 19 | 20 | 16 | 20 |
| Difference $(D = A - B)$ | 7 | −1 | 1 | 2 | 1 | 4 | 0 | 1 | 5 | 1 | 3 | 6 | 0 | 1 | 1 |

```
> t.test(Difference,alternative="greater")

One Sample t-test

data:   Difference
t = 3.5064, df = 14, p-value = 0.001745
alternative hypothesis: true mean is greater than 0
```

‖

## 8.5   Analysis of Variance

Analysis of variance (ANOVA) is a technique used to analyze the effect of
the levels of an independent categorical variable on a continuous dependent
response. It may be considered to be the natural extension of the test of
equality of two means from independent samples considered in Section 8.3. In
analysis of variance, we will test for the equality of several (more than two)
means. The different categories (or groups) may be considered to have been
generated by the different levels of the categorical variable referred to above.

The test proceeds by statistically analyzing whether the observed means
of the different categories of the independent variable are too different to have
occurred just by chance. If the observed group means do not differ significantly, then we infer that the independent variable did not have an effect
on the dependent variable and the group means are the same. Otherwise,
the inference is that the independent variable has an effect on the dependent
variable and the group means are different. In this case further comparisons
between the individual groups may be necessary to explore how this relationship controls the dependent variable in different categories of the independent
variable. The independent variable is often referred to in the analysis of variance literature as a *factor*. In our description we will only consider the case
of one independent variable, which is referred to in the literature as one-way
analysis of variance (one-way ANOVA). We should be aware, however, that
this procedure can be carried to higher levels where several factors (as well as
their interactions) may be simultaneously considered.

Suppose that this independent variable (factor) has $k$ categories. The factor
levels (categories) have been referred to by many other names in the literature, such as groups, treatments, populations, etc. It is customary to assume
that the distribution of the continuous dependent variable is normal for each
category, and in the $i$-th category this distribution has mean $\mu_i$ and variance
$\sigma^2$. Note that, although we model the means as being potentially different,
the variances are assumed to be equal. This is thus a direct generalization
of the set up considered in Section 8.3. We are interested in testing whether
the means for all the categories are the same. We assume that independent

random samples of the dependent response are available from each individual category. If the sample size for the $i$-th category is $n_i$, then our set up may be described as

$$X_{ij} = \mu_i + \epsilon_{ij}, \ j = 1, \ldots, n_i, \ i = 1, \ldots, k. \tag{8.11}$$

Here $X_{ij}$ is the response of the $j$-th observation in the $i$-th category, $\mu_i$ is the mean response for the $i$th category, and the $\epsilon_{ij}$ are independent normal variables with mean zero and variance $\sigma^2$, representing the part unexplained by the model. Our null hypothesis can then be expressed as

$$H_0 : \mu_1 = \mu_2 = \cdots = \mu_k, \tag{8.12}$$

which is to be tested against the alternative $H_A$ : not all the means are equal. Sometimes the set up in Equation (8.11) is alternatively expressed as

$$X_{ij} = \mu + \alpha_i + \epsilon_{ij}, \ j = 1, \ldots, n_i, \ i = 1, \ldots, k, \tag{8.13}$$

where $\mu$ now represents the overall mean, and $\alpha_i$ represents the deviation of the mean of the $i$-th category from the overall mean; we will refer to the $\alpha_i$s as the category effects. These constants must satisfy $\sum_{i=1}^{k} \alpha_i = 0$. In this set up our null hypothesis can be rewritten as

$$H_0 : \alpha_1 = \alpha_2 = \cdots = \alpha_k = 0 \tag{8.14}$$

against the alternative $H_A$ : not all the $\alpha_i$s are zero.

Let $n = \sum_{i=1}^{k} n_i$ represent the overall sample size. We will let $\bar{X}_i$, $i = 1, 2, \ldots, k$ and $\bar{X}$ represent the mean of the $i$-th category, $i = 1, 2, \ldots, k$ and the overall mean, respectively; these quantities are defined as

$$\bar{X}_i = \frac{1}{n_i} \sum_{j=1}^{n_i} X_{ij}, \quad \bar{X} = \frac{1}{n} \sum_{i=1}^{k} \sum_{j=1}^{n_i} X_{ij}. \tag{8.15}$$

Some simple algebra establishes that the total sum of squares of deviation of each observation from the overall mean may be decomposed as

$$\sum_{i=1}^{k} \sum_{j=1}^{n_i} (X_{ij} - \bar{X})^2 \ = \ \sum_{i=1}^{k} n_i (\bar{X}_i - \bar{X})^2 + \sum_{i=1}^{k} \sum_{j=1}^{n_i} (X_{ij} - \bar{X}_i)^2$$

$$SSTO \ = \ SSB \ + \ SSE. \tag{8.16}$$

Here $SSTO$ stands for the total sum of squares, $SSB$ stands for the between groups sum of squares (or sum of squares due to groups) and $SSE$ represents the sum of squares due to error (or sum of squares within groups). It may also be shown that $SSTO$, $SSB$ and $SSE$ are associated with $(n-1)$, $(k-1)$ and $(n-k)$ degrees of freedom, respectively. Like the sum of squares, the degrees of freedom are also additive, so that

$$df(SSTO) = n - 1 = (k - 1) + (n - k) = df(SSB) + df(SSE),$$

where $df(SSTO)$ represents the degrees of freedom of $SSTO$ and so on. The sum of squares $(SS)$ divided by the corresponding degrees of freedom generates the mean sum of squares $(MS)$. Thus

$$MSB = \frac{SSB}{k-1} \quad \text{and} \quad MSE = \frac{SSE}{n-k}.$$

It turns out that, when the null hypothesis of equality of all the means hold, the quantity

$$F = \frac{MSB}{MSE}$$

has an $F$ distribution with degrees of freedom $(k-1, n-k)$, so that a test for the null hypothesis can be constructed using this statistic. If the null is true, the value of the observed $F$ should be in the region that would be consistent with an $F(k-1, n-k)$ distribution. When the null hypothesis is false, the observed values of the $F$ statistic tend to be higher than those corresponding to the $F(k-1, n-k)$ variable. Thus the experimenter should reject the null hypothesis at level $\alpha$ when the observed $F$ exceeds $F_{1-\alpha}(k-1, n-k)$, the relevant quantile of the $F(k-1, n-k)$ distribution.

It is customary to present the results of an analysis of variance exercise in the form of the following table. The first column indicates the sources of variation. When analysis of variance is done using some software like R, the software will also produce an additional column providing the p-value of the test statistic.

**TABLE 8.1**
ANOVA table

| Source | $df$ | $SS$ | $MS$ | $F$ |
| --- | --- | --- | --- | --- |
| Group | $k-1$ | $SSB$ | $MSB = \frac{SSB}{(k-1)}$ | $\frac{MSB}{MSE}$ |
| Error | $n-k$ | $SSE$ | $MSE = \frac{SSE}{(n-k)}$ | |
| Total | $n-1$ | $SSTO$ | | |

The ANOVA method tests the null hypotheses that group means do not differ. It is NOT a test of differences in variances (in spite of the name), but rather assumes homogeneity of variances. Given that the procedure tests for the equality of means by analyzing the sum of squares, the name ANOVA is really a reasonable one, although it can perplex the first time reader.

The three main assumptions of ANOVA are that (i) samples should be independent, (ii) the distribution of the variable is normal for each category, and (iii) the variances of the categories are equal.

The three most important factors leading to significance in the ANOVA

*F*-test are (i) the magnitude of the difference of the means, (ii) the sample sizes in each group, and (iii) the magnitude of the variance. Larger sample sizes give more reliable information and even small differences in means may be significant if the sample sizes are large enough. Larger difference in means and a smaller variance can also lead to more significant results.

**Example 8.8.** (German Credit Data) The following table cross-classifies 1000 individuals in terms of their sex/marital status and purpose of credit. The categories for sex/marital status variable (coded as Sex.Marital.Status in R) are male divorced/single, male married/widowed and female. Similarly, the purpose for which the loan is sought has the classes new car, used car, home related and other.

```
                    Purpose
Sex.Marital.Status   1    2    3    4
                 1  27   89  128  116
                 2  70   85  189  204
                 3   6    7   47   32
```

Thus, when we are performing one-way ANOVA, there are two possible independent variables, (i) Sex.Marital.Status and (ii) Purpose. Also, there are two dependent variables, (i) amount of loan asked for and (ii) duration of loan. Thus it is possible to do $2 \times 2 = 4$ one-way ANOVA tests with this data. We present the R commands and outputs for each of these four tests in the following. For example, when the amount of loan (Credit.Amount) is the dependent variable and sex/marital status (Sex.Marital.Status) is the independent variable, there are $k = 3$ categories of the independent variable. We have $F_{0.95}(2, 997) = 3.004752$, so that we will reject the relevant null (of equality of means of the three categories) at level $\alpha = 0.05$ whenever the observed $F$ is larger than this value. The observed value of $F$, as provided in the R output below, is 16.19, so that the hypothesis is soundly rejected. In each of the other three cases also, the null hypothesis is rejected.

```
> a1 <- aov(Credit.Amount~Sex.Marital.Status, data=German.Credit.M1)
> a1
Call:
   aov(formula = Credit.Amount ~ Sex.Marital.Status,
 data = German.Credit.M1)

Terms:
                Sex.Marital.Status  Residuals
Sum of Squares          250342733  7709616837
Deg. of Freedom                 2          997

Residual standard error: 2780.794
```

Estimated effects may be unbalanced

```
> summary(a1)
                     Df    Sum Sq   Mean Sq F value   Pr(>F)
Sex.Marital.Status    2 2.503e+08 125171367   16.19 1.21e-07 ***
Residuals           997 7.710e+09   7732815
---
Signif. codes:  0 *** 0.001 ** 0.01 * 0.05 . 0.1   1

> a2 <- aov(Duration.of.Credit.Month ~ Sex.Marital.Status,
  data=German.Credit.M1)
> a2
Call:
   aov(formula = Duration.of.Credit.Month ~ Sex.Marital.Status,
   data = German.Credit.M1)

Terms:
              Sex.Marital.Status Residuals
Sum of Squares            2446.2  142823.4
Deg. of Freedom                2       997

Residual standard error: 11.96884
Estimated effects may be unbalanced

> summary(a2)
                     Df Sum Sq Mean Sq F value   Pr(>F)
Sex.Marital.Status    2   2446  1223.1   8.538 0.000211 ***
Residuals           997 142823   143.3
---
Signif. codes:  0 *** 0.001 ** 0.01 * 0.05 . 0.1   1

> a3 <- aov(Duration.of.Credit.Month ~ Purpose, data=German.Credit.M1)
> a3
Call:
   aov(formula = Duration.of.Credit.Month ~ Purpose,
 data = German.Credit.M1)

Terms:
                Purpose Residuals
Sum of Squares  3503.49 141766.10

Deg. of Freedom       3       996

Residual standard error: 11.93044
Estimated effects may be unbalanced

> summary(a3)
          Df Sum Sq Mean Sq F value   Pr(>F)
Purpose    3   3503  1167.8   8.205 2.13e-05 ***
```

```
Residuals    996 141766    142.3
---
Signif. codes:  0 *** 0.001 ** 0.01 * 0.05 . 0.1   1

> a4<- aov(Credit.Amount ~ Purpose, data=German.Credit.M1)
> a4
Call:
   aov(formula = Credit.Amount ~ Purpose, data = German.Credit.M1)

Terms:
                   Purpose  Residuals
Sum of Squares   660289896 7299669675
Deg. of Freedom          3        996

Residual standard error: 2707.21
Estimated effects may be unbalanced

> summary(a4)
             Df    Sum Sq   Mean Sq F value Pr(>F)
Purpose       3 6.603e+08 220096632   30.03 <2e-16 ***
Residuals   996 7.300e+09   7328986
---
Signif. codes:  0 *** 0.001 ** 0.01 * 0.05 . 0.1   1
```

‖

## 8.6    Chi-Square Tests

Some very important practical problems involve performing tests of hypothesis where the distributions of the test statistics are approximated by the chi-square distribution under the null. We will look at two such cases here.

(a) The first case involves performing tests of goodness-of-fit. Here we have a categorical variable, which can take values in a finite number of categories. The null hypothesis may specify probabilities of each of the categories (or cells), and our interest could be in testing whether the specified cell probabilities under the null describe the true distribution. More generally, the null hypothesis may simply give a structure on the probabilities without stating the exact values, in which case one needs to estimate one or more parameters to get the exact null probabilities.

(b) The second case involves tests of independence. In this case we have cross-classified tabular data, with the rows and columns representing the levels or categories of two different variables. We are interested in testing whether the two variables are independent. Tables representing cross-classified tabular data are sometimes called contingency tables.

## 8.6.1   Goodness-of-Fit Tests

Tests that deal with item (a) of Section 8.6 are called goodness-of-fit tests. Suppose we have have a die, and we want to know whether it is a fair die. That is, does each of the six faces have an equal chance of showing up when the die is rolled? If $p_i$ represents the probability that the $i$-th face will show up when rolled, the relevant null hypothesis for the goodness-of-fit test is

$$H_0 : p_1 = p_2 = \cdots = p_6 = \frac{1}{6},$$

which is to be tested against the alternative $H_A$ : not $H_0$. It is a goodness-of-fit test in the sense that it tries to determine whether the vector $(1/6, 1/6, \ldots, 1/6)$ provides a good fit to the data generated by an unknown probability vector.

A second example may be constructed with mutual funds which are typically subject to market risks. If we consider the categories Absolutely safe, Very safe, Somewhat safe, Not very safe and Not at all safe, the interest may be in exploring how dynamic the population risk perception for the different groups is. Suppose, from past data, we know that the proportion of investors rating mutual funds on safety is $p_1^0, p_2^0, \ldots, p_5^0$ for these five categories, $\sum_{i=1}^{5} p_i^0 = 1$. The experimenter wishes to know whether the present perception indicates a change from the past. In this case the experimenter will start by assuming that the past perceptions are still intact, and this is what the null hypothesis will represent. On the basis of present data, the experimenter will then test the null hypothesis

$$H_0 : p_1 = p_1^0, p_2 = p_2^0, \ldots, p_5 = p_5^0,$$

where $p_i$ is the true (unknown) value of the proportion of present investors rating the safety to be in the $i$-th category, and $p_i^0$ are the past (known) proportions for the same.

Yet another example may involve categories that are in the form of count data, representing values, $0, 1, 2, \ldots, k$. We may want to know, for example, if the probabilities of the categories have the structure specified by the Poisson probability model, so that the probability of the $i$-th cell is $p_i = e^{-\lambda}\lambda^i/i!$, $i = 0, 1, \ldots, k - 1$ for some known or unknown $\lambda$, for $i = 1, \ldots, k - 1$. In the $k$-th category we roll the entire remaining probability $Pr[X \geq k]$ as given by the Poisson law with parameter $\lambda$. When $\lambda$ is known, we can write down the hypothesis for the goodness-of-fit test exactly as in the two previous cases since each of the cell probabilities is now completely specified by the null hypothesis.

When $\lambda$ is unknown, we need to estimate it from the data. In this case the hypothesis does not specify the cell probabilities exactly. Now we will simply state the hypothesis as $H_0 : X \sim Poisson(\lambda)$ against $H_A$ : not $H_0$, where $\lambda$

needs to be estimated from the data. The usual practice is to estimate the Poisson mean parameter $\lambda$ with the sample mean. Thus we determine the exact cell probabilities using the $Poisson(\hat{\lambda})$ distribution, where $\hat{\lambda} = \bar{X}$ is the estimate obtained from the data. Note that, in this case, $\bar{X}$ is the maximum likelihood estimator of $\lambda$.

Suppose the null hypothesis is exactly specified. The goodness-of-fit test proceeds as follows. Let there be $k$ categories, and let the null specify the probabilities of the $k$ cells to be $p_1^0, p_2^0, \ldots, p_k^0$, respectively, $\sum_{i=1}^{k} p_i^0 = 1$. We choose a sample of size $n$ from the population of interest. Let $O_1, O_2, \ldots, O_k$ be the number of observations in the sample that belong to the first cell, second cell, ..., $k$-th cell, respectively; $\sum_{i=1}^{k} O_i = n$. On the other hand, assuming that the null hypothesis is true, the expected number of observations in these $k$ cells would be $np_1^0, np_2^0, \ldots, np_k^0$, respectively. When the null holds, the observed and the expected frequencies should be close. Denoting the expected frequencies as $E_1, E_2, \ldots, E_k$, our test statistic for testing the null hypothesis is given by the statistic

$$\chi^2 = \sum_{i=1}^{k} \frac{(O_i - E_i)^2}{E_i}. \tag{8.17}$$

If the null hypothesis is true, this statistic has an approximate $\chi^2$ distribution with degrees of freedom $(k-1)$. Small values of the statistic indicate that there is a good match between the observed and expected frequencies. We reject when the values of the statistic are very large, indicating that the observed data are poorly matched by the expected frequencies.

This needs further refinement when parameters are to be estimated, as in the Poisson model example given above where $\lambda$ is estimated with $\bar{X}$. In this case we use the same statistic as in Equation (8.17); however, the degrees of freedom has to be adjusted for the number of parameters estimated. Formally, if $s$ parameters have to be estimated, the chi-square statistic in Equation (8.17) has an asymptotic chi-square distribution with degrees of freedom $(k - s - 1)$, rather than just $(k-1)$; thus one degree of freedom is lost for every parameter estimated.

It is necessary that the expected frequencies of the cells should not be very small for the chi-square approximation to work satisfactorily. As a rule of thumb, we require that each expected frequency should be at least 5, and we pool adjacent cells in case this is not achieved for the current grouping.

**Example 8.9.** In today's competitive business market, the long-term success of electrical and electronics products depends not only on the quality of the product itself, but also on the level of service that the company provides for repair and maintenance. A major manufacturer of computers, printers and accessories conducted a very large nationwide survey in the United States, which followed up its customers for a period of two years after purchase and recorded their level of satisfaction for the after-sale service of the company. The satisfaction was measured in a five-point scale, ranging from Excellent to Poor. The nationwide proportions of the different categories based on the ratings of the individuals surveyed are as follows.

| Excellent | Very Good | Good | Fair | Poor |
|-----------|-----------|------|------|------|
| 15% | 22% | 32% | 16% | 15% |

The company opened its operations in India at a later date and carried out a similar customer satisfaction survey for an equivalent period on a smaller scale. A total of 480 customers were surveyed, and the first row of the following table presents the observed frequency of ratings in each of the five categories. The second row provides the expected frequencies in each class, if the category proportions of the US survey were the correct representations of the Indian scenario. The expected frequency for each class is obtained by multiplying the total frequency by the US proportion for that class. Thus the expected frequency of the class Excellent is obtained as $480 \times 0.15 = 72$ and so on.

| Excellent | Very Good | Good | Fair | Poor |
|-----------|-----------|------|------|------|
| 66 | 140 | 129 | 85 | 60 |
| 72.0 | 105.6 | 153.6 | 76.8 | 72.0 |

We want to determine if the Indian data follows a distributional pattern which matches the US proportions. Thus

$$H_0 : p_1 = 0.15, p_2 = 0.22, p_3 = 0.32, p_4 = 0.16, p_5 = 0.15$$

is the formal null hypothesis, to be tested against the alternative hypothesis $H_A :$ not $H_0$.

Our chi-square statistic is given by

$$\chi^2 = \frac{(66 - 72)^2}{72} + \frac{(140 - 105.6)^2}{105.6} + \cdots + \frac{(60 - 72)^2}{72} = 18.5214.$$

The asymptotic null distribution is $\chi^2(4)$, and at level of significance 0.05, we reject when the observed statistic exceeds $\chi^2_{0.95}(4) = 9.487729$. Our observed statistic is way above the critical value and the null hypothesis is rejected. As far as the sampled data is concerned, there is enough evidence to suggest that the distribution of customer satisfaction in India is different from the established distribution in the United States.

```
> chisq.test(c(66,140,129,85,60),p=c(.15,.22,.32,.16,.15))

Chi-squared test for given probabilities

data:  c(66, 140, 129, 85, 60)
X-squared = 18.5214, df = 4, p-value = 0.0009757
```

||

**Example 8.10.** It is believed that the pattern of the number of clients visiting an ATM machine in a prominent office location between the hours of 10 in the morning and 12 noon is well modeled by a Poisson distribution. In particular, we want to test whether the number of arrivals in any five-minute interval during this period has a $Poisson(3)$ distribution. Data is taken over 120 randomly chosen intervals of length five minutes during a particular month of the year. The corresponding expected frequencies are calculated under the $Poisson(3)$ distribution. The frequencies are displayed in the following table.

| Number of Customers | 0 | 1 | 2 | 3 | 4 | 5 | 6 | 7 | $\geq 8$ |
|---|---|---|---|---|---|---|---|---|---|
| Observed Frequency | 16 | 27 | 35 | 21 | 10 | 9 | 1 | 0 | 1 |
| Expected Frequency | 5.97 | 17.92 | 26.88 | 26.88 | 20.16 | 12.10 | 6.05 | 2.59 | 1.43 |

However, the expected frequencies of the last two classes are smaller than 5; the expected frequency remains smaller than 5 when we merge these two classes. So we merge the last three classes and the last class is interpreted as Number of customers $\geq 6$; this class has an observed frequency of 2 and an expected frequency of 10.07. The merged table is given below. We then perform a chi-square goodness-of-fit test with this data using seven classes.

| Number of Customers | 0 | 1 | 2 | 3 | 4 | 5 | $\geq 6$ |
|---|---|---|---|---|---|---|---|
| Observed Frequency | 16 | 27 | 35 | 21 | 10 | 9 | 2 |
| Expected Frequency | 5.97 | 17.92 | 26.88 | 26.88 | 20.16 | 12.10 | 10.07 |

The hypothesis being tested can be expressed as $H_0 : X \sim Poisson(3)$, to be tested against the alternative $H_A :$ not $H_0$. The $\chi^2$ statistic is

$$\chi^2 = \frac{(16 - 5.97)^2}{5.97} + \frac{(27 - 17.92)^2}{17.92} + \cdots + \frac{(2 - 10.07)^2}{10.07} = 37.5417.$$

More details are given in the R output below. The test leads to solid rejection at $\alpha = 0.05$. Clearly, the $Poisson(3)$ distribution does not adequately represent this dataset. For reference, the quantile of the $\chi^2(6)$ distribution corresponding to $q = 0.95$ is 12.59159.

```
> pstar=c(dpois(c(0:5),3),1-ppois(5,3))
> chisq.test(c(16,27,35,21,10,9,2),p=pstar)

Chi-squared test for given probabilities

data:  c(16, 27, 35, 21, 10, 9, 2)
X-squared = 37.5417, df = 6, p-value = 1.381e-06
```

This does not, however, lead to the conclusion that the Poisson distribution itself is an inaccurate model for the data in this case. Perhaps there is another value of the parameter $\lambda$ for which the expected frequencies will fit the data. We estimate the mean parameter of the Poisson distribution from the data. Using the sample mean $\bar{X}$ (the maximum likelihood estimator of $\lambda$) to estimate the value of $\lambda$, we find that, in this case, the estimate is $\hat{\lambda} = \bar{X} = 2.158333$. Deriving the expected frequencies under the *Poisson*(2.158333) model, we get the expected and observed frequencies described in the following table.

| Number of Customers | 0 | 1 | 2 | 3 | 4 | 5 | 6 | 7 | $\geq 8$ |
|---|---|---|---|---|---|---|---|---|---|
| Observed Frequency | 16 | 27 | 35 | 21 | 10 | 9 | 1 | 0 | 1 |
| Expected Frequency | 13.86 | 29.93 | 32.29 | 23.22 | 12.53 | 5.41 | 1.95 | 0.60 | 0.16 |

In order to meet the requirement of having all expected frequencies greater than 5, we merge the last four cells. The relevant table is then as given below.

| Number of Customers | 0 | 1 | 2 | 3 | 4 | $\geq 5$ |
|---|---|---|---|---|---|---|
| Observed Frequency | 16 | 27 | 35 | 21 | 10 | 11 |
| Expected Frequency | 13.86 | 29.92 | 32.29 | 23.22 | 12.53 | 8.16 |

Direct calculation gives a $\chi^2$ statistic of 2.5543. As $\chi^2_{0.95}(4) = 9.487729$, we cannot reject the null hypothesis. The p-value for this test is 0.3651. We have used 4 degrees of freedom in these calculations. ‖

## 8.6.2 Tests of Independence

We use the following contingency table from the German Credit Data to motivate the problem.

| | Purpose | | | | Total |
|---|---|---|---|---|---|
| Sex/ | 27 | 89 | 128 | 116 | 360 |
| Marital | 70 | 85 | 189 | 204 | 548 |
| Status | 6 | 7 | 47 | 32 | 92 |
| Total | 103 | 181 | 364 | 352 | 1000 |

A total of 1000 people who have applied for loans have been cross-classified according to their sex/marital status and purpose of loan. See the description of the data in Section 8.5. This has generated the $3 \times 4$ table given above. Are the two variables independent? We have already looked, in Section 6.4.2, at the definition of independence of random variables when the joint (bivariate) distribution of the two random variables is given. However, in this case, we do not know the distribution exactly, but have a classification based on a sample from the joint bivariate population. Does the sample behave as if it were generated from a joint distribution where the components are independent?

The contingency table of expected frequencies under independence is given below. The frequency in each cell is the product of the corresponding row and margin totals of the original table divided by the overall total frequency (1000).

| | Purpose | | | | Total |
|---|---|---|---|---|---|
| Sex/ | 37.08 | 65.16 | 131.04 | 126.72 | 360.00 |
| Marital | 56.44 | 99.19 | 199.47 | 192.90 | 548.00 |
| Status | 9.48 | 16.65 | 33.49 | 32.38 | 92.00 |
| Total | 103.00 | 181.00 | 364.00 | 352.00 | 1000.00 |

For a $k \times l$ contingency table, we denote the observed frequencies by $O_{ij}$, $i = 1, \ldots, k$, and $j = 1, 2, \ldots, l$. Let the corresponding expected frequencies under independence be $E_{ij}$. Then the $\chi^2$ test statistic for independence is

$$\chi^2 = \sum_{i=1}^{k} \sum_{j=1}^{l} \frac{(O_{ij} - E_{ij})^2}{E_{ij}}. \tag{8.18}$$

When the null hypothesis of independence is true, the $\chi^2$ statistic has an asymptotic $\chi^2((k-1) \times (l-1))$ distribution. The null hypothesis

$H_0$ : The row variable and the column variable are independent

is rejected at level $\alpha$ when the observed statistic is exceeds $\chi^2_{1-\alpha}((k-1) \times (l-1))$.

For the present data, $k = 3$ and $l = 4$, so that the statistic has $2 \times 3 = 6$ degrees of freedom. The observed value is

$$\chi^2 = \frac{(27 - 37.8)^2}{37.8} + \frac{(89 - 65.16)^2}{65.16} + \cdots + \frac{(32 - 32.38)^2}{32.38} = 31.24.$$

We have $\chi^2_{0.95}(6) = 12.5916$, so that the null hypothesis is soundly rejected. There is enough evidence in the data to suggest that the row and the column variables (sex/marital status and purpose) are not independent at level of significance 0.05.

The relevant R output is appended below.

```
> a=c(27,89,128,116,70,85,189,204,6,7,47,32)
> A=matrix(a,nrow=3,ncol=4,byrow=T)
> chisq.test(A)

Pearson's Chi-squared test

data:  A
X-squared = 31.2402, df = 6, p-value = 2.281e-05
```

## 8.7   Inference about Proportions

So far we have considered point estimation, hypothesis testing and interval estimation in relation to the mean parameters of one or more continuous variables. Sometimes the theoretical proportions of a categorical variable are also of importance to us, and we may want to estimate or perform tests of hypothesis about these unknown parameters.

A bank may want to know, when it is giving out loans to different applicants for various purposes, what proportion of the loan recipients are likely to default during the repayment process. A store manager may want to know what proportion of times a particular employee makes a sale as customers walk in. A coffee brand company may want to know its true market share. On the other hand, sometimes we may want to do inference about the difference of two proportions. Between two major soda companies, such as Coca-Cola and Pepsi-Cola, which one commands a greater proportion of loyal clientele? Has the nationwide proportion of individuals who support a particular presidential candidate changed after a nationally televised presidential debate? These and many other important questions of practical interest can be investigated using techniques of inference about one or two population proportions.

### 8.7.1   Inference about a Single Proportion

Consider the Coffee Brand Selection Data that we have used extensively throughout this book (see, e.g., Case Study 3.3). What is the true market share proportion of the coffee brand Aldi (including both its variants)?

To answer this, we will treat brand selection as a discrete random variable where each time a packet is bought it is Aldi or one of its variants with an (unknown) probability $p$. We assume that the sample is an i.i.d. random sample. Suppose our data is coded as $X_1, X_2, \ldots, X_n$, where $X_i$ is equal to 1 if the $i$-th packet sold in the sample is an Aldi, and 0 otherwise. In relation to this data, we take up the following inference problems.

(a) Point estimation: The sample proportion $\bar{X} = \frac{1}{n} \sum_{i=1}^{n} X_i$ is our estimator of the unknown true proportion $p$. As an estimator of $p$, it is also often referred to as $\hat{p}$. Our description in Section 8.8.1 indicates that this is also the maximum likelihood estimator of the parameter $p$. For the market share of Aldi, this leads to the estimate $\hat{p} = 29,718/163,625 = 0.1816226$. (See the R output below.) [Note that the number 163,625 is larger than the number 130,986 reported earlier (e.g., Chapter 3). The reason for this discrepancy is that the figure 130,986 is the number of instances of purchase. Since some instances may include the purchase of more than one packet, the total number of packets purchased is more than 130,986.]

(b) Confidence interval: It can be shown that the estimator $\hat{p}$ of the population proportion $p$ has, for a large sample size, an approximate normal distribution with mean $p$ and variance $p(1-p)/n$. Using the central limit theorem, the quantity

$$Z = \frac{\hat{p} - p}{\sqrt{p(1-p)/n}}$$

has an asymptotic standard normal distribution. This can then be used as the pivotal element in our analysis. However, the standard deviation $\sqrt{p(1-p)/n}$ of $\hat{p}$ is not computable as $p$ is unknown. We substitute the unknown $p$ with $\hat{p}$, which leads to $SE(\hat{p}) = \sqrt{\hat{p}(1-\hat{p})/n}$, the standard error of $\hat{p}$. Using

$$Z = \frac{\hat{p} - p}{\sqrt{\hat{p}(1-\hat{p})/n}}$$

as an approximate $N(0,1)$ variable, and employing the readjustments of the inequalities that we have done many times in this chapter, we get the $100(1-\alpha)\%$ confidence interval of $p$ to be

$$\hat{p} \pm Z_{1-\alpha/2} SE(\hat{p}).$$

For $\alpha = 0.05$, we have $Z_{1-\alpha/2} = 1.959964$, and the standard error is calculated to be 0.000953. Thus the 95% confidence interval for the market share of Aldi is

$$0.1816226 \pm 1.959964 \times 0.000953 = (0.1798, 0.1835).$$

(c) Hypothesis testing: Now consider testing the null hypothesis $H_0 : p = 0.2$ against the alternative hypothesis $H_A : p < 0.2$. In general, when we test the null hypothesis $H_0 : p = p_0$, our test statistic is

$$Z = \frac{\hat{p} - p_0}{\sqrt{p_0(1-p_0)/n}},$$

which has an asymptotic $N(0,1)$ distribution under the null. Note that, in this case, the standard error estimate is $\sqrt{p_0(1-p_0)/n}$, which is different from the standard error in case (b), where we had set up the confidence interval. The reason is that, in this case, we have a tentative idea about the unknown parameter as given by the null hypothesis. Our observed test statistic is

$$Z = \frac{0.1816 - 0.2}{\sqrt{0.2 \times 0.8/163,625}} = -18.58.$$

At the 5% level of significance, we reject the null when the observed $Z$ is smaller than $-1.644854$. In this case the null hypothesis is overwhelmingly rejected.

In the following, the R output dealing with these questions in respect of this data is provided.

```
> prop.test(29718, 163625, p=0.2)

        1-sample proportions test with continuity correction

data:  29718 out of 163625, null probability 0.2
X-squared = 345.2652, df = 1, p-value < 2.2e-16
alternative hypothesis: true p is not equal to 0.2
95 percent confidence interval:
0.1797590 0.1835012
sample estimates:
       p
0.1816226


> prop.test(29718, 163625, p=0.2, correct=F)

        1-sample proportions test without continuity correction

data:  29718 out of 163625, null probability 0.2
X-squared = 345.38, df = 1, p-value < 2.2e-16
alternative hypothesis: true p is not equal to 0.2
95 percent confidence interval:
 0.1797621 0.1834981
sample estimates:
       p
0.1816226
```

Note that, in the R output, the statistic used is $Z^2$, which has an approximate $\chi^2(1)$ distribution under the null. One rejects at level $\alpha$, when the observed $Z^2$ is greater than $\chi^2_{1-\alpha}(1)$. Since the statistic is squared in this case, this test can only detect the magnitude of deviation, and not the direction of deviation. Thus the $\chi^2$ statistic given in the above outputs essentially perform the test against the two sided alternative $H_A^* : p \neq 0.2$, rather than the less than type alternative $H_A : p < 0.2$ considered in item (c). Note also that the output lists an additional case where the test statistic is computed with continuity correction, leading to a slightly different value of the test statistic and p-value.

## 8.7.2    Inference about Two Proportions

Our interest now is in comparing two different proportions. We continue with the Coffee Brand Selection Data to illustrate this. Suppose we want to compare the proportion of market share of Eduscho coffee in the two lowest economic groups (having income less than 1499 DM and between 1500 and 3000 DM, respectively). Let the sample sizes and the population proportions for these two groups be denoted by $n_i$ and $p_i$, respectively, $i = 1, 2$.

In the first group 5337 Eduscho varieties were bought out of the total of 23,927 packets sold. In the second group the corresponding numbers are 5236 out of 26,068. Thus $n_1 = 23,927$ and $n_2 = 26,068$.

(a) Point estimation: To estimate the difference of the population proportions, we still use the sample version $\hat{p}_1 - \hat{p}_2$. This is the maximum likelihood estimate of the difference. In this instance this estimate is

$$\frac{5377}{23,927} - \frac{5236}{26,068} = 0.020413.$$

(b) Confidence interval: Here we want to set up a $100(1 - \alpha)\%$ confidence interval for $p_1 - p_2$. Now

$$Z = \frac{\hat{p}_1 - \hat{p}_2 - (p_1 - p_2)}{\sqrt{\frac{p_1(1-p_1)}{n_1} + \frac{p_2(1-p_2)}{n_2}}}$$

has an approximate $N(0,1)$ distribution. As both $p_1$ and $p_2$ are unknown, we replace them with their estimates $\hat{p}_1$ and $\hat{p}_2$, respectively, in the denominator, which gives the pivotal element

$$Z = \frac{\hat{p}_1 - \hat{p}_2 - (p_1 - p_2)}{\sqrt{\frac{\hat{p}_1(1-\hat{p}_1)}{n_1} + \frac{\hat{p}_2(1-\hat{p}_2)}{n_2}}}.$$

When $n_1$ and $n_2$ are large (as they are in this case), the above has an approximate $N(0,1)$ distribution. Then, using standard manipulations, the $100(1 - \alpha)\%$ confidence interval for $p_1 - p_2$ is

$$(\hat{p}_1 - \hat{p}_2) \pm z_{1-\alpha/2}\sqrt{\frac{\hat{p}_1(1 - \hat{p}_1)}{n_1} + \frac{\hat{p}_2(1 - \hat{p}_2)}{n_2}}.$$

The actual value of the confidence interval is given in the R output below for the $\alpha = 0.05$ case.

(c) Hypothesis testing: Now we test for $H_0 : p_1 = p_2$ against the not equal to alternative. In this case the standard error of $(\hat{p}_1 - \hat{p}_2)$ is slightly different, since now we know that the proportions are the same under the null. Thus the standard error to be used is

$$\sqrt{\hat{p}(1 - \hat{p})\left(\frac{1}{n_1} + \frac{1}{n_2}\right)},$$

where $\hat{p}$ is the pooled estimate of the common (but unknown) value of $p$. In this case $\hat{p} = (5337 + 5236)/(23,937 + 26,068) = 0.2114$. The statistic

$$Z = \frac{\hat{p}_1 - \hat{p}_2}{\sqrt{\hat{p}(1 - \hat{p})\left(\frac{1}{n_1} + \frac{1}{n_2}\right)}}$$

has an asymptotic $N(0,1)$ distribution under the null. Consequently, $Z^2$ has an asymptotic $\chi^2(1)$ distribution under the null. We reject when the observed $\chi^2$ value is greater than $\chi^2_{1-\alpha}(1)$. When $\alpha = 0.05$, this critical value is 3.84146. The observed value of the statistic is much higher (see the R output below), and the null hypothesis is overwhelmingly rejected.

```
> x1 <- c(5337,23937)
> x2 <- c(5236,26068)
> x <- rbind(x1,x2)
> prop.test(x)

        2-sample test for equality of
        proportions with continuity correction

data:   x
X-squared = 23.675, df = 1, p-value =
1.141e-06
alternative hypothesis: two.sided
95 percent confidence interval:
 0.008961623 0.021136336
sample estimates:
   prop 1    prop 2
0.1823119 0.1672630

> prop.test(x, correct=F)

        2-sample test for equality of proportions
        without continuity correction

data:   x
X-squared = 23.7793, df = 1, p-value = 1.08e-06
alternative hypothesis: two.sided
95 percent confidence interval:
 0.008994675 0.021103284
sample estimates:
   prop 1    prop 2
0.1823119 0.1672630
```

## 8.8   Appendix

### 8.8.1   Maximum Likelihood Estimator

Point estimation is one of the important components of statistical inference. In this chapter we have been partial to interval estimation and hypothesis testing, and have given relatively less attention to point estimation. In all the estimation questions we have faced so far, we have estimated the population mean with the sample mean and estimated the population variance (in the normal case) with the sample variance.

While it is generally reasonable to estimate the population mean with the sample mean, it is not necessarily the best estimator for this purpose, and using the sample mean to estimate the population mean blindly is in some sense a naive approach. Good estimation methods should be based on suitable optimality criteria. Of course, it may well be that such criteria will generate the sample mean as the estimator of choice for the population mean, but this needs to be demonstrated.

In this connection we will briefly describe the method of *maximum likelihood estimation*. This is by far the most popular method of estimation in statistics. We explain this method below, starting with a demonstration for the Bernoulli case.

Suppose that we have a *Bernoulli(p)* distribution. The parameter $p$ is unknown, and we need to estimate it. Suppose we have an i.i.d. random sample $X_1, X_2, \ldots, X_{10}$ from this distribution. We get to see the outcomes, record each one of them as being 0 or 1 (i.e., as a success or a failure), and on the basis of this we want to estimate the unknown parameter $p$. To be specific, suppose our Success and Failure sequence is SSFSSFSSSF, so that there are seven successes and three failures. If we use the sample mean to estimate $p$, we would get the estimate $\hat{p} = 7/10$. Should it be the estimate of our choice? Why?

If the unknown success parameter is $p$, the probability of observing the above sequence is $p^7(1-p)^3$. One could approach the estimation problem in the following manner. The true parameter is unknown, but a particular sequence of outcomes has been observed. How likely would the observed sequence be under different parameter values? In the following we present a table which gives us the probabilities of observing the sample in question for different values of the unknown parameter $p$.

| Value of $p$ | Probability of the obtained sample |
|:---:|:---:|
| 0.1 | $7.29 \times 10^{-8}$ |
| 0.3 | $7.50 \times 10^{-5}$ |
| 0.5 | 0.000977 |
| 0.7 | 0.002224 |
| 0.9 | 0.000478 |

While we have recorded the probabilities for only a few values of $p$, it gives us the general idea. There are some values which are extremely unlikely to have generated this particular sample. Common sense says that our estimate $\hat{p}$ of $p$ should not be one of those values. There are some other values for which this probability is much more reasonable. The maximum likelihood philosophy chooses the estimator to be that value of the parameter for which this sample has the highest *likelihood* of occurring. The formal procedure in this case would be to choose that value of $p$ for which the objective function $p^7(1-p)^3$, the probability of the given sample as a function of $p$, is maximum. In this case, the likelihood represents an actual probability. This objective function, or the likelihood, quantifies the relative chance of occurrence of this particular sample as a function of the unknown parameter (in this case $p$). Our problem then reduces to a maximization problem – choose the value of $p \in [0,1]$ for which this probability is the maximum. Taking the derivative of the objective function with respect to $p$, we get the estimating equation to be

$$\frac{7}{p} - \frac{3}{1-p} = 0.$$

The solution to this equation is $\hat{p} = 7/10$, which is the same as the sample mean. So the sample mean does emerge as the *maximum likelihood estimate* in this case. A plot of the likelihood for this data is given in Figure 8.4, which demonstrates that the likelihood smoothly increases until it reaches a peak at $p = 0.7$, and then drops again.

In general, the estimator which maximizes the likelihood is called the *maximum likelihood estimator*. Given an i.i.d. random sample from a density $f$, the likelihood of the unknown parameter is

$$f(x_1)f(x_2)\ldots f(x_n), \tag{8.19}$$

where $x_1, x_2, \ldots, x_n$ are the actual observed values of the random variables. As the density $f$ is also a function of the parameter, so is the likelihood in Equation (8.19), and, given the data, the likelihood is a function of the unknown parameter only. The maximum likelihood estimator is then obtained by maximizing the likelihood with respect to the parameter. Note that the likelihood in Equation (8.19) does not represent a probability in continuous models, but the idea of maximization to obtain the estimator with the highest likelihood as in the discrete case should not be difficult to understand.

For the Poisson distribution, given $x_1, x_2, \ldots, x_n$, the observed values of an i.i.d. random sample, the maximum likelihood estimate of the Poisson parameter $\lambda$ is found to be $\hat{\lambda} = \bar{x}$, the observed sample mean. When we view the observations $X_1, X_2, \ldots, X_n$ as random, the maximum likelihood estimator of $\lambda$ is $\bar{X}$.

Similarly, for the $N(\mu, \sigma^2)$ distribution, the maximum likelihood estimators of $\mu$ and $\sigma^2$ are

$$\hat{\mu} = \bar{X} = \frac{1}{n}\sum_{i=1}^{n} X_i \text{ and } \hat{\sigma}^2 = \frac{1}{n}\sum_{i=1}^{n}(X_i - \bar{X})^2. \tag{8.20}$$

**The Bernoulli likelihood**

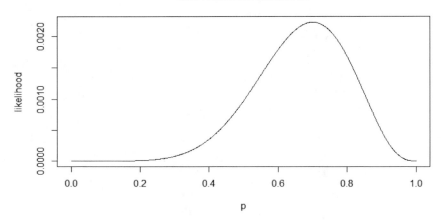

**FIGURE 8.4**
The likelihood function for the Bernoulli model.

Thus the maximum likelihood estimator of the variance parameter has a denominator of $n$ and is different from the usual unbiased estimator in Equation (7.12). However, the estimators $s^2$ and $\hat{\sigma}^2$ get closer and closer to each other with increasing $n$.

We will use the abbreviation MLE to represent the maximum likelihood estimator or the maximum likelihood estimate, depending on the context.

### 8.8.2 Levene's Test for Equality of Variances

In Examples 8.5 and 8.6 we have done a test for equality of two means. As we have seen, this test is based on the assumption that the sample variances are equal. In a real situation it may have to be checked to verify if this assumption is reasonable. We will not provide the technical details, but R provides a test for equality of variances called Levene's test. The relevant output in relation to Examples 8.5 and 8.6 is presented below. The p-value of 0.2934 shows that there is no evidence of violation of the equal variances assumption.

```
> leveneTest(y,group)
Levene's Test for Homogeneity of Variance (center = median)
      Df F value Pr(>F)
group  1  1.1287 0.2934
      48
```

### 8.8.3   Unbiasedness of the Pooled Estimator

In Section 8.3, we used the pooled estimator $s_p^2$ as an estimator for the un-known common variance $\sigma^2$. Here we demonstrate that this is an unbiased estimator for $\sigma^2$. We know that

$$\frac{(n_1-1)s_1^2}{\sigma^2} \sim \chi^2(n_1-1), \text{ and } \frac{(n_2-1)s_2^2}{\sigma^2} \sim \chi^2(n_2-1),$$

where $s_1^2$ and $s_2^2$ are the sample variances of the two populations. From our results in Section 7.13, both $s_1^2$ and $s_2^2$ are unbiased estimators of $\sigma^2$. Thus

$$E\left[\frac{(n_1-1)s_1^2+(n_2-1)s_2^2}{n_1+n_2-2}\right] = \frac{(n_1-1)\sigma^2+(n_2-1)\sigma^2}{n_1+n_2-2} = \frac{n_1+n_2-2}{n_1+n_2-2}\sigma^2 = \sigma^2.$$

## 8.9   Suggested Further Reading

The book by DeGroot and Schervish (2012) is our favorite for additional reading on material covered in this chapter. We also recommend Walpole et al. (2010) and Akritas (2014) to complement the discussion in this chapter. Among more basic texts that cover statistical inference, Pal and Sarkar (2009), Anderson et al. (2014), Levin and Rubin (2011), Black (2012) and Mann (2013) represent some of the books which are well liked. Casella and Berger (1990) is also an excellent resource for topics in inference, although at a much higher level than what is considered in this book.

In a departure from usual bookish presentations, the book by Agresti and Franklin (2009) provides several practical angles of looking at data and do-ing inference, and is very different from most other textbooks dealing with inference.

# 9

## Regression for Predictive Model Building

A business of any size needs to make important decisions all the time. Starting from the neighborhood tea stall to Microsoft and Google, millions of businesses are making billions of decisions every minute and each one of these is having an effect on their performance. If the decision is correct, then the business is likely to make a profit; if not, it is likely to incur a loss. The objective of any business is to make a profit; otherwise it will not be able to survive. Therefore every business strives to make better decisions. How does a business make a decision? How does the neighborhood tea stall make a decision? The proprietor perhaps knows all his clients and their habits and prepares accordingly. How do the businesses that do not know their clients personally take decisions regarding stocks maintained, manpower planning, shipping arrangements and another thousand and one situations?

Many businesses, even today, are run by "gut feeling," or, more formally, by domain knowledge. If it is a long-standing business of good repute and small size, the behavior of the clients over the years is more or less stable and that knowledge is leveraged upon. However, this process is not efficient and will definitely not work if the business is large. When there is a huge amount of data or when a business is growing or when new avenues are chartered and new products are being marketed, such ad hoc decision making will not work. In these days of data explosion, transaction data should be used to understand the health of the business and give it a direction. For improved decision making, an analytical process should be applied to extract intelligent information from the data.

Businesses want to take faster and better decisions compared to their competitors. So they would like to get a fairly good idea regarding what is expected to happen in the future. Focus on predictive analytics is the future of business. There are many different methods of predicting the future, of which two are the most important: (i) regression, where a response is predicted on the basis of one or more predictors which are supposed to be known, and (ii) time series analysis where the historical behavior of the variable of interest is modeled for the future. In this chapter we concentrate on regression and in Chapter 12, the time series method is studied. All these methods which attempt to estimate the status of the business for the future come under the umbrella of predictive analytics.

## 9.1   Simple Linear Regression

Often such situations arise where the value of one variable is estimated on the basis of one or more related variables. Consider, for example, the salary of employees. It is known that salary depends on a person's education level, experience and age, among other things. If the HR department has this information, it will have a good idea about what salary to offer to a new employee. The procedure that relates a response to one or more predictors is known as the regression procedure. If there is only one predictor and the dependence structure is linear, the method is known as *simple linear regression*; with more than one predictor, the method is called *multiple linear regression*. In its most general form, the response and the predictors can be continuous or categorical variables. However, when the response is not a continuous variable, a different procedure is mandated. To start with, we keep the regression problem limited to a continuous response and one continuous predictor. We will enrich the model as we go along.

**Example 9.1.** (Auto Data) One of the major considerations for an automobile buyer is a car's miles per gallon (MPG) performance. The following data is extracted from the StatLib data library, which is maintained at the Carnegie Mellon University. The current version of the data is available in the R (version 3.1.1) ISLR library under the name Auto. It has 392 observations and 8 different characteristics on each car. Recall that the same data was used for illustration in Chapter 4 (Statistical Graphics and Visual Analytics) and elsewhere in this book. It is known that the horsepower of a car has an impact on its MPG efficiency. Here we will consider the problem of predicting MPG based on a car's horsepower.

Let us first understand the segment of the data that is important to us in this problem. There are 392 units or observations in the sample, and each observation is characterized by a pair of variables, namely, MPG and Horsepower. This is a bivariate sample and the pairing is inviolate. A scatterplot of the observations is shown in Figure 9.1, which has Horsepower on the $X$-axis and MPG on the $Y$-axis. It is conventional to display the response (here MPG) on the $Y$-axis and the predictor (here Horsepower) on the $X$-axis. The concept of interdependence of variables was discussed in Chapters 3, 6 and 7. However, there we did not focus on the response-predictor relationship and hence the treatment was different.

The scatterplot indicates that there is a relationship between MPG and Horsepower. There is a decreasing trend in the data; if Horsepower (predictor) increases, MPG (response) generally decreases. This is an example of *negative association* between the predictor and the response.                                    ‖

One measurement of the strength of association is the correlation coefficient. In Chapters 3 and 6 the correlation coefficient was discussed at some

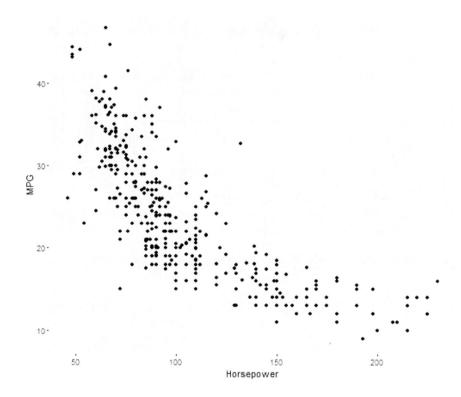

**FIGURE 9.1**
Scatterplot of MPG versus Horsepower for Auto Data.

length. It is a number between −1 and + 1, and the nearer it is to 1 in magnitude, the stronger is the association. For the above data the correlation coefficient between Horsepower and MPG is −77.8%, indicating a strong association between them. Also, the sign of the correlation is negative, which is expected, as the graph indicates that MPG decreases with higher Horsepower.

However, the correlation measure, by itself, is not enough to answer questions like the following: What is the (average) MPG of a car when the car's Horsepower is 150 (or 200)? The observed value of average MPG at a given Horsepower is not used as the estimated MPG. This is a naive estimate and does not borrow strength from the fact that MPG is a decreasing function of Horsepower. To estimate a response depending on a predictor, the relation between the variables is exploited through the method of simple linear regression.

The simple linear regression model assumes that the expected value or mean of the response $Y$, which is a random variable, depends on the observed value $x$ of the predictor $X$, through a linear function as $E(Y) = \alpha + \beta x$ where $\alpha$ and $\beta$ are the intercept and the slope of the straight line, respectively. Technically, this expectation is the conditional expectation of the response $Y$, given $X = x$.

Since $Y$ is a random variable, the above relationship can also be expressed as $Y = \alpha + \beta x + \epsilon$ where $\epsilon$ denotes the associated error term. For a random sample of size $n$, the relationship in the $i$-th pair may be expressed as $Y_i = \alpha + \beta x_i + \epsilon_i$ for all $i = 1, \ldots, n$. The error term is required to accommodate the variability. Assuming that the errors follow a normal distribution with equal variance, the following two conditions are equivalent: $Y_i \sim N(\alpha + \beta x_i, \sigma^2)$ and $\epsilon_i \sim N(0, \sigma^2)$. This is the simplest form of regression, and, if the numerical values of $\alpha$ and $\beta$ were known, the full dependence structure of the response on the predictor would have been known.

Let us try to understand why regression works. Suppose that there are only three values (levels) $x_1$, $x_2$ and $x_3$ of the predictor $X$. At each of these three levels of $X$, several responses are observed. Because of the variability of these observations, there is a level-specific distribution of the response $Y$ at each value of $X$. If the predictor influences the response, then mean of the response $Y$ will be a function of the level of the predictor. If the variability of the response does not depend on the level of the predictor, and if the mean response is linear in the value of the predictor, then linear regression will be able to provide a good estimator of the response. This scenario is described by the equation $E(Y) = \alpha + \beta X$, where, implicitly, $E(Y)$ is the conditional expectation of the response given the predictor. More explicitly, at the three levels of the predictor, the mean responses are given by $E(Y|x_1) = \alpha + \beta x_1$, $E(Y|x_2) = \alpha + \beta x_2$ and $E(Y|x_3) = \alpha + \beta x_3$. The focus of the regression is on estimating the expected response $E(Y|x)$ at a given level of the predictor, which depends on $\alpha$ and $\beta$. These are known as the regression parameters and are estimated based on the sample observations.

It is to be noted that the means of the response distribution, which are completely known if the levels of the predictor are known, lie on the straight line but the individual observations from the distributions may not lie on the straight line. This is because the observations are realizations from the response distribution and will be different from the mean of the distribution. Had all realizations of the random variable been the same, the variability of the distribution would be zero and the random variable would be a constant! Due to the inherent variability in the response distribution at each level of the predictor, it is possible to have a pair of observations $(x_1, y_1)$ and $(x_2, y_2)$ for which $x_1 < x_2$ but $y_1 > y_2$, even when $X$ and $Y$ have a positive association. In reality it is possible to find two workers where the first one has 12 years of education and makes 15.50 dollars per hour while the second one has 16 years of education but makes only 9.30 dollars per hour. The difference between the population mean and the realized value of the random variable may be denoted by $\epsilon$ and is called the error, even though no concept of incorrectness

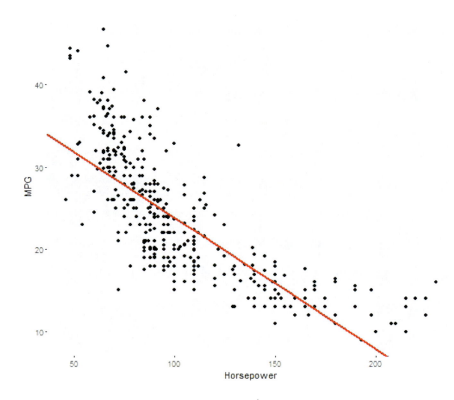

**FIGURE 9.2**
Regression of MPG on Horsepower.

is associated with it. Since an individual value can be different from the mean, the regression equation for the $i$-th case can also be written as $Y_i = \alpha + \beta x_i + \epsilon_i$.

Note that $\alpha$ and $\beta$ are population parameters and hence they need to be estimated. There are many methods of estimation of the regression parameters, but the most common one is called the *least squares* method of estimation. This method minimizes the sum of the squared differences between the estimated and observed values of the response. We will denote the estimates of $\alpha$ and $\beta$, computed from the sample, as $a$ and $b$, respectively. Hence the estimated regression equation is $\hat{y}_i = a + bx_i$, where $\hat{y}_i$ is the estimated mean response when the value of the predictor is $x_i$; it is also called the fitted value of the response at this level of the predictor. When $a$ and $b$ have been computed, the estimated or fitted values of the response at each level of the predictor can be obtained by plugging in the relevant quantities in the above formula.

**Example 9.1.** (Continued) Applying the lm() functionality of R, regression parameter estimates come out to be $\hat{\alpha} = a = 39.93$ and $\hat{\beta} = b = -0.16$. Hence the regression equation of MPG on Horsepower is

$$\widehat{\text{MPG}} = 39.93 - 0.16 \times \text{Horsepower}.$$

Thus the estimated mean value $\widehat{\text{MPG}}$ of MPG can be obtained by plugging in the value of the Horsepower variable on the right hand side of the above equation. If the Horsepower of a car is 130 (e.g., Chevrolet Chevelle Malibu), then its expected MPG will be $(39.93 - 0.16 \times 130) = 19.13$. Similarly, if the Horsepower of a car is 200 (e.g., Ford Maverick), then its expected MPG is 7.93.

The straight line in Figure 9.2 shows the estimated linear regression line. The estimated regression line, or the fitted line, as it is more commonly known, passes through the general mass of the observations. However, not all observations lie on the regression line, nor are they expected to do so. Even though not all observed points lie on the fitted regression line, it denotes the functional relationship between Horsepower and (expected) MPG. The difference between the observed and the estimated response is known as the residual. Residuals represent important diagnostic features of a regression and will be studied in detail later.

The coefficient of the predictor in the regression equation is the slope of the regression line. It measures the change in the (expected) response for one unit change in the predictor. In the above example, for one unit increase in Horsepower, MPG is expected to decrease by 0.16 units. Converting it into commonly measurable units, one concludes that for every 10 unit increase in Horsepower, a car goes 1.6 miles less per gallon. The intercept is the value of the response when the predictor is equal to zero. However, the intercept may not be always be readily interpretable. For example, a car will never have zero Horsepower! ∥

Formally, the simple linear regression model is estimated under the following assumptions:

- The true relationship between the mean response and the predictor is linear.

- The errors of the regression model are independently and identically distributed, having the $N(0, \sigma^2)$ distribution.

In a general regression set up, our data contains $n$ pairs of observed values $(x_1, y_1), (x_2, y_2), \ldots, (x_n, y_n)$. Suppose $\bar{x}$ and $\bar{y}$ are the sample averages of $X$ and $Y$, respectively, and $S_{xx} = \sum_{i=1}^{n}(x_i - \bar{x})^2$, $S_{yy} = \sum_{i=1}^{n}(y_i - \bar{y})^2$ and $S_{xy} = \sum_{i=1}^{n}(x_i - \bar{x})(y_i - \bar{y})$. $S_{xx}$ and $S_{yy}$ are also known as the sum of squares of $X$ and the sum of squares of $Y$, respectively, and $S_{xy}$ is known as the sum of products of $X$ and $Y$. The least square estimates of the regression parameters are given by

$$\hat{\beta} = b = S_{xy}/S_{xx} \quad \text{and} \quad \hat{\alpha} = a = \bar{y} - b\bar{x}.$$

Recall that the association between the predictor $X$ and the response $Y$ can be measured by the correlation coefficient. Hence it is but natural that the sample correlation coefficient $r$ and the regression slope $b$ should have some relationship. If $b > 0$, for every unit increase in $X$, $Y$ increases by $b$ units. If $b < 0$, for every unit increase in $X$, $Y$ decreases by $|b|$ units. For $b > 0$, $Y$ and $X$ are positively correlated and for $b < 0$, $Y$ and $X$ are negatively correlated. Furthermore,

$$r = \frac{S_{xy}}{\sqrt{S_{xx}}\sqrt{S_{yy}}} = b\frac{\sqrt{S_{xx}}}{\sqrt{S_{yy}}},$$

where the positive square roots are used. Thus the sign of $r$ is the same as the sign of $b$. Note also that $S_{xx}$ and $S_{yy}$ are the numerators in the formula of sample variances of $X$ and $Y$.

There are a few misconceptions associated with regression applications that need to be dispelled. One popular misconception is that, once the regression equation is obtained, it can be used for any pairs of $X$ and $Y$ values. That certainly is not true. A regression equation works well within the scope of the regression. Essentially this means that we should restrict ourselves to predicting future values of the response within the range of the predictor values that have been used in the modeling exercise. If any extrapolation is to be done, it should be done in a restricted area around the range of the predictor. Outside the scope of the regression, the nature of dependence between $Y$ and $X$ may not remain the same. Another caveat needs to be considered with the appropriateness of the model in mind. The regression line, or any business model, is data-driven and data-dependent. If the data changes, so does the model. It is always prudent to monitor the model against the course of the business and update the model continuously. Updating in such cases indicates a change in the nature of the model or simply a reevaluation of the model parameters.

Even when the assumed model is appropriate for the data, the predicted and observed values of the response will generally not be identical. Corresponding to the $i$-th observation, $r_i = y_i - \hat{y}_i$ is the $i$-th residual. Note that the residuals may be positive or negative and they may be taken as estimates of the unknown population errors. Observations farther from the regression line (in the vertical direction) have larger residuals compared to observations closer to the regression line. It can be shown mathematically that $\sum_{i=1}^{n} r_i = 0$ for a least squares regression line. An overall measure of closeness of the estimated response to the actual values of the response may be defined as the Sum of Squares of Errors, $SSE = \sum_{i=1}^{n} r_i^2 = \sum_{i=1}^{n} (y_i - \hat{y}_i)^2$. An overall measure of variance in the response is defined as the Mean Square Error $MSE = SSE/(n-2)$. Assuming that $\epsilon_i \sim N(0, \sigma^2)$, an estimate of the error variance is given by the $MSE$ and the standard deviation of error distribution is obtained as $\sqrt{MSE}$. For a regression model to perform well, $MSE$ must be small.

## 9.1.1 Regression ANOVA

Variability is an integral part of all data, and model building is an attempt to explain variability. From our discussion in the previous section, it is clear that, at each level of the predictor, the response $Y$ has a different distribution. The difference is due to the mean of the response distribution, which is a linear function of the predictor and changes if the value of the predictor changes. For a moment let us not consider the dependence of $Y$ on $X$. The total sum of squares of deviation among the observed values of $Y$ (around their mean) is defined as

$$SSTO = S_{yy} = \sum_{i=1}^{n}(y_i - \bar{y})^2.$$

Recall that this is nothing but $(n-1)$ times the sample variance of $Y$. Given a set of observations, $SSTO$ is a constant regardless of the existence or the nature of the dependence of $Y$ on $X$. Regression model fitting is an attempt to partition the total sum of squares into two additive parts: sum of squares due to regression and sum of squares due to error.

$$SSTO = SSR + SSE = \sum_{i=1}^{n}(y_i - \bar{y})^2 = \sum_{i=1}^{n}(\hat{y}_i - \bar{y})^2 + \sum_{i=1}^{n}(y_i - \hat{y}_i)^2.$$

The sum of squares due to regression is the explained part of the variability that occurs due to the fact that the observations $y_i, i = 1, 2, \ldots, n$ are expected to be different since each of them may represent a different population. Sum of squares due to error is the unexplained part of variability that is inherent in the distribution. Naturally, a regression model works well if the explained part of variability is high compared to the unexplained part of variability. This leads us to an overall measure of regression effectiveness called the *Coefficient of Determination* $R^2$, which measures the proportion of total sum of squares that is explained by regression. This measure is given by

$$R^2 = \frac{SSR}{SSTO} = \frac{SSTO - SSE}{SSTO} = 1 - \frac{SSE}{SSTO}.$$

The higher the $R^2$, the better is the fit of the regression model. The maximum possible value of $R^2$ is 100%, which occurs only if all the observed points lie on a straight line. On the other hand, $R^2$ is 0 if there is absolutely no structure in the data. For a real-life dataset, neither 0 nor 100% values of $R^2$ is ever expected. Nevertheless, a high enough value of $R^2$ indicates that the model under consideration is able to satisfactorily explain the variability in the data. $R^2$ can also be used to compare relative performance of two simple linear regression models with different predictors. The regression model for predicting MPG based on Horsepower has an $R^2$ of 60%. This means that 60% of the total variability existing in the MPG is explained by Horsepower. For simple linear regression, the coefficient of determination equals $r^2$, the squared correlation coefficient between $X$ and $Y$.

## 9.1.2 Inference for Simple Linear Regression

As in any other statistical problem, regression analysis is done on a sample, but it must provide a glimpse into the parent population. When the sample mean or the sample standard deviation are computed from a random sample, their numerical values are often used to estimate the corresponding population parameters, namely, the population mean and the population standard deviation. Sample statistics are random variables since they are based on random samples drawn from the population. By the same logic, the estimators of the regression slope and intercept are also sample statistics; for a large enough sample size, they may be assumed to follow normal distributions with means equal to the corresponding population parameters. (If the original assumptions of linearity of response and normality of errors are satisfied, the estimators of slope and intercept have exact normal distributions even for small sample sizes.) Specifically, $E(\hat{\alpha}) = \alpha$ and $E(\hat{\beta}) = \beta$. Their variances have more complicated forms (not shown here) and depend on the error variance $\sigma^2$. Since $\sigma^2$ is unknown and is estimated by the $MSE$, the inference problems concerning regression parameters are handled by a $t$ distribution with $(n-2)$ degrees of freedom, $n$ being the sample size.

The most important inference problem is to test whether there is indeed a regression of $Y$ on $X$. If there is no regression of $Y$ on $X$, then the regression coefficient of $Y$ on $X$ is 0, which implies that there is no dependence of $Y$ on $X$ and all $Y_i$ have a common mean. Analytically speaking, the hypothesis of interest is $H_0 : \beta = 0$ versus $H_A : \beta \neq 0$. To test this hypothesis, the test statistic is computed using sample data and its p-value is determined. (We will not get into the actual form of the test statistic, but it turns out to be a scaled version of the slope estimate $b$.) At a prespecified level of significance, commonly 5%, $H_0$ is rejected if the p-value of the test is less than 5%. Rejection of the null hypothesis indicates the likely existence of regression of $Y$ on $X$.

**Example 9.1.** (Continued) The following output is available as a result of regressing MPG on Horsepower in the Auto Data.

```
Coefficients:
             Estimate  Std. Error  t-value   Pr(>|t|)
(Intercept) 39.935861   0.717499     55.66   <2e-16 ***
Horsepower  -0.157845   0.006446    -24.49   <2e-16 ***
---
Signif.codes:  0 '***' 0.001 '**' 0.01 '*' 0.05 '.' 0.1 ' ' 1
Residual standard error: 4.906 on 390 degrees of freedom
Multiple R-squared:  0.6059, Adjusted R-squared:  0.6049
F-statistic: 599.7 on 1 and 390 DF,  p-value: < 2.2e-16
```

The regression estimates are given along with their standard errors, value of the $t$ statistics and the corresponding p-values. For the slope coefficient of Horsepower, the $t$ statistic has a very high absolute value of 24.49 with a highly significant p-value. Hence the null hypothesis of no regression is rejected.

It is also possible to construct a confidence interval for the population regression slope $\beta$. With a 95% level of confidence, this interval is given by

$$b \pm t_{.975}(n-2) \times SE(b),$$

where $SE(b)$ is the standard error of $b$ and $t_{.975}(n-2)$ is the quantile of the $t$ distribution with $(n-2)$ degrees of freedom corresponding to $q = 0.975$. From the above output, the 95% C.I. for the true regression slope is $(-0.157845 \pm 1.97 \times 0.006446)$ or $(-0.1451, -0.1705)$. Hence we may say, with 95% confidence, that the true regression slope lies between $-0.1451$ and $-0.1705$. Since this interval does not contain the value 0, $H_0 : \beta = 0$ will be rejected in favor of $H_A : \beta \neq 0$. Note also that, for very large degrees of freedom, the $t$ distribution is identical to the standard normal distribution for all practical purposes. Hence the confidence interval for the regression slope may also be computed using the quantiles of the $Z$ distribution.                                    ‖

It is important to note here that even a highly significant regression coefficient may not be able to explain a very high proportion of variability in the model. It is true that, if the regression coefficient is not significantly different from zero, then there is little effect of regression, and, therefore, the numerical value of the coefficient of determination $R^2$ will not be high. But the opposite is not necessarily true. A non-zero slope is not enough by itself to explain most of the variability in the response. A regression model has high predictive power only if $R^2$ has a high value.

### 9.1.3    Predicted Values, Confidence and Prediction Intervals

To understand the full implication of the predicted values of a regression model, several concepts behind regression must be revisited. Recall that $\hat{y}_i = a + bx_i$ is the predicted value of the response as given by the least squares regression model. Recall also that, at a given level of the predictor $X = x_i$, $E(Y_i|x_i) = \mu_i = \alpha + \beta x_i$. If there is a regression of $Y$ on $X$, then the mean of the distribution of $Y_i$ is a function of $X$ and depends on the level of $X$. Hence $\hat{y}_i$, the predicted value, is an estimate of the population mean $\mu_i$.

**Example 9.1.** (Continued) In the Auto Data, at certain levels of Horsepower there are multiple observations of the response, whereas at certain other levels there are only single observations. In Figure 9.3, four levels of the predictor are displayed. Multiple cars are found with Horsepower 100 and 105; on the other hand, with Horsepower 102 and 103 each, only one car is found. The estimated MPG values are given in Table 9.1.

The estimated values of MPG are the estimated means of the distribution of MPG (response) at different values of Horsepower. Using the estimated values of MPG, it is possible to determine the confidence interval for the true population mean of MPG at a given level of Horsepower. The 95% confidence limits for certain levels of Horsepower are given above. If the Horsepower is

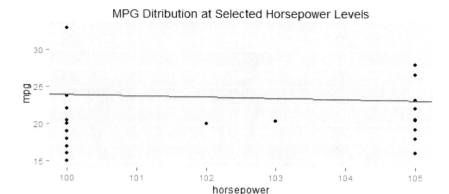

**FIGURE 9.3**
MPG distribution.

**TABLE 9.1**
Confidence and prediction intervals

| Horsepower | Estimated MPG | 95% Confidence Interval for Mean Response | 95% Prediction Interval for a New Observation |
|---|---|---|---|
| 100 | 24.15 | [23.66, 24.64] | [14.49, 33.81] |
| 102 | 23.85 | [23.34, 24.32] | [14.18, 33.49] |
| 103 | 23.68 | [23.19, 24.16] | [14.02, 33.33] |
| 105 | 23.36 | [22.87, 23.85] | [13.70, 33.02] |

100, then with 95% confidence it may be said that the MPG performance of a car will be between 23.66 and 24.64.

Suppose now an analyst is interested in knowing the range within which a *new observation* from the response distribution at a given level of the predictor is expected to lie. This is a completely different proposition than constructing a confidence interval for a population parameter. A new observation from the response distribution is a realization of a random variable; hence a confidence interval cannot be constructed. The point estimate of the new observation at a given level of the predictor is equal to the corresponding fitted value. However, the prediction interval is wider, sometimes considerably, than the confidence interval since it takes into account the variability inherent in the distribution of response, which is denoted by $\sigma^2$, in addition to the variability due to estimation. We may say, therefore, that, for a car with Horsepower 100, mean MPG performance is expected to be between 23.66 and 24.64 while a particular car with Horsepower 100 may have its MPG between 14.5 and 33.8.
‖

### 9.1.4 Regression Assumptions and Model Diagnostics

Although least squares estimation of regression model parameters does not require any assumption, the performance of a regression model is based on underlying assumptions regarding the data. Model building is an exercise in representing the reality in a simpler form. The closer to reality the model is, the better it is expected to perform. A regression model is developed based on certain assumptions about the data, and, post model building, it is important to check the validity of those assumptions. For simple linear regression, the following assumptions are made for all $i = 1, \ldots, n$:

1. Linearity: $E(Y_i|x_i) = \alpha + \beta x_i$ or, equivalently, $Y_i = \alpha + \beta x_i + \epsilon_i$, where $\epsilon_i$ is the error component associated with the $i$-th observation.

2. Independence and homoscedasticity: All observations are independently distributed with variance $\sigma^2$.

3. Normality: $Y_i$ follows a normal distribution.

The statements "$Y_i$ follows a normal distribution" and "$\epsilon_i$ follows a normal distribution" are equivalent, which can be shown mathematically. While $E(Y_i|x_i) = \alpha + \beta x_i$, $E(\epsilon_i) = 0$. Both random variables have the same variance $\sigma^2$. Among the assumptions, linearity, independence and homoscedasticity are very important. Normality is also important for the regression inference results to hold, but in real life very few datasets may be regarded as following an exact normal distribution, even though many will follow an approximate normal distribution. Generally, the inference procedure on regression is not very seriously affected by mild departures from normality. However, if the data distribution shows substantial departure from normality, analysts can always opt for other methods such as bootstrap inference procedures instead of normality-based inference. Departures from the other assumptions have more serious repercussions on the model building procedure.

When the true relationship is nonlinear, using a straight line relationship will give rise to biased estimates. Bias is any systematic departure from the norm (see Section 7.8.3). If the linearity assumption is invalid, the predicted values will deviate from the true values in a systematic pattern. If error variances are not constant, the confidence or prediction intervals will all be wrong. However, the principal consequence of non-constancy of error variance is deeper than that. Regression ANOVA assumes all variances are the same and estimates them with the $MSE$. If the error variances are different at different levels of the predictor, then $MSE$ is not a valid estimate of any parameter.

It is therefore very important to validate the regression assumptions, and if the assumptions are not tenable, to take measures to rectify them. The principal tool to check regression assumptions is the residual plot, where residuals are plotted against the fitted values. Since the regression procedure ensures that $\sum_{i=1}^{n} r_i = 0$, if all assumptions are met, the residual plot should look like

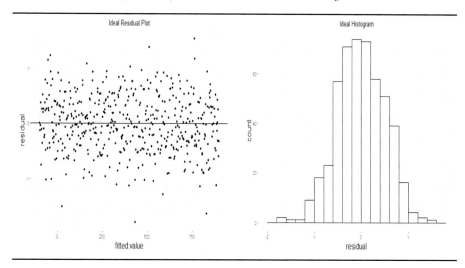

**FIGURE 9.4**
Distribution of residuals:I.

a uniform band around the $r = 0$ line. The histogram of residuals should be symmetric around the 0 value. See Figure 9.4.

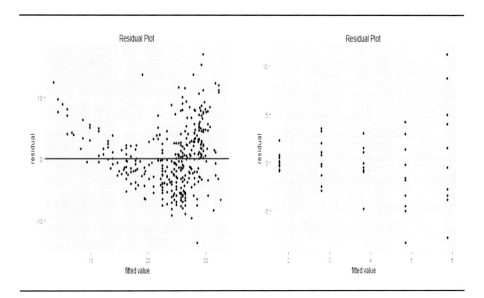

**FIGURE 9.5**
Distribution of residuals: II.

Let us investigate residual plots which exhibit systematic departure from the ideal form observed in Figure 9.4. Different types of departures indicate different problems. If the residuals follow a curved pattern (left panel of Figure 9.5), it may indicate a departure from linearity. If the residual plot shows a funnel-like shape (right panel of Figure 9.5), with a wider mouth on the right or the left, it may indicate heteroscedasticity in the data. In this case the heteroscedasticity occurs through an increase in the error variance as the value of the predictor increases.

If the assumption of independence is violated, i.e., if subsequent observation pairs are not independent, then an index plot of the residuals will show a distinctive wave pattern. An index plot is an order plot where residuals are plotted against the order of the observation instead of against the predicted values. Figure 9.6 provides an example of such a pattern.

Normality of a distribution is usually tested through a quantile-quantile (Q-Q) plot where quantiles computed from the data are plotted against quantiles of the theoretical distribution. If the observations truly follow a normal distribution, then the Q-Q plot will be as in Figure 9.7. The straight line makes a 45 degree angle with the $X$-axis and passes through the origin. If the theoretical and observed quantiles coincide and lie exactly on the reference

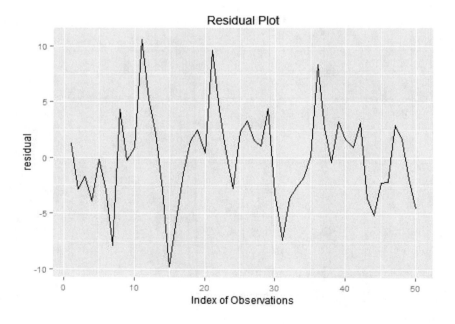

**FIGURE 9.6**
Distribution of residuals: III.

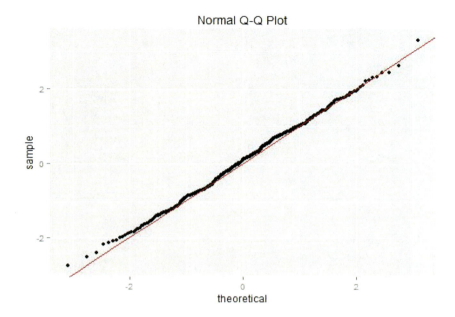

**FIGURE 9.7**
Normal Q-Q plot.

line, it indicates perfect normality. However, in reality, this will be observed very rarely. The left panel in Figure 9.8 shows a residual plot which is not even close to normal, while its histogram, given in the right panel of the same figure, also corroborates its non-normal nature.

Of the above assumptions, normality is the least important. Many of the regression inference procedures are moderately robust to departure from normality. If the departure is very significant, alternative procedures like bootstrap exist to determine the confidence intervals. But steps need to be taken to rectify the other departures.

## 9.1.5 Outliers and Leverage Points

Residual plots of regression are also used to identify outliers and leverage points. Note that boxplots identify only univariate outliers. Detection of outliers is very important because a small number of outlying observations which are very different from the bulk of the data may heavily influence the inference process. For example, if the data distribution has a long tail, the mean is expected to be significantly larger than the median and hence may not be

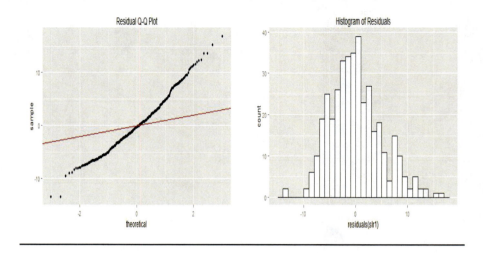

**FIGURE 9.8**
Distribution of residuals: IV.

a true representative value of the central tendency. However, in the regression context, univariate outliers may or may not exert a large influence on the regression line. It is important to note here that a "good" regression line will pass through the main body of the data, and all data points will have equal "influence" on the regression line. These concepts will be explained later in detail. For now, let us look at Figure 9.9.

There are two distinct regression lines through the data – the red line is the result of a model fit without the three points around the coordinates (40, 0); the green line is the result of a model fit when all points are included in the regression. The three points at the lower right hand corner are not outliers, if either the predictor level or the response is considered separately. However, they are certainly *influential points* in the sense that only 3 points out of a total of 53 are able to substantially change the intercept and slope of the regression line.

Alternatively, consider a point (100, 322). This point is certainly an outlier in both the predictor and the response, but it is not an influential point. Inclusion or exclusion of this point will not produce any appreciable change in the regression line in terms of the estimated values of the slope or the intercept. In fitting a regression line, the outliers, in one or the other variable, do not necessarily provide a cause for concern. It is the influential points that are to be identified and appropriately neutralized. Often the influential points are high leverage points; these points are outliers in the $X$ direction. Residual plots and various functions of the residuals are used to identify the leverage points. These points can potentially have very large residuals.

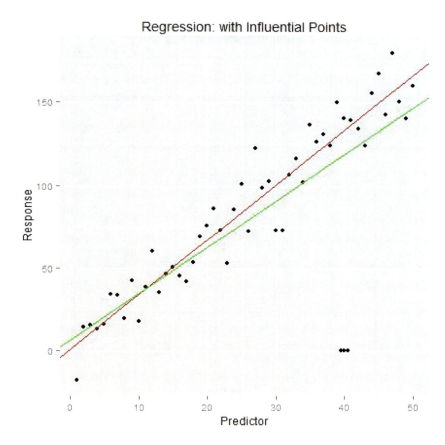

**FIGURE 9.9**

Example of influential points.

The following provides a summary of the discussion given above.

(a) Influential points cause a large change in the fitted regression line.

(b) Observations which are geometrically well separated from the majority of the data (in either direction) may be called outliers. Sometimes observations may not be outlying in the $X$ or $Y$ direction individually, but the $(X, Y)$ pair may be an outlier. Outliers may or may not be influential points.

(c) High leverage points are such points which have extreme values of the predictor variable (so they are outliers in the $X$ direction).

(d) Very often influential points are both outliers and high leverage points. But observations (sometimes a number of observations) may be influential without having high leverage.

## 9.1.6 Transformation of Variables

Transformation of variables is sometimes recommended as a panacea for improving regression model assumptions as well as taking care of the influential points. It is true that in many situations transformations are useful in improving the regression model fit, e.g., incorporating curvature in the model and stabilizing the variance. However, transformations are to be used with caution, as they may also introduce problems in the model.

Recall that the response is assumed to be normally distributed. If this assumption is valid, then any nonlinear transformation on $Y$ will destroy the normality. Transformation on $Y$ is recommended when the variance is non-constant and distribution is non-normal, in addition to having nonlinearity in the data. Transformation on the predictor is recommended if the variance seems to be constant. However, transformation of variables is as much an art as a science! Several guidelines are provided here but alternatives must be tried out to finalize which transformation, if any, would lead to a better model fit. It is also important to recognize that the conclusions drawn using transformed data may not always transfer neatly to the original scale. While that is not a reason to rule out any transformation, it is necessary to consider interpretability while using transformations.

If the scatterplot shows an increasing trend but the shape is concave, a $\log(X)$ or $\sqrt{X}$ transformation is recommended. If the increasing trend follows a convex pattern, a squared transformation might work. If the scatterplot is decreasing but convex, an inverse transformation $1/X$ on $X$ might linearize the relation. The best practice is to try out different alternatives and then decide which transformation is the most appropriate.

To determine an appropriate transformation on $Y$, often a family of power transformations, called the Box–Cox family of transformations, is used. Transformed $Y$ is $Y^\lambda$, where $\lambda$ is a parameter to be determined from the data. The normal error regression model with a Box–Cox transformation is $Y_i^\lambda = \alpha + \beta x_i + \epsilon_i$. The value of $\lambda$ that maximizes the likelihood function is the recommended transformation.

R has a function called `boxcox()` in the MASS library that computes the value of the likelihood over a range of $\lambda$ and plots it. It is possible to visually identify an appropriate $\lambda$ through that procedure. Figure 9.10 shows the maximum likelihood function for the regression model of MPG on Horsepower using the Auto Data. It is clear that the likelihood function is maximized for $\lambda = -0.5$. Hence the recommended strategy in this case, without any other consideration like interpretability, is to regress $Z = 1/\sqrt{\text{MPG}}$ on Horsepower. The results of the regression are given below:

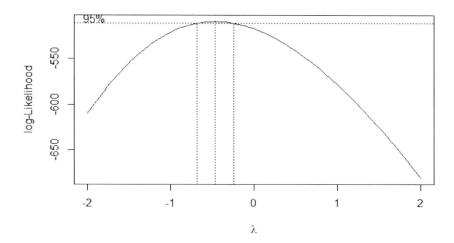

**FIGURE 9.10**
Determining the power of the transformation.

```
Call:
lm(formula = (1/sqrt(MPG)) ~ Horsepower)

Residuals:
     Min        1Q     Median        3Q       Max
-0.067822 -0.012913 -0.001744  0.012773  0.069129
Coefficients:
             Estimate Std. Error t value Pr(>|t|)
(Intercept) 1.304e-01  2.890e-03   45.11   <2e-16 ***
Horsepower  8.149e-04  2.597e-05   31.38   <2e-16 ***
---
Signif.codes:  0 **0.001 *0.01 0.05 0.1 1
Residual standard error: 0.01976 on 390 degrees of freedom
Multiple R-squared:  0.7163,Adjusted R-squared:  0.7156
F-statistic: 984.9 on 1 and 390 DF,  p-value: < 2.2e-16
```

The multiple $R^2$ has improved from 60% to 71%. Also there is a significant improvement in the residual plot (see Figure 9.11). The curvature in $Y$ has been appropriately taken care of.

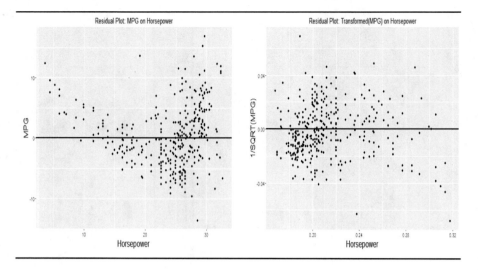

**FIGURE 9.11**
Distribution of residuals: V.

## 9.2   Multiple Linear Regression

Very rarely is a response aptly modeled by a single predictor. Several predictors are generally needed to explain the variability within the response. When more than one predictor is used in the regression equation, it is called a multiple regression problem. Continuing with a car's MPG efficiency (see Example 9.1), it is known that, along with Horsepower, Cylinders (number of cylinders), Weight, Acceleration, etc., may also have an impact on MPG. Multiple regression deals with the joint effect of all predictors on the response. The observations are multivariate, i.e., on each unit one observes a single response but several predictors. The multiple regression equation is written as

$$Y_i = \alpha + \beta_1 x_{1i} + \beta_2 x_{2i} + \cdots + \beta_k x_{ki} + \epsilon_i$$

where $Y$ is the response and $(X_1, X_2, \ldots, X_k)$ are the predictors. One important condition for multiple regression parameters to be uniquely estimable is that the number of observations $n$ has to be larger than the number of parameters to be estimated. If there are $k$ predictors, then the number of parameters to be estimated is $p = k + 1$. There are special situations where $n < p$, but, for big data analytics, such situations are less frequent.

Continuing with the MPG efficiency of cars, let us include Weight as a second predictor in the model along with Horsepower, without any transformation of the response. The following result is observed.

```
Coefficients:
              Estimate    Std. Error    t value    Pr(>|t|)
(Intercept) 45.6402108    0.7931958     57.540     < 2e-16  ***
Horsepower  -0.0473029    0.0110851     -4.267     2.49e-05 ***
Weight      -0.0057942    0.0005023    -11.535     < 2e-16  ***
---
Signif.codes:  0 *** 0.001 ** 0.01 * 0.05 . 0.1   1
Residual standard error: 4.24 on 389 degrees of freedom
Multiple R-squared:  0.7064,Adjusted R-squared:  0.7049
F-statistic: 467.9 on 2 and 389 DF,  p-value: < 2.2e-16
```

The regression equation may be written as

$$\widehat{\text{MPG}} = 45.64 - 0.05 \times \text{Horsepower} - 0.01 \times \text{Weight}.$$

Even though the values of the regression coefficients are small, they are highly significant, as corroborated by the associated p-values. The regression coefficients signify that, for every unit increase in Horsepower, the expected MPG of a car goes down by 0.05 units, when, Weight of the car is kept at a constant level. When the Horsepower of a car is kept at a constant level, for every unit increase of Weight, a car's expected MPG goes down by 0.01 unit. For the Chevrolet Chevelle Malibu with Horsepower 130 and Weight 3504, the expected MPG is 19.19, whereas for the Buick Estate SW with Horsepower 225 and Weight 3086, the expected MPG is 17.11.

The multiple regression equation is not an aggregation of two simple regression equations using Horsepower or Weight individually. Simple regression of MPG on Horsepower yields the relation

$$\widehat{\text{MPG}} = 39.93 - 0.16 \times \text{Horsepower}$$

and simple regression of MPG on Weight yields the relation

$$\widehat{\text{MPG}} = 46.22 - 0.007 \times \text{Weight}.$$

Individually Horsepower is able to explain 60% of the total variability in MPG while Weight is able to explain 69% of the variability. However, Horsepower and Weight together are able to explain 71%. How much more of the total variability in the response is explained by addition of an extra predictor in the model depends on the correlation among the predictors as well as the correlation between the response and the predictors.

Recall that total variability in the response is $SSTO = \sum_{i=1}^{n}(y_i - \bar{y})^2$, which does not depend on the number of predictors in the model. Given a dataset, $SSTO$ is a constant. For the Auto Data $SSTO = 23,819$. A regression equation is developed with the objective of apportioning $SSTO$ into two parts: $SSR$ and $SSE$. The higher the $SSR$, the better the performance of the model in describing the dependence of the response on the predictors. Inclusion of an additional predictor in the model is justified only if it is able to take a significant part of $SSE$ and add it to $SSR$, which in turn will increase the value of $R^2$.

Whether inclusion of an additional predictor will significantly increase the value of $R^2$ depends on the correlation between the existing predictors in the model as well as the correlation between the new predictor and the response. Continuing with the Auto Data, note that the correlation between Horsepower and Weight is 86%, the correlation between MPG and Weight is −83% and the correlation between MPG and Horsepower is −77%. Since Horsepower and Weight have a high correlation, the addition of Weight AFTER Horsepower does not increase the value of $R^2$ to a large extent.

## 9.3    ANOVA for Multiple Linear Regression

Consider a multiple regression equation in the most general form as

$$Y_i = \alpha + \beta_1 x_{1i} + \beta_2 x_{2i} + \cdots + \beta_k x_{ki} + \epsilon_i$$

where $Y$ is the response and $(X_1, X_2, \ldots, X_k)$ are the predictors. The number of parameters to be estimated is $p = k + 1$. Table 9.2 presents the analysis of variance for multiple linear regression.

**TABLE 9.2**
ANOVA table

| Sources | $df$ | $SS$ | $MS$ | $F$ |
|---|---|---|---|---|
| Regression | $k = p - 1$ | $\sum_{i=1}^{n}(\hat{y}_i - \bar{y})^2$ | $MSR = \frac{SSR}{(p-1)}$ | $\frac{MSR}{MSE}$ |
| Error | $n - p$ | $\sum_{i=1}^{n}(y_i - \hat{y}_i)^2$ | $MSE = \frac{SSE}{(n-p)}$ | |
| Total | $n - 1$ | $\sum_{i=1}^{n}(y_i - \bar{y})^2$ | | |

To test the null hypothesis $H_0 : \beta_1 = \beta_2 = \cdots = \beta_k = 0$ against the alternative $H_A :$ At least one $\beta_j \neq 0$, the statistic $MSR/MSE$ is used. This follows an $F$ distribution with $(p - 1, n - p)$ degrees of freedom. The null hypothesis that all regression coefficients are simultaneously zero indicates that there is no regression effect; the response does not depend on any of the predictors. The alternative hypothesis that the response depends on at least one predictor is given by the condition that there is at least one $j$ for which $\beta_j \neq 0$. The rejection region for the null hypothesis is on the right hand side, i.e., if the observed value of

$$F = \frac{MSR}{MSE}$$

is larger than the quantile $F_{1-\alpha}(p-1, n-p)$ of the indicated $F$ distribution, then the null hypothesis is rejected. Rejection of the null hypothesis indicates that there exists a regression of the response on at least one of the predictors, but it does not specify which predictor (or which group of predictors) is significant. To identify significant predictors, the individual null hypothesis $H_0 : \beta_j = 0$ against $H_A : \beta_j \neq 0$ is to be tested for all $j = 1, 2, \ldots, k$.

**Example 9.2.** (Auto Data) In this dataset, MPG may be regressed on the following predictors: Cylinders (number of cylinders), Displacement, Horsepower, Weight, Acceleration and Year.

```
Coefficients:
                Estimate   Std. Error   t value   Pr(>|t|)
(Intercept)   -1.454e+01   4.764e+00    -3.051    0.00244 **
Cylinders     -3.299e-01   3.321e-01    -0.993    0.32122
Displacement   7.678e-03   7.358e-03     1.044    0.29733
Horsepower    -3.914e-04   1.384e-02    -0.028    0.97745
Weight        -6.795e-03   6.700e-04   -10.141    < 2e-16 ***
Acceleration   8.527e-02   1.020e-01     0.836    0.40383
Year           7.534e-01   5.262e-02    14.318    < 2e-16 ***
---
Signif.codes:  0 *** 0.001 ** 0.01 * 0.05 . 0.1   1
Residual standard error: 3.435 on 385 degrees of freedom+
Multiple R-squared:  0.8093,Adjusted R-squared:  0.8063+
F-statistic: 272.2 on 6 and 385 DF,  p-value: < 2.2e-16
```

From the p-values associated with the coefficients, it is clear that Cylinders, Displacement, Horsepower and Acceleration are not significant; only Weight and Year are highly significant.

```
Analysis of Variance Table
Response: MPG
              Df  SumSq Mean Sq   F value    Pr(>F)
Cylinders      1 14403.1 14403.1 1220.5070 < 2.2e-16 ***
Displacement   1  1073.3  1073.3   90.9544 < 2.2e-16 ***
Horsepower     1   403.4   403.4   34.1845 1.07e-08 ***
Weight         1   975.7   975.7   82.6822 < 2.2e-16 ***
Acceleration   1     1.0     1.0    0.0819   0.7749
Year           1  2419.1  2419.1  204.9945 < 2.2e-16 ***
Residuals    385  4543.3    11.8
---
Signif.codes:  0 *** 0.001 ** 0.01 * 0.05 . 0.1   1
```

The ANOVA table shows the sequential sum of squares for each predictor, in the order they are included in the model, and the residual sum of squares ($SSE$). In the above model $SSE = 4543.3$ with $n - p = 392 - 7 = 385$ df. The regression sum of squares $SSR = SSTO - SSE = 23,819 - 4543.3 = 19,275.7$, which is given in the ANOVA table in a sequential form corresponding to each predictor in the model. Adding all the sequential sums of squares for each single degree of freedom, $SSR$ is obtained. The general format given is

$$
\begin{aligned}
&SS(\text{Cylinders}) + SS(\text{Displacement}|\text{Cylinders}) \\
&+SS(\text{Horsepower}|\text{Cylinders, Displacement}) \\
&+SS(\text{Weight}|\text{Cylinders, Displacement, Horsepower}) \\
&+SS(\text{Acceleration}|\text{Cylinders, Displacement, Horsepower, Weight}) \\
&+SS(\text{Year}|\text{Cylinders, Displacement, Horsepower, Weight, Acceleration}) \\
=\ &SSR.
\end{aligned}
$$

It is important to understand the sequential nature of the sum of squares vis-a-vis the regression coefficients and their significance in a multiple linear regression. The regression coefficients are also known as partial regression coefficients, which deal with the partial effect of a certain predictor on the response, after adjusting for the effects of all other predictors, regardless of the order in which the predictors enter the model. The sequential sums of squares depend on the order in which the predictors are specified in the model statement, though the values of $SSR$ and $SSE$ do not depend on the order.

For the same example with Auto Data, consider the following model in R.

```
Call:
lm(formula = MPG ~ Horsepower + Weight + Acceleration
   + Cylinders + Displacement + Year).
```

The corresponding ANOVA table and sequential $SS$ are given below.

```
Analysis of Variance Table
Response: MPG
```

| | Df | SumSq | Mean Sq | F value | Pr(>F) | |
|---|---|---|---|---|---|---|
| Horsepower | 1 | 14433.1 | 14433.1 | 1223.0487 | | |
| Weight | 1 | 2392.1 | 2392.1 | 202.7023 | <2e-16 | *** |
| Acceleration | 1 | 0.0 | 0.0 | 0.0004 | 0.9835 | |
| Cylinders | 1 | 31.4 | 31.4 | 2.6584 | 0.1038 | |
| Displacement | 1 | 0.0 | 0.0 | 0.0001 | 0.9910 | |
| Year | 1 | 2419.1 | 2419.1 | 204.9945 | <2e-16 | *** |
| Residuals | 385 | 4543.3 | 11.8 | | | |

```
---
Signif.codes:  0 *** 0.001 ** 0.01 * 0.05 . 0.1   1
```

As the order in which predictors enter the regression model has changed, the significance level of the sequential SS has also changed, albeit the value and significance of the regression coefficients will not change. This apparent contradictory nature of sequential SS and significance of regression coefficients is essential to understand how to come up with interpretable models while building the model automatically.

The degree to which the order of entry in the model impacts the sequential $SS$ depends on the correlation among the predictors. Ideally, if all the predictors are uncorrelated, the order of entry in the model does not matter. For the Auto Data, the correlation matrix among the predictors is given below. Going by the rule of thumb that a correlation above 70% is high, the variable Cylinders has high correlation with Displacement, Horsepower and Weight; Displacement is highly correlated with Horsepower and Weight and Horsepower has high correlation with Weight. Such interdependence is the reason that the significance of sequential $SS$ changes drastically depending on the order in which the variables are put in the model.

|  | Cylinders | Displacement | Horsepower | Weight | Acceleration | Year |
|---|---|---|---|---|---|---|
| Cylinders | 1.00 | 0.95 | 0.84 | 0.90 | −0.50 | −0.35 |
| Displacement | 0.95 | 1.00 | 0.90 | 0.93 | −0.54 | −0.37 |
| Horsepower | 0.84 | 0.90 | 1.00 | 0.86 | −0.69 | −0.42 |
| Weight | 0.90 | 0.93 | 0.86 | 1.00 | −0.42 | −0.31 |
| Acceleration | −0.50 | −0.54 | −0.69 | −0.42 | 1.00 | 0.29 |
| Year | −0.35 | −0.37 | −0.42 | −0.31 | 0.29 | 1.00 |

‖

## 9.3.1 Multicollinearity and Variance Inflation Factor

In the example involving Auto Data, it is clear that the predictors that are highly correlated are responsible for certain non-uniqueness in the model, and the order of specification becomes an important determinate for a predictor to be included in the model. The term *multicollinearity* in regression refers to a linear relationship among two or more predictors. When two variables are nearly linearly related, their correlation is high. However, pairwise correlations fail to detect scenarios where three or more variables are linearly related. Along with effects already considered, multicollinearity has other ramifications, including inaccuracy in regression parameter estimates, inflated standard errors of regression coefficients and false non-significance, which in turn leads to degradation in model predictability. Hence detection and elimination of multicollinearity is of primary importance.

One simple way of detecting multicollinearity is to consider all pairwise scatterplots, or, equivalently, compute all pairwise correlations. However, for big data, where the number of predictor variables may run into the hundreds, this is not possible. Also as explained above, pairwise correlations fail to detect linearity among three or more predictors. The most useful tool to detect multicollinearity is the *variance inflation factor*, defined as $\mathrm{VIF}_j = 1/(1 - R_j^2)$, where $R_j^2$ is the coefficient of multiple determination obtained by regressing $X_j$ on the other $k-1$ predictors. A variance inflation factor equal to or close to 1 is desirable. A rule of thumb is that, if $1 < \mathrm{VIF}_j < 5$, then there is moderate multicollinearity, and for $VIF_j > 5$, there is strong multicollinearity. In effect, a model having VIFs close to 1 is acceptable. If high VIFs are observed, the model may be improved by removing one or more predictors suspected of having linear dependence, either from a statistical point of view or from business considerations or domain knowledge. If no such easy way out is available, or in spite of removing a few predictors multicollinearity among the rest remains high, it is prudent to remove one variable at a time from the model, starting with the predictor having the highest VIF. Usually it drives down the VIF for the other predictors also.

The multiple linear regression with all predictors of MPG leads to the following VIFs.

| Predictor | Cylinders | Displacement | Horsepower | Weight | Acceleration | Year |
|---|---|---|---|---|---|---|
| VIF | 10.63 | 19.64 | 9.40 | 10.73 | 2.63 | 1.24 |

It is clear that multicollinearity is a very big problem for this data. Displacement has the highest VIF and hence it is removed from the model. For the revised model, VIFs are as given below.

| Predictor | Cylinders | Horsepower | Weight | Acceleration | Year |
|---|---|---|---|---|---|
| VIF | 5.86 | 8.87 | 8.64 | 2.59 | 1.24 |

Removing Horsepower and Cylinders in sequence, the model arrived at is the following.

| Predictor | Weight | Acceleration | Year |
|---|---|---|---|
| VIF | 1.27 | 1.25 | 1.14 |

Note that Weight, Acceleration and Year do not have very high pairwise correlation among themselves. Also note that this three-predictor model has an $R^2$ value of 81%, which is practically equal to the multiple $R^2$ of the original six-predictor model. Hence addition of the other three predictors could not explain more variability than what has already been explained by Weight, Acceleration and Year. The fact that multicollinearity may lead to false non-significance is also manifested in the example. In the six-predictor model, only Weight and Year had a significant p-value for the regression coefficients; Acceleration had a p-value of 40%.

Removal of predictors with high VIF may lead to significant reduction in $R^2$.

**Example 9.3.** (Strength of Concrete) The data is abstracted from the Machine Learning Repository of UC Irvine, and the relevant URL is
`https://archive.ics.uci.edu/ml/datasets/Concrete+Compressive+Strength`.
We consider the compressive strength of concrete (coded as Strength), which is a nonlinear function of Age and ingredients such as Cement, Blast.Furnace.Slag, Fly.Ash, Water, Superplasticize, Coarse.Aggregate and Fine.Aggregate. The results of this regression are given below.

```
Coefficients:
                     Estimate Std. Error t value Pr(>|t|)
(Intercept)        -23.331214  26.58550   -0.878 0.380372
Cement               0.119804   0.008489  14.113  < 2e-16 ***
Blast.Furnace.Slag   0.103866   0.010136  10.247  < 2e-16 ***
Fly.Ash              0.087934   0.012583   6.988 5.02e-12 ***
Water               -0.149918   0.040177  -3.731 0.000201 ***
Superplasticize      0.292225   0.093424   3.128 0.001810 **
Coarse.Aggregate     0.018086   0.009392   1.926 0.054425 .
Fine.Aggregate       0.020190   0.010702   1.887 0.059491 .
Age                  0.114222   0.005427  21.046  < 2e-16 ***
---
Signif.codes:  0 *** 0.001 ** 0.01 * 0.05 . 0.1   1
Residual standard error: 10.4 on 1021 degrees of freedom
Multiple R-squared:  0.6155,Adjusted R-squared:  0.6125
F-statistic: 204.3 on 8 and 1021 DF,  p-value: < 2.2e-16
```

The corresponding VIFs are shown in the following table.

| Predictor | Cement | Blast.Furnace.Slag | Fly.Ash | Water |
|---|---|---|---|---|
| VIF | 7.49 | 7.28 | 6.17 | 7.00 |
| Predictor | Superplasticize | Coarse.Aggregate | Fine.Aggregate | Age |
| VIF | 2.96 | 5.07 | 7.01 | 1.12 |

Removing the predictor Cement with the highest VIF leads to the following results.

```
Coefficients:
                     Estimate Std. Error t value Pr(>|t|)
(Intercept)        301.532014  14.533501  20.747  < 2e-16 ***
Blast.Furnace.Slag  -0.023371   0.005061  -4.618 4.37e-06 ***
Fly.Ash             -0.068234   0.006546 -10.424  < 2e-16 ***
Water               -0.549056   0.031181 -17.609  < 2e-16 ***
Superplasticize      0.279643   0.102076   2.740  0.00626 **
Coarse.Aggregate    -0.085002   0.006451 -13.176  < 2e-16 ***
Fine.Aggregate      -0.109407   0.006005 -18.220  < 2e-16 ***
Age                  0.109522   0.005919  18.504  < 2e-16 ***
---
Signif.codes:  0 *** 0.001 ** 0.01 * 0.05 . 0.1   1
Residual standard error: 11.36 on 1022 degrees of freedom
Multiple R-squared:  0.5405,Adjusted R-squared:  0.5374
F-statistic: 171.7 on 7 and 1022 DF,  p-value: < 2.2e-16
```

The removal of Cement has reduced other VIFs, as seen in the next table, even though not all of them are close to 1. The removal of Cement has also impacted the significance level of the other predictors, and all predictors are now significant. However, the multiple $R^2$ value has reduced to 54%, while in the previous model it was above 61%. It also happens that Cement does not have a very high correlation with any of the predictors; the highest correlation is between Cement and Fly.Ash (−40%).

| Predictor | Blast.Furnace.Slag | Fly.Ash | Water | Superplasticize |
|-----------|--------------------|---------|-------|-----------------|
| VIF       | 1.52               | 1.40    | 3.53  | 2.96            |
| Predictor | Coarse.Aggregate   | Fine.Aggregate | Age |  |
| VIF       | 2.00               | 1.84    | 1.11  |                 |

## 9.4  Hypotheses of Interest in Multiple Linear Regression

In the simple linear regression model, there is only a single predictor and hence the important hypothesis to test is $H_0 : \beta = 0$ against $H_A : \beta \neq 0$. This is the equivalent of testing $H_0$: There is a regression of $Y$ on $X$ against the alternative $H_A$: There is no regression effect. In the case of multiple linear regression many different types of hypotheses could be of interest.

Consider a multiple regression model

$$Y = \alpha + \beta_1 x_1 + \beta_2 x_2 + \cdots + \beta_k x_k + \epsilon.$$

A multivariate sample of size $n$ is taken and all inference problems are based on these $n$ observations. To determine whether $X_j$, $j = 1, 2, \ldots, k$ is a useful predictor in the model, the hypothesis $H_0 : \beta_j = 0$ versus $H_A : \beta_j \neq 0$ is to be tested. If the null hypothesis is true, then a change in the value of $X_j$ would not affect the value of the response, given that the other $k - 1$ predictors are included in the model. The test statistic for the null hypothesis is the associated $t$ value, and a conclusion is drawn based on the p-value given by R. In the following example with Auto data, after correcting for VIF, the three predictors Weight, Acceleration and Year were considered for MPG.

```
Call:
lm(formula = MPG ~ Weight + Acceleration + Year)

Coefficients:
              Estimate Std. Error t value Pr(>|t|)
(Intercept) -14.936555   4.055512  -3.683 0.000263 ***
Weight       -0.006554   0.000230 -28.502  < 2e-16 ***
Acceleration  0.066359   0.070361   0.943 0.346204
Year          0.748446   0.050366  14.860  < 2e-16 ***
---
```

```
Signif.codes:   0 *** 0.001 ** 0.01 * 0.05 . 0.1    1
Residual standard error: 3.428 on 388 degrees of freedom
Multiple R-squared:  0.8086,Adjusted R-squared:   0.8071
F-statistic: 546.5 on 3 and 388 DF,  p-value: < 2.2e-16
```

For each predictor, its sample estimate, standard error, corresponding $t$ value and p-values are given. For Weight and Year, p-values are small, indicating that these predictors are useful for MPG, while Acceleration has a p-value of 35%, indicating this predictor may be dropped from the model.

The null hypothesis $H_0$: There is no regression of $Y$ on any of the predictors is equivalent to testing $H_0 : \beta_1 = \beta_2 = \beta_3 = \cdots = \beta_k = 0$ against the alternative $H_A$: At least one $\beta_j \neq 0$. An $F$ statistic is used to test this hypothesis, which is defined as

$$F = \frac{SSR/(p-1)}{SSE/(n-p)} \sim F(p-1, n-p).$$

For Auto Data and the three-variable model, numerical value of the $F$ statistic is 546.5, which, with 3 and 388 degrees of freedom, has a very significant p-value. Hence the null hypothesis is rejected in favor of the alternative that all regression coefficients are not simultaneously 0.

Another important type of testing procedure is the General Linear $F$ test. This is a technique to test multiple null hypotheses simultaneously. An alternative way to look at multiple linear hypotheses is to apply constraints to the linear model in order to simplify. In the example of Compressive Strength of Cement (Example 9.3), the components Course.Aggregate (CA) and Fine.Aggregate (FA) are both significant, and estimates of regression coefficients are numerically close, with comparable standard errors. Hence the null hypothesis that both regression coefficients are equal is a sensible one. To test $H_0 : \beta_{CA} = \beta_{FA}$ against the alternative $H_A : \beta_{CA} \neq \beta_{FA}$, the General Linear $F$ test is applicable.

The General Linear $F$ test depends on two concepts – the full model, or the model with the least number of restrictions, and the reduced model, or the model with a higher number of restrictions, which is nested (i.e., fully contained) within the full model. A sequence of nested or hierarchical models is obtained by successively applying more and more restrictions. If there are $k$ predictors in the model, the most general model is $M_1$ with all $k$ predictors in the model.

$$M_1 : Y = \alpha + \beta_1 x_1 + \beta_2 x_2 + \beta_3 x_3 + \cdots + \beta_k x_k + \epsilon.$$

A model which has only $k-1$ predictors is nested within $M_1$. Consider the models

$$M_2 : Y = \alpha + \beta_2 x_2 + \beta_3 x_3 + \cdots + \beta_k x_k + \epsilon$$

and

$$M_3 : Y = \alpha + \beta_1 x_1 + \beta_2 x_2 + \cdots + \beta_{k-1} x_{k-1} + \epsilon.$$

Both $M_2$ and $M_3$ are nested within $M_1$. But $M_2$ and $M_3$ are not hierarchical, since neither of them can be constructed from the other by applying a restriction such as $\beta_j = 0$ for some $j$. The General Linear $F$ test is appropriate for testing model reduction only in a hierarchical fashion. The null and the alternative hypotheses considered are $H_0$: Reduced model is adequate versus $H_A$: Full model is appropriate. The statistic is defined as

$$F = \frac{([SSE(R) - SSE(F)]/[df E(R) - df E(F)])}{(SSE(F))/(df E(F))},$$

where $SSE(R)$ and $SSE(F)$ are error sums of squares for the reduced and full model, respectively, and $df E(R)$ and $df E(F)$ are the corresponding degrees of freedom. This follows an $F$ distribution with $(df E(R) - df E(F), df E(F))$. If a constraint is added to a model, the model is further simplified and as a consequence $SSE$ increases. The General Linear $F$ statistic tests whether the resultant increase in $SSE$ is significant. If it is, the null hypothesis is rejected and the experimenter is led to the decision that the model suggested by the null hypothesis is not adequate for the given data.

**Example 9.4.** (Strength of Concrete) Compressive strength of concrete (coded as Strength) may be written as a linear function of the predictors as

$$\widehat{\text{Strength}} \;=\; 301.53 - 0.02 \times \text{Blast.Furnace.Slag} - 0.07 \times \text{Fly.Ash}$$
$$-0.55 \times \text{Water} + 0.28 \times \text{Superplasticize}$$
$$-0.09 \times \text{Course.Aggregate} - 0.11 \times \text{Fine.Aggregate} + 0.11 \times \text{Age}.$$

This may be considered as the full model. The hypothesis of interest is

$$H_0 : \beta_{CA} = \beta_{FA} \text{ against } H_A : \beta_{CA} \neq \beta_{FA}.$$

The reduced model may then be fit with the predictor Aggregate (which is the sum of Coarse.Aggregate and Fine.Aggregate) along with the other five predictors in the model.

```
Call:
lm(formula = Strength ~ Blast.Furnace.Slag + Fly.Ash + Water +
Superplasticize + Aggregate + Age, data = Concrete_Data)
```

The fit produced by the above model is given in the next table. The $SSE$(FullModel) is 131,952 with 1022 degrees of freedom, while the $SSE$(ReducedModel) is 133,935 with 1023 degrees of freedom. Hence the value of the General Linear $F$ statistic is $(133,935 - 131,952)/129 = 15.37$. The p-value is highly significant, implying that Coarse.Aggregate and Fine.Aggregate do not have equal impact on compressive strength of cement.

```
Coefficients:
                     Estimate Std. Error t value Pr(>|t|)
(Intercept)        309.764283  14.481490  21.390  < 2e-16 ***
Blast.Furnace.Slag  -0.022966   0.005095  -4.507 7.32e-06 ***
Fly.Ash             -0.066059   0.006568 -10.058  < 2e-16 ***
Water               -0.561426   0.031238 -17.973  < 2e-16 ***
Superplasticize      0.164935   0.098474   1.675   0.0943 .
Aggregate           -0.098954   0.005417 -18.266  < 2e-16 ***
Age                  0.110976   0.005949  18.656  < 2e-16 ***
---
Signif.codes:  0 *** 0.001 ** 0.01 * 0.05 . 0.1   1
```

$\parallel$

The General Linear $F$ test is a versatile statistic to take care of many different types of situations. It is widely applicable as long as the models are nested. We will have other applications of the General Linear $F$ test later.

### 9.4.1   Categorical Predictors

So far we have discussed situations when both the response and the predictors are continuous variables. Let us now consider the situation where one or more predictors are categorical. Categorical predictors may be nominal, where the numerical values associated with the categories are for labeling purposes only. On the other hand, values assumed by ordinal predictors have a hierarchical relationship. In the Auto Data, the predictor Origin is a nominal variable, assuming values 1 for cars of American origin, 2 for cars of European origin and 3 for cars of Japanese origin. Naturally, the values taken by this variable do not have any hierarchical meaning.

To include a nominal variable with $L$ levels in the model, a set of $L - 1$ indicator variables needs to be introduced, where the $l$-th indicator variable will take the value 1 if the observation has the $l$-th level of the predictor. In Auto Data, since Origin has three levels, two indicator variables are to be introduced.

| Variables in Auto Data | | | Indicator Variables | |
|---|---|---|---|---|
| Observation | Name | Origin | Indicator 1 | Indicator 2 |
| 1 | Chevrolet Chevelle Malibu | 1 | 0 | 0 |
| 2 | BMW 2002 | 2 | 1 | 0 |
| 3 | Datsun pl510 | 3 | 0 | 1 |

Indicator variable 1 takes the value 1 only if Origin $= 2$, i.e., only if the car is European and indicator variable 2 takes the value 1 only if Origin $= 3$, i.e., only if the car is Japanese. Cars with Origin $= 1$, i.e., American cars, are by default taken to be baseline, corresponding to which no separate variable is introduced. This is necessary because otherwise the regression coefficients will not be estimable. Against the baseline, the other levels of the categorical variable will be compared. In R the lowest level of the categorical variable is taken as the baseline by default; however, it can be changed. Other software may treat the highest numbered level as the baseline. By no means is 0-1 coding the only way to deal with categorical predictors. But we have not considered any alternative coding since the final result is independent of coding.

**Example 9.5.** (Auto Data) The inclusion of Origin as a categorical predictor of MPG in addition to Weight and Year gives rise to the following model:

```
Call:
lm(formula = MPG ~ Weight + Year + Origin)

Coefficients:
             Estimate Std. Error t value Pr(>|t|)
(Intercept) -1.831e+01  4.017e+00  -4.557 6.96e-06 ***
Weight      -5.887e-03  2.599e-04 -22.647  < 2e-16 ***
Year         7.698e-01  4.867e-02  15.818  < 2e-16 ***
Origin2      1.976e+00  5.180e-01   3.815 0.000158 ***
Origin3      2.215e+00  5.188e-01   4.268 2.48e-05 ***
---
Signif.codes:  0 *** 0.001 ** 0.01 * 0.05 . 0.1   1
Residual standard error: 3.337 on 387 degrees of freedom
Multiple R-squared:  0.819,Adjusted R-squared:  0.8172
F-statistic: 437.9 on 4 and 387 DF,  p-value: < 2.2e-16
```

The regression coefficients corresponding to the two levels of the categorical variable Origin provide comparison with the baseline level. Compared to the baseline of American cars, European cars show 1.97 units of improvement in MPG performance when Weight and Year stay the same. Compared to American cars Japanese cars show 2.21 units of improvement in MPG performance when Weight and Year stay the same. Compared to the baseline, both levels are significantly different, but that does not necessarily indicate that there is significant difference between the Origin 2 and Origin 3. Categorical predictors give rise to a system of linear equations, since, for alternative levels of the categorical predictors, only the corresponding indicator variable contributes. In this case the regression equations are

• Origin = American

$$\widehat{MPG} = -1.83 - 0.006 \times \text{Weight} + 0.77 \times \text{Year}$$

• Origin = European

$$\widehat{MPG} = -1.83 - 0.006 \times \text{Weight} + 0.77 \times \text{Year} + 1.97$$

• Origin = Japanese

$$\widehat{MPG} = -1.83 - 0.006 \times \text{Weight} + 0.77 \times \text{Year} + 2.21$$

The only difference among the equations in the system is in the intercepts. In this case it may be said that European cars have a better baseline performance of MPG compared to American cars, and Japanese cars have the best baseline performance among the three. To determine whether the categorical predictor with $L$ levels is significant, the hypothesis to test is $H_0 : \beta_1 = \beta_2 = \ldots = \beta_{L-1} = 0$ against the alternative $H_A$: that at least one $\beta_j \neq 0$, where $\beta_j$ is the regression coefficient corresponding to the $j$-th indicator variable. The test statistic is a General Linear $F$ statistic which compares the full model with the set of $L - 1$ indicator variables against the reduced model where all indicator variables are absent. In the Auto Data example $SSE(F) = 4310.4$ with 387 df while $SSF(R) = 4569.0$ with 389 df. Hence the $F$ statistic equals 11.64 with 2 and 387 df and has a p-value less than 0.001.

In this example, both levels of the categorical variable are significantly different from the baseline. However, if it happens that one or more levels of the predictor are not significantly different from the baseline, but other levels are, it is possible that the General Linear $F$ statistic is rejected and the predictor cannot be removed from the model. ‖

**Example 9.6.** (German Credit Data) Consider Credit.Amount to be dependent on Duration.of.Credit.Month (loan duration), Purpose (of loan) and Sex.Marital.Status of applicants. While Duration is continuous, Purpose and Sex.Marital.Status are both categorical. Purpose has 4 levels (1: new car, 2: used car, 3: home related, 4: other). Sex.Marital.Status has 3 levels (1: male divorced/single, 2: male married/widowed, 3: female), arrived at by collapsing the original levels to avoid the problem of having too few observations in some cells of the table. A multiple regression model is fitted to the data.

```
lm(formula = Credit.Amount ~ Duration.of.Credit.Month + Purpose +
    Sex.Marital.Status)

Coefficients:
                          Estimate Std. Error t value Pr(>|t|)
(Intercept)               1654.128    272.704   6.066 1.87e-09 ***
Duration.of.Credit.Month   137.742      5.674  24.277  < 2e-16 ***
Purpose2                 -1333.089    266.348  -5.005 6.60e-07 ***
Purpose3                 -1896.193    240.057  -7.899 7.43e-15 ***
Purpose4                 -1156.025    239.555  -4.826 1.61e-06 ***
Sex.Marital.Status2        235.840    145.741   1.618   0.1059
Sex.Marital.Status3       -575.291    250.276  -2.299   0.0217 *
---
Signif.codes:  0 *** 0.001 ** 0.01 * 0.05 . 0.1   1
Residual standard error: 2122 on 993 degrees of freedom
Multiple R-squared:  0.4384, Adjusted R-squared:  0.435
F-statistic: 129.2 on 6 and 993 DF,  p-value: < 2.2e-16
```

All levels of Purpose seem to be significant compared to the baseline level new car. However, compared to the baseline level of male divorced/single, the level male married/widowed does not seem to be significant, but female is. Non-significance in one or more levels of the categorical predictor does not mean that the indicator variables corresponding to those levels can be dropped from the model. The indicator variables corresponding to a categorical predictor act together as a system and are to be included in the model in totality or dropped from the model altogether. Whether the Sex.Marital.Status predictor can be dropped from the model depends on the outcome of the General Linear $F$ test. We have $SSE(F) = 4,470,377,810$ with 993 df while $SSE(R) = 4,524,209,641$ with 995 df. The value of the $F$ statistic is 5.97 with a p-value of 0.002 at 2 and 993 degrees of freedom. Hence we may conclude that the predictor Sex.Marital.Status cannot be dropped from the model. ∥

## 9.5 Interaction

Interaction in multiple linear regression occurs if the impact of one predictor variable $X_1$ on the response $Y$ varies depending on the levels of a second predictor $X_2$. Suppose for a group of on-line purchasers it is seen that there is a distinct difference in the behavioral pattern of females and males. For every extra minute spent on an e-commerce site, the incremental increase in amount spent is different between males and females. In such a case we will consider that there is an interaction effect between the predictors Gender and Time Spent. Most commonly, the interaction effect between $X_1$ and $X_2$ is represented in the model as a product term $X_1X_2$.

Consider, again, the regression of MPG on Weight, Year and Origin. Assuming that there is no interaction in the model, three different linear equations, one for each country of origin, are obtained. The three equations differ among themselves in the numerical value of the intercept. But the effects of Weight and Year on MPG would be the same regardless of Origin. If, however, an interaction term of Origin and Weight is included in the model, the dependence of MPG on Weight will change according to the Origin of a car. Similarly, if an interaction term with Year and Origin is included in the model, then the dependence of MPG on Year will vary according to Origin. The model with interaction terms may be written as

$$E(\text{MPG}) = \alpha + \beta_1\text{Weight} + \beta_2\text{Year} + \beta_3\text{Origin}$$
$$+\beta_4\text{Weight} \times \text{Origin} + \beta_5\text{Year} \times \text{Origin}$$

```
Call:
lm(formula = MPG ~ Weight + Year + Origin + Weight * Origin +
Year * Origin)
```

The fit produced by the above model is presented in the next table.

```
Coefficients:
                Estimate Std. Error t value Pr(>|t|)
(Intercept)    -9.439e+00  4.997e+00  -1.889  0.05964 .
Weight         -5.653e-03  2.764e-04 -20.450  < 2e-16 ***
Year            6.421e-01  6.007e-02  10.690  < 2e-16 ***
Origin2        -3.219e+01  9.855e+00  -3.267  0.00119 **
Origin3        -1.209e+01  9.296e+00  -1.301  0.19409
Weight:Origin2 -2.663e-03  8.390e-04  -3.174  0.00162 **
Weight:Origin3 -5.579e-03  1.144e-03  -4.878 1.57e-06 ***
Year:Origin2    5.402e-01  1.287e-01   4.197 3.36e-05 ***
Year:Origin3    3.513e-01  1.145e-01   3.069  0.00230 **
---
Signif.codes:  0 *** 0.001 ** 0.01 * 0.05 . 0.1   1
Residual standard error: 3.138 on 383 degrees of freedom
Multiple R-squared:  0.8417,Adjusted R-squared:  0.8384
F-statistic: 254.5 on 8 and 383 DF,  p-value: < 2.2e-16
```

Note that the two interaction terms in the model improve the multiple $R^2$ from 82% to 84%.

- Origin = American

$$\widehat{\text{MPG}} = -9.44 - 0.006 \times \text{Weight} + 0.64 \times \text{Year}$$

- Origin = European

$$
\begin{aligned}
\widehat{\text{MPG}} \;=\; & -9.44 - 0.006 \times \text{Weight} + 0.64 \times \text{Year} - 32.15 - 0.003 \times \text{Weight} \\
& +0.54 \times \text{Year} \\
\;=\; & -41.59 - 0.009 \times \text{Weight} + 1.18 \times \text{Year}
\end{aligned}
$$

- Origin = Japanese

$$
\begin{aligned}
\widehat{\text{MPG}} \;=\; & -9.44 - 0.006 \times \text{Weight} + 0.64 \times \text{Year} - 12.09 - 0.006 \times \text{Weight} \\
& +0.35 \times \text{Year} \\
\;=\; & -21.53 - 0.012 \times \text{Weight} + 0.99 \times \text{Year}
\end{aligned}
$$

The regression coefficients corresponding to the terms Weight × Origin and Year × Origin are responsible for changing the coefficients for Weight and Year. This implies that, for cars of different origins, the impact of Weight and Year on MPG is different. For American cars, if year increases by one, MPG increases by 0.64 units, while for European cars it increases by 1.18 units and for Japanese cars it increases by 0.99 units. If plotted on a three-dimensional graph, the three equations give rise to three intersecting planes. Similarly, an interaction term can be defined for two continuous variables also. In the model above, another interaction term Weight × Year could have been added. Interpretation of this interaction is also similar, i.e., dependence of

MPG on Weight changes each year. However, it is difficult to visualize such an interaction geometrically.

A product term with two predictors is called a two-factor interaction. Likewise three-factor interactions can also be defined with product terms for three predictors. Even higher-order interactions are also possible but are very rarely used because of the difficulty in interpretability. In a model, an interaction is included only if the main effects or the components of the interaction are significant. It may seem counterintuitive, but it is possible to have significant interactions without the main effects being significant. In such a case it is necessary that the lower-order terms are not dropped from the model.

## 9.6 Regression Diagnostics

A concern already identified in the residual analysis of regression is the identification of influential points. Ideally, each sample point will exert an equal amount of influence on the regression line. If the regression line happens to be too dependent on one or a few of the data points, possibly removed from the main concentration of the data, then these points are known as influential points for that model. Influential points may lead to a model with suboptimal predictive ability. Several statistics are used to measure the influence of each point on the regression line. All of these statistics are functions of residuals $r_i$, defined as the difference between the observed and the estimated response. For the $i$-th data point, $r_i = y_i - \hat{y}_i$, which is also a surrogate measurement of the error $\epsilon_i$ attached to the $i$-th observation. The error component in a regression model is never measurable, but the residuals always are.

To properly understand the source and impact of each data point on various aspects of linear regression, it is important to get acquainted with the hat matrix $H$. In matrix notation, we can write

$$\hat{Y} = HY,$$

where $H$ is a square matrix with the element $h_{ij}$ in the $i$-th row and $j$-th column defined as

$$h_{ij} = \frac{1}{n} + [(x_i - \bar{x})(x_j - \bar{x})]/[\sum_{k=1}^{n}(x_k - \bar{x})^2],$$

where $x_i$ is a vector of values of all predictor variables corresponding to the $i$-th observation and $x_j$ is a vector of values of all predictor variables corresponding to the $j$-th observation. The $i$-th diagonal element $h_{ii}$ of the hat matrix $H$ is known as the leverage of the $i$-th data point. Through the value of $h_{ii}$, each observation influences the predicted value. It may be shown that $0 < h_{ii} < 1$ and $\sum_{i=1}^{n} h_{ii} = p$, where $p = k + 1$ and $k$ is the number of predictors in the model. If all observations have equal influence on the regression equation, then all diagonal elements are equal to $p/n$. A rule of thumb suggests that a point would be defined as a high leverage point if $h_{ii} > 3p/n$. It is important to note that high leverage points are not necessarily bad observations. Since they exert more influence on the regression line than other points, it is a good strategy to do a further scrutiny and ensure that these are legitimate observations.

Assuming that the regression model errors are normally distributed with mean 0 and variance $\sigma^2$, the standardized error term associated with the $i$-th observation is defined as $\epsilon^* = \epsilon_i/\sigma$. However, neither $\epsilon_i$ nor $\sigma$ is known and both are to be estimated. The mean square error $MSE$ is an unbiased estimate of $\sigma^2$ and the $i$-th residual $r_i$ may be used as a surrogate for $\epsilon_i$. The studentized residual $t_i$ is defined as

$$ t_i = \frac{r_i}{(1 - h_{ii})\sqrt{MSE}}. $$

Studentized residuals have equal variance with numerical value 1. An observation may be considered an outlier if $|t_i| > 3$. If the sample size is small and one observation is believed to be very different from the rest, then it would have an influence on the computation of the $MSE$. To counter that, in computing the $i$-th studentized residual, $MSE$ is computed by eliminating the $i$-th observation from the model and fitting the model with the remaining $n-1$ observations. This generates the studentized deleted residual. But when sample size is large, there will not be any significant difference between the ordinary and the studentized deleted residuals.

Another quantification of influence of observations is through the difference between two estimated responses, one in which all observations are used to fit the regression model and the other where the $i$-th observation is deleted. This difference is simply called DFITs, difference in fits. Let $\hat{y}_i$ be the estimated value of the $i$-th response and $\hat{y}_{i(i)}$ be the value of the $i$-th response when the $i$-th observation is removed from the model. A scaled difference of these two quantities is known as DFIT$_i$, and, if $|\text{DFIT}_i| > 1$, then the $i$-th observation is considered an influential point.

While DFIT$_i$ measures the influence of the $i$-th observation on the $i$-th predicted value, Cook's distance (Cook's D) measures how much the $i$-th point influences the complete set of predicted values, in aggregation. Cook's D is computed for each observation, and a large value indicates that the corresponding observation has large influence. Usually, no cut-off is used for identification, but if one or more points are seen to have relatively large values compared to the others, they may be identified as outliers. Visualization of Cook's D generally provides such identification.

The difference in beta (DFBETA) criterion measures the standardized change in a regression coefficient when the $i$-th observation is omitted from the model. A rule of thumb suggests a value larger than $2/\sqrt{n}$ for DFBETA$_k$ indicates that the $i$-th observation has undue influence in estimating $\beta_k$.

Leverage points or outliers may not necessarily be "bad" points or erroneous points or points carrying no potential information. On the contrary, in many cases, these points, being different from the rest, contribute significantly to the overall understanding of the business process. The presence of outliers may be indicative of two different groups of observations mixed together. In such cases the groups need to be identified and treated differently. It is very important that the outliers or leverage points are examined carefully. Suppose a regression model is built on airlines passengers. If business class and first class passengers are mixed in a group of economy class passengers, the former group will come out to be outliers. Instead of deleting them from the dataset and building a model on the rest, the two groups are to be separated and different models need to be built. Throwing out the revenue-earning group will affect any business decision made by the airline.

It is to be noted that different methods of identification of influential observations may identify different sets of points as influential. There may be significant overlap among the sets, but often the sets will not be identical. A decision then needs to be taken which points are to be separated from the main bulk of the data. There is no clear-cut rule in this respect, and the decision needs to be taken based on common sense and domain knowledge.

**Example 9.7.** (Strength of Concrete) Consider the regression model of Concrete Strength on the predictors as

$$\widehat{\text{Strength}} = 301.53 - 0.02 \times \text{Blast.Furnace.Slag} - 0.07 \times \text{Fly.Ash}$$
$$+0.28 \times \text{Superplasticize} - 0.09 \times \text{Coarse.Aggregate}$$
$$-0.11 \times \text{Fine.Aggregate}.$$

The corresponding residual plot is given in Figure 9.12.

Even though the plot seems well behaved overall, it is always a good idea to check for outliers and leverage points and take a decision regarding them. See Figures 9.13 and 9.14. From these graphs it is clear that different tools of measuring the influence of observations are identifying different sets of points as most influential. The hat matrix identifies the highest number of points as high leverage values. On the other hand, only two observations have studentized residuals outside of the $\pm 3$ limit. These two observations are the 225th and the 382nd observations and they need to be looked at closely. While DFITs do not identify any point to be highly influential, several points show up on Cook's distance plot. The largest two are the 225th and 611th observations. A natural question that arises at this juncture is about the treatment of the observations identified as influential. Do we remove them and run a regression on the cleaned data? For this dataset there are over 1000 observations and removing all three identified points makes little difference to the sample size. We note that, by removing all three points, regression $R^2$ improves from 54% to 55%, not a significant change. A robust statistician might consider data deletion to be a naive idea, but more advanced robust techniques of dealing with outliers are beyond the scope of this book.

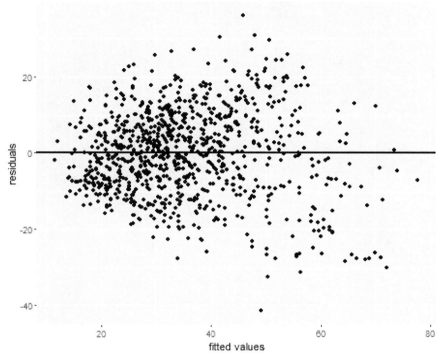

**FIGURE 9.12**
Distribution of residuals: VI.

## 9.7 Regression Model Building

In this age of Big Data and automatic data capture, any dataset typically contains many variables. Even after initial screening for quality and missing data and removing variables that do not contain useful information, there may still remain in the dataset a large number of predictors. Many of these predictors may also be highly correlated. It is important that the final recommended regression model has a limited number of predictors. A regression model with numerous predictors is difficult to understand and work with. We have already explained that correlated predictors give rise to multicollinearity and thereby

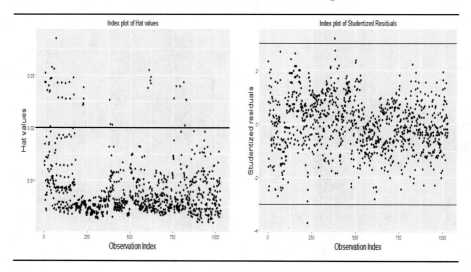

**FIGURE 9.13**
Identification of influential points: I.

the significance of model parameters goes astray. Further, inclusion of too many predictors in a model may lead to overfitting and reduce the predictive power of the model. Hence a few good regression models need to be identified for further intensive studies. Note also that uses of regression models vary. From that perspective too no single "best" model may be available. Choosing a subset of predictors must be done in such a way that no important predictor is left out. That will lead to a biased model. For example, in predicting sales of a particular brand of toothpaste, prices of close competitors are to be included in the model.

Identification of one or more subsets of predictor variables is one of the most challenging problems in regression analysis. All analytical software applications have the capability to run automatic procedures to recommend "good" models. However, an automatically run model may not be the most useful model if the predictors included in the model are not under the researcher's control. Take, for example, crop yield. If an automatic model search procedure recommends a model built on rainfall and temperature, that model will hardly be of use, as those two predictors are not under human control. For marketing research, if sales of a product is shown to depend on the GDP of a country, inflation and unemployment, that model is also of limited use. The insight may still be there, but it will not be actionable.

Recall that fitting all available predictors in a regression model and dropping predictors showing non-significant p-values is an incorrect strategy for model building. Regression coefficients in the model measure effects of corresponding predictors, given all other variables are included in the model.

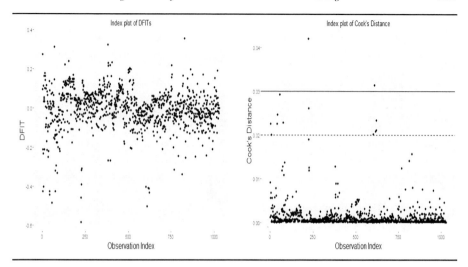

**FIGURE 9.14**
Identification of influential points: II.

Hence, this may lead to the elimination of important but correlated variables. Here we will discuss two different model-building strategies – in one a model is built step by step, either by adding or deleting one variable at a time, and in the other one, best regression models with the given number of predictors are compared and several competing working models are recommended.

### 9.7.1 Stepwise Procedure—Forward Selection and Backward Elimination

Possibly the most common automatic search procedure is the forward selection (FS) method where at each stage one predictor is included in the model and the procedure stops when no more variables may enter.

FS starts with no predictor in the model. The first predictor to enter the model is the one that has the highest correlation with the response $Y$ and is significant at a predetermined level $\alpha$. This is known as the entry criterion or $\alpha$-to-enter, and its value is recommended to be kept at a higher level, e.g., 20% or 25%, compared to standard values. The objective is to build a model possibly with a few extra variables and later eliminate them if necessary. Suppose the first variable to enter the model is $X_{(1)}$.

Next, all the remaining $n-1$ predictors are examined to see the addition of which increases the model $R^2$ the most, keeping $X_{(1)}$ in the model. Suppose the candidate variable is $X_{(2)}$. If $X_{(2)}$ is found significant at that stage at the $\alpha$-to-enter level, then $X_{(2)}$ is included in the model. At subsequent stages the same procedure is repeated. The procedure stops when, at any stage, no other variable is found to fulfill the entry criterion. The FS procedure does an initial screening when the number of predictors is very large and provides a working model.

The backward elimination (BE) procedure is the exact opposite of this. This procedure starts with all predictors in the model and the first predictor to be eliminated is the one removal of which decreases $R^2$ the least and is not significant at a suitable $\alpha$-to-remove level. Typically, the BE procedure keeps too many predictors in the model and does not eliminate enough. Hence it is not much preferred by analysts.

There is a third procedure which is a combination of FS and BE and is known as the stepwise procedure. The intended direction of the procedure is forward and it starts with no predictor in the model. After adding a predictor at each stage, each of the other predictors is checked for redundancy. In the forward step the best available predictor is included in the model if found significant at the $\alpha$-to-enter level. In the backward step the weakest predictor is removed from the model if not found significant at the $\alpha$-to-remove level. Two criteria are defined for model building, $\alpha$-to-enter and $\alpha$-to-remove. It is to be noted that $\alpha$-to-remove has to be equal or greater than $\alpha$-to-enter, otherwise the procedure will go into a loop and the same predictor will continue to enter the model in one step and will be removed in the next.

The order in which a predictor enters the model does not reflect its importance in prediction. Note that both the entry and exit of predictors are conditional on the variables that are already included in the model. It is possible that domain knowledge and other business considerations dictate inclusion of one or more predictors in the model. Such constraints can be admitted in model building but that may lead to a completely different model compared to the situation when the model is built automatically without any intervention.

### 9.7.2  All Possible Regressions

This technique considers all possible subsets of predictors to identify several good regression models. If there are $k$ predictors in the data, this technique considers all subsets of size 1, all subsets of size 2, all subsets of size 3 and so on. Naturally, with even a moderate number of predictors, this is a computationally intensive technique. Most of the analytical software shows the best three to five models at every level. But it is possible to look at all regression models for a given number of predictors. This procedure is also known as the best subset procedure.

Models identified by the "All Possible Regression" method cannot be compared using $R^2$ since more predictors in the model automatically increases its value. Even if predictors are not significant, $R^2$ will always increase, or at least stay the same, but it will never decrease. To take into consideration the number of predictors in the model, several alternative measures of model comparison are used.

- Adjusted $R^2$: It is defined as

$$1 - \frac{(n-1)MSE}{SSTO}.$$

The main difference between $R^2$ and adjusted $R^2$ is that the latter starts to decrease when insignificant variables are included in the model. Using $MSE$ instead of $SSE$ in the numerator acts as a penalty for including more variables than are optimally necessary.

- $\sqrt{MSE}$: Recall that the $MSE$ is an estimator for the error variance $\sigma^2$. A smaller value of error variance is expected to lead to a better model. Two otherwise equivalent models may be compared through their $\sqrt{MSE}$ values.

- Information criteria: Another tool to compare models is through their information content. The following are two important criteria for model comparison, where $p$ is the number of parameters to be estimated.

  (a) Akaike's information criterion: $n\log(SSE) - n\log(n) + 2p$.
  (b) Bayesian information criterion: $n\log(SSE) - n\log(n) + p\log(n)$.
      Akaike's information criterion (AIC) and the Bayesian information criterion (BIC) are similar in spirit. The primary difference is that the BIC penalizes a model more for an additional number of predictors.

- Mallow's $C_p$ criterion: At the core of the calculation is a comparison of the candidate model being considered and the full model, i.e., the model with all predictors. If a model works well, then the numerical value of $C_p$ is expected to be around $p$. For a potentially good model, $C_p \le p$. Hence models that have very large values of $C_p$ are not considered. On the other hand, if a model is biased, $C_p$ may have very small values or may even be 0. Hence models with extremely large or extremely small values of $C_p$ are not considered.

- Prediction sum of squares $PRESS_p$: This is used to assess the predictive ability of a model, which may be quite different, and generally lower, than the model's ability to fit. Typically, statistics which quantify the degree of model fit are used to evaluate the extent of similarity of the expected values and the observed values, when the same data has been used to develop the model; this relates to the model's ability to fit. Prediction is used to determine how well the model is performing when a set of new observations, which has not been used to develop the model, is being predicted. Denoting $\hat{Y}_{i(i)}$ as the estimated value of the $i$-th response when predictors for the $i$-th observation have not been used to estimate regression parameters,

$$PRESS_p = \sum_{i=1}^{n}(Y_i - \hat{Y}_{i(i)})^2.$$

In general, a smaller value of $PRESS_p$ indicates better predictive ability.

$PRESS_p$ can also be used to calculate the predicted $R^2$ ($R^2_{Pred}$) as $1 - PRESS_p/SSTO$, which is generally more intuitive than $PRESS_p$. $R^2$ and $R^2_{Pred}$ are related in nature and their numerical values are similar but they are not exactly the same, since the former is computed on the training set (the part of the data used to build the model) and the latter on the test set (the part of the data used to validate the model). One would expect $R^2$ to be high compared to $R^2_{Pred}$, but an exorbitantly high $R^2$ compared to $R^2_{Pred}$ indicates overfitting, i.e., the model developed is too close to the observations used and is not expected to work well for other sets of data.

As mentioned before, model building is as much an art as science. Even though several useful criteria are available to check for model adequacy, each of these addresses a different characteristic of the data. Unless the data is unusually well behaved, the choice of the model may depend on the criterion used. In fact, it is possible that several candidate models emerge as good working models and there may not be a single best model among them. In such cases, the choice of the model should not depend on statistical motivations only. Ease of maintenance, business applicability, client preference etc., need to be taken into consideration.

## 9.8 Other Regression Techniques

Linear regression is just one of many useful and frequently used regression techniques. There is a tendency to believe that this is the only regression technique available, but by no means is that the case. Regression involves two main steps. The first step is identification of the nature of the dependence and the second step is the estimation of the parameters. In this chapter we have dealt with linear dependence and a particular method of estimation, known as the least squares method. We have also confined ourselves to a continuous response. Within that realm, there are many alternatives for both steps.

In linear regression the dependence of the response on the predictors is linear. However, the nature of the dependence may be polynomial or involve other nonlinear functions. In the example involving Auto Data, inclusion of a second-degree term of Horsepower will drastically improve the fit, as seen in Figure 9.15.

A polynomial regression of degree 2 is essentially a linear regression with a squared term of one or more of the predictors in the model. Similarly, it is possible to have a nonlinear dependence of the form $E(Y) = (\theta_1 x)/(x + \theta_2)$. This is an example of nonlinear regression.

Whether linear or nonlinear, all situations discussed so far have parametric

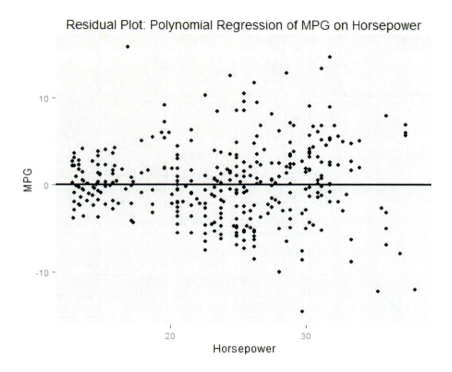

**FIGURE 9.15**
Distribution of residuals: VII.

forms, i.e., the nature of the dependence of the response on the predictors can be explicitly expressed through a functional form. An alternative model is nonparametric regression where neither the conditional mean response nor the error distribution is parametrically prespecified. In this case the response is estimated at a target predictor value obtained by taking a local weighted average of the predictors around it.

An alternative method of estimation of regression parameters is known as robust regression. These are especially useful to treat outliers. Many of the robust regression procedures are such that, if observations exert undue influence on the regression line, the procedure automatically discounts those observations. Consider, for example, the least median of squares estimator where, instead of minimizing the sum of squared residuals, the median of squared residuals is minimized. Another alternative estimation method minimizes the sum of absolute deviations, or various functions of scaled absolute deviations.

R is capable of handling many of the nonlinear and robust regression procedures. Notwithstanding that, few of these regression procedures are actually applied in big data analytics. Admitting that these procedures may be cumbersome to apply and computationally time intensive, the time probably has come to move beyond the realm of linear regression.

## 9.9    Logistic Regression

While the predictors may be discrete or continuous, so far we have assumed the response to be continuous only. However, there may be many occasions when a discrete response needs to be predicted. Banks often have to take a decision regarding whether to approve a loan to an applicant. Telecommunication companies are always appraising customers based on their churn status. In both cases the response is binary: Yes or No. In the healthcare industry it is often of interest to determine whether exposure to several risk factors contributes to disease development at various hierarchical levels, such as mild, moderate or severe. Multiple linear regression is not a valid technique here. Recall that one of the assumptions of multiple regression is that the response follows a normal distribution. If the response takes on two values, Yes or No, usually coded as 1 or 0, that assumption is completely violated. For a binary response the expected response at a given level of predictor is

$$E(Y|x) = 1 \times Pr(Y = 1|x) + 0 \times Pr(Y = 0|x) = Pr(Y = 1|x).$$

Hence regression of $Y$ on $X$ reduces to a problem of regressing the probability of the response taking the value 1. Since probability is bounded by 0 and 1, a linear regression function is not applicable. A special regression function is necessary which is bounded by 0 and 1 for any value of the predictors. To achieve that, an appropriate transformation on $Pr(Y = 1)$ is used to linearize it. Several such transformations are used in practice, but the most commonly used one is called a logistic transformation.

Before we are in a position to describe logistic transformation, we need to talk about odds and log odds of success. Logistic distribution is closely related to binomial distribution. A revisit to this important concept may not be out of order. When an experiment is performed with only two outcomes, where the probability of each outcome remains the same for each repetition of the experiment, the total number of successes is said to follow a binomial distribution. The binomial distribution (defined in Section 6.2.1) is characterized by the total number of times the experiment is repeated (usually denoted by $n$) and the probability of success (usually denoted by $p$). Recall that the two outcomes of the experiment are denoted by success and failure. These terms are just labels and have nothing to do with success or failure per se. Suppose a bank decides to approve a loan to each applicant with a probability of 20%. This is an experiment, and the probability of success (loan approval) remains the same for each applicant. If, on a single day, 30 applicants walk through the door, the number of experiments is 30. The total number of loans approved on that day would follow a binomial distribution with parameters $n = 30$ and $p = 0.2$. The odds of success is defined as the ratio of the probability of success and the probability of failure. We have

$$\text{odds of success} = \frac{Pr(Y = 1)}{Pr(Y = 0)} = \frac{Pr(Y = 1)}{1 - Pr(Y = 0)} = \frac{\pi}{1 - \pi},$$

with $\pi = Pr(Y = 1)$, 1 denoting success. The log the quantity presented in the last equation is known as the log odds of success. Now consider a situation where the probability of loan approval depends on one or more characteristics of the applicant. For simplicity, suppose the approval rate for a male applicant is 30% while that for a female applicant is 18%. The success probability depends on the applicant's gender, and, log odds of success, in turn, also depends on the single predictor, gender. In logistic regression, the log odds of success or the logit of success is expressed as a linear function of the predictors. The logistic regression function is given as

$$\log\left(\frac{Pr(Y = 1|x)}{1 - Pr(Y = 1|x)}\right) = \alpha + \beta_1 x_1 + \beta_2 x_2 + \cdots + \beta_k x_k,$$

when $k$ predictors are involved.

Note that there is no explicit error term associated with the right hand side of the equation. This is because, unlike multiple linear regression, the model predicts the probability of the response being a success (taking the value 1), rather than the value of the response. Another major departure from multiple linear regression is that scatterplots do not provide any indication of dependence of the response on any predictor.

Consider the data on creditworthiness among a sample of 1000 German loan applicants. The response is creditworthiness, i.e., whether the applicant is considered a good risk and his/her loan application is approved. One principal determiner of approval is the amount of loan asked for. Since the response can take the values 0 or 1 only, scatterplots are non-informative (see Figure 9.16).

## 9.10 Interpreting the Logistic Regression Model

Logistic regression is part of a larger group of models known as the generalized linear model (GLM). One of the principal characteristics of a GLM is that $E(Y|x)$ is an invertible function of $\eta = \alpha + \beta_1 x_1 + \beta_2 x_2 + \cdots + \beta_k x_k$. In other words, $g(E(Y|x)) = \eta$, where the function $g$ is known as a link function. For a logistic model, the link function is the logit function. Other common link functions are the probit and the complementary log-log function, which are not described here. In the case of a linear model, the function $g(\cdot)$ is

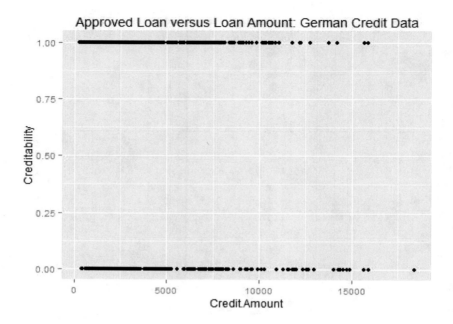

**FIGURE 9.16**
Scatterplot for binary response.

the identity function. Further, in the case of a linear model, the response is assumed to follow a normal distribution, but in the case of a GLM the response may have other discrete distributions like binomial or Poisson or continuous distributions such as Weibull, etc.

**Example 9.8.** (German Credit Data) Let us investigate the dependence of Creditability on Credit.Amount.

```
Call:
glm(formula = Creditability ~ Credit.Amount, family = binomial,
data = German.Credit.M1)

Coefficients:
                Estimate Std. Error z value Pr(>|z|)
(Intercept)    1.229e+00  1.083e-01   11.348  < 2e-16 ***
Credit.Amount -1.119e-04  2.355e-05   -4.751 2.02e-06 ***
---
Signif.codes:  0 *** 0.001 ** 0.01 * 0.05 . 0.1   1
```

(Dispersion parameter for binomial family taken to be 1)

    Null deviance: 1221.7  on 999  degrees of freedom
Residual deviance: 1199.1  on 998  degrees of freedom
AIC: 1203.1

Number of Fisher Scoring iterations: 4

The logistic regression equation is given as

$$\log\left(\frac{Pr(Y=1)}{1-Pr(Y=1)}\right) = 1.23 - 0.0001 \times \text{Credit.Amount}$$

After a simple exponentiation procedure to get rid of the logarithm, odds of loan approval is

$$\frac{Pr(Y=1)}{1-Pr(Y=1)} = \exp(1.23 - 0.0001 \times \text{Credit.Amount}).$$

Another simple algebraic manipulation provides

$$Pr(Y=1) = \frac{\exp(1.23 - 0.0001 \times \text{Credit.Amount})}{1 + \exp(1.23 - 0.0001 \times \text{Credit.Amount})}.$$

Hence the probability of a loan approval has been expressed as a function of Credit.Amount. The predicted value of the logit function or the log odds and the estimated probability of success (loan approval) are directly available from the model. A few of representative loan amounts and their corresponding predicted log odds and probability of approval are given in Table 9.3.    ‖

**TABLE 9.3**
Credit amount table

| Credit.Amount (in Deutsche Marks) | Predicted Log Odds of Success | Predicted Probability of Success |
|:---:|:---:|:---:|
| 250 | 1.2014 | 0.77 |
| 6331 | 0.5209 | 0.62 |
| 8086 | 0.3246 | 0.58 |
| 14,421 | −0.3842 | 0.40 |
| 18,424 | −0.8321 | 0.30 |

Logistic regression is linear in the log odds, but it is not linear for $Pr(Y = 1) = E(Y)$. It is an S-shaped function having asymptotes at 0 and 1 (see Figure 9.17), i.e., for very small values of success probability or very large values of success probability, the logistic regression function is parallel to the X-axis.

This is because probability is bounded by 0 and 1; whatever the value of the predictor, estimated success probability will remain between these two values. It is clear from Figure 9.17 that, for a negative regression coefficient, the logistic function is an inverted S whereas, if the coefficient has a positive sign, the logistic function is the other way around. The sharpness of the curve depends on the numerical value of the regression coefficient. The logistic regression function of loan approval on credit amount is very flat and does not depart too much from a linear relationship.

The S-shape of the logistic regression curve implies that the rate of change in the predicted probability is not constant over the range of the predictor variable. Toward the middle, the rate of change is higher compared to the two ends when the function becomes flat and parallel to the Y-axis. Toward the extreme end, when the value of the estimated probability has attained 0 or 1, the logistic regression function will not show any change whatsoever.

## 9.11    Inference for the Logistic Regression Model

The above example contains only a single predictor, but logistic regression may have multiple predictors as well. Consider Creditability as the response and Credit.Amount, Duration.of.Credit.Month, Length.of.Current.Employment

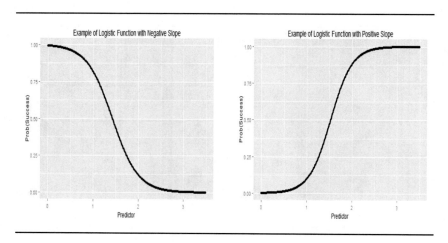

**FIGURE 9.17**
Shape of the logistic function.

and Purpose (of loan) as the predictors. The output of the GLM fit is given below.

```
Coefficients:
                               Estimate Std. Error z value Pr(>|z|)
(Intercept)                   2.199e+00  3.603e-01   6.103 1.04e-09 ***
Credit.Amount                -3.456e-05  3.272e-05  -1.057 0.290732
Duration.of.Credit.Month     -3.837e-02  7.562e-03  -5.075 3.88e-07 ***
Length.of.Current.Employment  2.658e-01  6.773e-02   3.925 8.67e-05 ***
Purpose2                     -1.158e+00  3.253e-01  -3.560 0.000371 ***
Purpose3                     -9.053e-01  3.087e-01  -2.933 0.003359 **
Purpose4                     -1.334e+00  3.010e-01  -4.432 9.33e-06 ***
---
Signif.codes:  0 *** 0.001 ** 0.01 * 0.05 . 0.1   1
```

Credit.Amount and Duration.of.Credit.Month are continuous predictors. Length.of.Current.Employment is an ordinal variable with four levels and Purpose (of loan) is a nominal variable with levels 1: New car, 2: Used car, 3: Home related and 4: Others.

Significance of the predictors is determined as in the case of linear regression. From the above table it is clear that, after Duration.of.Credit.Month, Length.of.Current.Employment and Purpose, Credit.Amount is not required to be included in the model. The significance test is done using a normal test, instead of a $t$-test. This is because the underlying model is binomial, and, when sample size is large, significance testing for binomial proportions follows a normal distribution. It is also to be noted that, for a large sample size, the $t$ distribution also reduces to a standard normal distribution.

The model for Creditability is

$$\text{logit(Creditability)} \; = \; 2.20 - 0.00003 \times \text{Credit.Amount}$$
$$-0.04 \times \text{Duration.of.Credit.Month}$$
$$+0.27 \times \text{Length.of.Current.Employment}$$
$$-1.16 \times \text{Purpose2}$$
$$-0.90 \times \text{Purpose3} - 1.33 \times \text{Purpose4}$$

Given a particular covariate combination, $\log\left(\frac{Pr(Y=1)}{1-Pr(Y=1)}\right)$ is estimated directly. Consider a particular combination of covariates as Credit.Amount = 1049, Duration.of.Credit.Month = 18, Length.of.Current.Employment = 1 and Purpose = 2 (used car). The predicted value of the logit at this combination is 0.58, and thus $Pr(Y = 1)$ is estimated to be 0.64. For one unit change in each covariate, the change in the logit is equal to the coefficient of that covariate, while the change in the probability of success is equal to the exponential of the coefficient. In the above example suppose other predictors remain the same but Purpose is 3 (home related) instead of 2 (used car). The numerical value of the logit is 0.82 and $Pr(Y = 1) = 0.69$. The difference between logit(Purpose = 2) and logit(Purpose = 3) is the difference between $\beta$(Purpose = 2) and $\beta$(Purpose = 3). Hence the log odds ratio of Creditability at Purpose = 2 versus Purpose = 3 is exponential $(\beta_2 - \beta_3)$, where $\beta_i$ corresponds to coefficient for Purpose = $i$. Because the log odds ratio has a direct interpretation through the regression coefficients, logistic regression is very popular compared to other generalized linear models.

For a given covariate combination, the expected value of logit(Creditability) and $Pr$(Creditability = 1) are easily estimated. Using the standard error obtained in the prediction process, upper and lower confidence intervals are also found. But no direct $F$-test for overall model fit is applicable for logistic regression.

## 9.12    Goodness-of-Fit for the Logistic Regression Model

After fitting a model to the data, it is natural to check whether the model is adequate for the data. In the case of a linear regression, the coefficient of multiple determination $R^2$ provides such a measure. In the case of a logistic regression, no such direct comparison of $SSTO$ and $SSE$ is recommended, even if it is possible to construct it. Instead, logistic regression models are compared using deviance statistics and Akaike's information criterion (AIC).

The parameters of a logistic regression are estimated through maximum likelihood estimation. If there is no model to fit, maximum likelihood estimates of the probability of success at any level of the predictor is the sample proportion, if the predictor pattern is repeated. Under the recommended model, the model parameters are used to estimate the probability of success at every level of the covariate, whether the level is repeated or not. If the model under consideration is

$$\log\left(\frac{Pr(Y = 1|x)}{1 - Pr(Y = 1|x)}\right) = \alpha + \beta_1 x_1 + \beta_2 x_2 + \cdots + \beta_k x_k,$$

then

$$Pr(Y = 1|x) = \frac{\exp(\alpha + \beta_1 x_1 + \beta_2 x_2 + \cdots + \beta_k x_k)}{(1 + \exp(\alpha + \beta_1 x_1 + \beta_2 x_2 + \cdots + \beta_k x_k))}.$$

An iterative procedure called the Fisher scoring algorithm is used to estimate the model parameters. Usually, convergence of the iterative algorithm is not a problem, but it is not guaranteed. The likelihood function summarizes the information that the data provides about the unknown parameters of the model. The value of the likelihood function, when the parameters are replaced by their MLEs, is the numerical value of the likelihood under the current model. Let it be denoted by $L_M$. The value of the likelihood function is the highest when no model is assumed on the success probabilities. Let this be denoted by $L_F$. The difference of these two likelihood functions indicates how good the model fit is. Formally, $D = -2[\log L_C - \log L_F]$ so that a large value of $D$ indicates that the current model is a poor fit. The underlying principle is that, if the model is good for the data, the value of the likelihood function is not too small compared to the maximum value possible.

The null deviance provided by R is the deviance statistic for an intercept-only model, and the deviance for the model under consideration is given by the residual deviance. If the model under consideration is expected to be an improvement over the intercept-only model, then the deviance statistic must show considerable reduction in its numerical value. For the German Credit Data example the null deviance is 1221.7 and the residual deviance is 1199.1, when Credit.Amount is included in the model, thereby providing a reduction of 22.6 in the value of the deviance and loss of 1 degree of freedom. Consider now an alternative model where the length of time the loan is applied for (Duration.of.Credit.Month) is included in the model. There will be no change in the null deviance, but the residual deviance is now 1177.1, providing a reduction of 44.6 in the value of the deviance and loss of 1 degree of freedom. The second model is a better one in explaining the variability. Just as in the case of multiple linear regression, more than one predictor can be included in a logistic regression model. Following is an example of an analysis of deviance table for German Credit Data where Credit.Amount, Duration.of.Credit.Month, Length.of.Current.Employment and Purpose (of loan) have been included.

```
Call:
glm(formula = Creditability ~ Credit.Amount + Duration.of.Credit.Month
+ Length.of.Current.Employment + Purpose, family = binomial,
data = German.Credit.M1)

Coefficients:
                               Estimate Std. Error z value Pr(>|z|)
(Intercept)                   2.199e+00  3.603e-01   6.103 1.04e-09 ***
Credit.Amount                -3.456e-05  3.272e-05  -1.057 0.290732
Duration.of.Credit.Month     -3.837e-02  7.562e-03  -5.075 3.88e-07 ***
Length.of.Current.Employment  2.658e-01  6.773e-02   3.925 8.67e-05 ***
Purpose2                     -1.158e+00  3.253e-01  -3.560 0.000371 ***
Purpose3                     -9.053e-01  3.087e-01  -2.933 0.003359 **
Purpose4                     -1.334e+00  3.010e-01  -4.432 9.33e-06 ***
---
Signif.codes:  0 *** 0.001 ** 0.01 * 0.05 . 0.1   1

(Dispersion parameter for binomial family taken to be 1)

    Null deviance: 1221.7  on 999  degrees of freedom
Residual deviance: 1133.2  on 993  degrees of freedom
AIC: 1147.2

Number of Fisher Scoring iterations: 4
```

```
> anova(GDlr3)
```

Analysis of Deviance Table

Model: binomial, link: logit

Response: Creditability

Terms added sequentially (first to last)

|  | Df | Deviance | Resid. Df | Resid. Dev |
|---|---|---|---|---|
| NULL |  |  | 999 | 1221.7 |
| Credit.Amount | 1 | 22.665 | 998 | 1199.1 |
| Duration.of.Credit.Month | 1 | 22.511 | 997 | 1176.5 |
| Length.of.Current.Employment | 1 | 18.221 | 996 | 1158.3 |
| Purpose | 3 | 25.166 | 993 | 1133.2 |

The analysis of deviance table shows reduction in deviance after each predictor has been added sequentially. Note that Purpose is a nominal variable with four levels and therefore has 3 degrees of freedom. After addition of all four predictors, the total reduction in deviance is 88.5 while 6 degrees of freedom are lost.

Since there is no concept of adjusted $R^2$ for logistic regression, it is difficult to check the redundancy of a predictor. From the $z$-table of significance of predictors it needs to be inferred. In this example Credit.Amount seems to be redundant in the presence of the other predictors in the model. An alternative model is fitted as follows.

Analysis of Deviance Table

Model: binomial, link: logit

Response: Creditability

Terms added sequentially (first to last)

|  | Df | Deviance | Resid. Df | Resid. Dev |
|---|---|---|---|---|
| NULL |  |  | 999 | 1221.7 |
| Duration.of.Credit.Month | 1 | 44.615 | 998 | 1177.1 |
| Length.of.Current.Employment | 1 | 18.455 | 997 | 1158.7 |
| Purpose | 3 | 24.381 | 994 | 1134.3 |

It is clear that the reduction of deviance in the absence of Credit.Amount is only slightly less (87.4).

Another measure of the goodness-of-fit of a logistic model is the AIC. It is also based on deviance of a model, and, like multiple linear regression, it also penalizes a more complex model. Its behavior is similar to that of adjusted $R^2$, but the number itself does not provide any measure of model fit. AIC is meaningful in a comparative sense. If there are several candidate models, nested or not, the lowest value of AIC indicates the best among them. It is a useful criterion for comparing model fit, but is not interpretable in an absolute sense.

## 9.13   Hosmer–Lemeshow Statistics

Possibly the most intuitive and commonly used statistic to test the fit of a logistic model is the Hosmer–Lemeshow goodness-of-fit (GoF) statistic. The deviance test for the logistic regression model fit works well if unique profiles of the predictor variable combinations are well defined. If predictor variables are all categorical variables, then the number unique profiles is limited. However, if there are one or more continuous (or almost continuous) variables among the predictors, then the number of unique profiles is too large. The deviance test is based on the proportion of successes in each unique profile combination of the predictors. With one or more continuous variables, not only the number of unique profiles is too large, the proportion of success in each profile may be 0 or 1, as there may be single observations in each of the profiles.

A goodness-of-fit statistic is a measure of closeness between two distributions – one of them is the actual distribution and the other is specified under the current model assumption. The statistic is used to test whether the specified model holds. The GoF statistic is computed over several classes (groups) and follows a $\chi^2$ distribution with degrees of freedom equal to the difference of the number of parameters of the two models. See Section 8.6.1 for a description of the Pearson's goodness-of-fit statistic under the multinomial model.

The Hosmer–Lemeshow statistic tests the hypothesis $H_0$ : Current model fits the data versus $H_A$ : Current model does not fit well. Observations are grouped according to the estimated values of success probability under $H_0$, i.e., the estimated probabilities that are available through the model fit. It is recommended that 10 equal-sized groups be used, since empirically it has been found to give good results. The number of observations per group, the sum of the observed responses and sum of predicted responses in each group are computed. The statistic computed is

$$G_2 HL = \sum_{j=1}^{10} \frac{(O_j - E_j)^2}{(E_j(1 - E_j/n_j))}$$

which follows $\chi_8^2$. The rejection region is always on the right hand side. If the value of the test statistic is too large, then the null hypothesis is rejected. For the German Credit Data example, for the model with the variables Duration.of.Credit.Month, Length.of.Current.Employment and Purpose, the fitted values and observed values are close enough so that the p-value of the test statistic is 15%. Hence this model is acceptable for the data.

```
Hosmer and Lemeshow goodness of fit (GOF) test

data:  Creditability, GDlr4$fitted.values
X-squared = 12.0945, df = 8, p-value = 0.147
```

## 9.14   Classification Table and Receiver Operating Curve

The output of a logistic regression is the probability of success at a given level of predictor. However, ultimately one would require a prediction of success or failure. Let us continue with the German Credit Data example. When a new loan application comes in it is not enough to know that the probability of the applicant not defaulting is 65%. The loan manager needs to take a decision whether it is advisable to extend a loan to this applicant. Therefore the probability of success needs to be mapped to a 0–1 decision rule using a threshold. If the predicted success probability is greater than or equal to the threshold, then the decision is 1 or success; if it is less than the threshold, the decision is 0 or failure. A classification table may be created as below:

|                                          | Observed Success | Observed Failure |
|------------------------------------------|:----------------:|:----------------:|
| Predicted Success (above threshold)      | a                | b                |
| Predicted Failure (below threshold       | c                | d                |

The sum total of all the cells equals the sample size. A different threshold level would give a different classification table. The optimum level of threshold is where the misclassification probability is minimized. The misclassification probability is $(b + c)/(a + b + c + d)$. For a different level of threshold this value will change. Two related quantities are sensitivity or true positive rate defined as $a/(a + c)$ and specificity, or true negative rate defined as $d/(b + d)$. High sensitivity and high specificity indicate a better fit. But sensitivity and specificity cannot be maximized simultaneously.

Extending the concept of a $2 \times 2$ table, instead of selecting one particular threshold, one can examine the full range of thresholds from 0% to 100%. For each given value of threshold, one classification table can be constructed and a pair of sensitivity and specificity values may be obtained. The curve obtained by plotting sensitivity against (1–specificity) is known as the receiver operating curve (ROC). For the ROC, the area under the curve (AUC) is an overall measure of the fit of the logistic regression model. The AUCs for two competing models are shown in Figure 9.18.

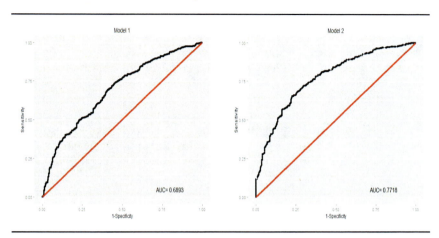

**FIGURE 9.18**
Area under the ROC curve.

Model 1 contains

{Credit.Amount, Duration.of.Credit.Month,
Length.of.Current.Employment, Purpose}

whereas Model 2 contains

{Duration.of.Credit.Month, Length.of.Current.Employment
Purpose, Account.Balance, Installment.Percent}.

For Model 1, AIC is 1147.2 and for Model 2, AIC is 1038.2. Comparatively, Model 2 is better as far as AIC is concerned. In Figure 9.18, the diagonal line indicates random classification, i.e., if no model is used, then each applicant will have an equal probability of being classified as a success or a failure. Hence, the area under the curve is 50%. Any model, if working, is expected to do better than 50%. The discriminatory power of Model 1 is 68%, but the discriminatory power of Model 2 is 77%.

## 9.15 Suggested Further Reading

Regression is a vast topic and there probably does not exist any area where it is not applied. For a comprehensive treatment of theory and application in regression, Montgomery et al. (2012), Draper and Smith (1998), Weisberg

(2013) and Kutner et al. (2003) may be used. Among other books on regression, Chatterjee and Hadi (2012), Pardoe (2012) and Ryan (2009) are useful resources. All regression books contain multiple data analytic examples and case studies involving datasets of various sizes and complexity. Regression is a subject that is best learned through application; hence exposure to varied datasets and analytical techniques always helps. For detection of outliers and leverage points, Cook and Weisberg (1982) is one of the most useful resources, even though most of the regression texts mentioned above also spends time on this important aspect. For rudimentary understanding of alternative regression techniques, such as robust and nonparametric regression, Rousseeuw and Leroy (2003), Takezawa (2005) and Davino et al. (2013) may be consulted. For logistic regression, Hosmer et al. (2013) and Kleinbaum and Klein (2010) may be consulted. Additionally, Collett (1991) and Harrell (2015) discuss other related concepts. Linear regression, robust and nonparametric regression as well as bootstrap applications in regression are also discussed in Kundu and Basu (2004). A data mining approach to regression including model development and validation can be found in James et al. (2014).

# 10

## Decision Trees

In the previous chapter we discussed fairly sophisticated regression models with different types of responses, continuous and binary, and different types of predictors, continuous, binary, nominal and ordinal. We also touched upon the fact that regression models may not always be linear. Polynomial regression and interactions among the predictors may make the model very complicated. If there are hundreds of predictors, then variable selection and incorporation of interactions may not always work out well, primarily due to volume and complexity. In such situations decision-rule-based models are used to split the sample into homogeneous groups and thereby come up with a prediction. Tree-based models are also classification models, but the main difference between these and ordinary classification models is that here the sample is split successively according to answers to questions like whether $X_1 \geq a$. All observations with $X_1 \geq a$ are classified into one group and the rest are classified into a second group. Typically, the split is binary, but there are procedures when split may be multiple. Tree-based models are easily interpretable and they are able to take into account many variables automatically, using only those which are most important. Tree-based models are also useful to identify interactions, which may later be used in regression models.

In medical diagnostics, tree-based methods have many applications. A common goal of many clinical researches is to develop a reliable set of rules to identify new patients into hierarchical risk levels. Based on data already in the database, and based on the clinical symptoms and ultimate outcome (death or survival) of the patients, a new patient may be classified into high-risk and low-risk categories. Decision trees have very high applicability in industrial recommender systems. A recommender system recommends items or products to prospective buyers based on their purchase pattern. In many such cases, especially if the store does not have any information on the customer other than their previous purchases, or in the case of online stores, prospective buyers' choices and click-stream responses, regression or any other model building procedures will not be applicable. Decision trees are applicable in marketing to identify homogeneous groups for target marketing. Decision trees segment the predictor space into several similar regions and use the mean of the continuous response or mode of categorical response in each separate region for prediction.

## 10.1 Algorithm for Tree-Based Methods

*Classification and regression tree* (CART) is a recursive binary splitting algorithm introduced by Breiman et al. (1984). Even though there have been many improvements on this method, the core algorithm is still identical to that of CART. Consider the set of all predictors, continuous and categorical, together. CART does not make any differentiation between categorical and continuous predictors. CART is based on nodes and split. The full sample is known as the *root node*. At each instance of split, a variable and its level is selected, so that purity at each child node is the highest possible at that level. The basic idea of tree growing is to choose a split among all possible splits at each node so that the resultant child nodes are the purest. At each split the sample space is partitioned according to one predictor. Only univariate splits are considered, so that the partitions are parallel to the axes. If $X$ is a categorical variable with $I$ levels, then the number of splits possible is $2^{(I-1)} - 1$. This is because each possible subset is considered as a candidate split. If $X$ is an ordinal variable or a continuous variable with $K$ distinct values in the sample, then the number of splits possible is $K - 1$. For each continuous or ordinal variable, the split is of the form $X \leq x_0$; if, for an observation, this condition is satisfied, the observation falls in the left child node, otherwise it is included in the right child node. The best split is the one that maximizes the splitting criterion. Depending on the splitting criterion, splits may be different. Several commonly used splitting criteria are discussed. For every node, the same algorithm is followed until a stopping rule is reached.

A stopping rule may be defined in different ways. If a node is pure, i.e., value of the response in each node is identical, then that node is a terminal node and it will not be split any more. If all observations in a node have identical values on the predictor combination, whatever the value of the response, that node is also a terminal node. In addition to these natural stopping rules, user-defined stopping rules may be provided, where nodes will not be split if maximum tree depth is achieved, or, if the node contains less than the user-defined minimum number of observations, the node will not be split.

A tree starts with all observations in the root node. Successive splitting divides each node into two child nodes. Splitting stops when the conditions mentioned in the stopping rule are met. The nodes that are not split any more are called the terminal nodes or leaf nodes. There may be a number of intermediate nodes in the path to reach the leaf nodes from the root node. Each and every observation falls through the intermediate nodes to one child node only. The total number of observations in the child nodes is equal to the sample size. The path to each child node is also non-ambiguous, i.e., each observation will reach the destined node through one path only. Each path indicates a rule of classification for the observations.

If the response is categorical, then in each terminal node, the predicted value of response is the modal value in that node. If the response is continuous, the predicted value is the average response.

## 10.2 Impurity Measures

Impurity in the context of decision trees means the mix of observations belonging to different classes in a single node. If all observations in a node belong to the same class, the node is pure and impurity is zero. Impurity is highest if there are an equal number of observations belonging to each class in a node.

Suppose in the sample there are $K$ classes and $p_j$ is the proportion of the $j$-th class, $p_j \geq 0$ and $\sum p_j = 1$. Formally, the impurity function $\Phi(\cdot)$ is defined as a function on $(p_1, p_2, p_3, \ldots, p_K)$ so that $\Phi(\cdot)$ is maximum when all probabilities are equal and minimum when only $p_j = 1$ and other $K - 1$ proportions are all 0. At a node $t$, impurity measure $i(t)$ is defined as

$$i(t) = \Phi(p(1|t)p(2|t)\ldots p(K|t))$$

where $p(j|t)$ is the proportion of class $j$ given a point is in node $t$. The aggregated impurity measure of a tree $T$ is given by

$$I(T) = \sum_{t \in T} i(t)p(t)$$

where $p(t)$ is the proportion of observations in terminal node $t$.

Two common impurity functions for classification are

1. Gini index: $\sum_{j=1}^{K} p_j(1 - p_j) = 1 - \sum_{j=1}^{K} p_j^2$.

2. Entropy: $\sum_{j=1}^{K} p_j \log \frac{1}{p_j}$ (if $p_j = 0$, then a limiting value is used.)

For a regression tree, the least squares deviance is used as the splitting criterion.

**Example 10.1.** (Auto Data) This is an application of a regression tree. The objective here is to predict the MPG performance of different cars, information for which is available in Auto Data. In R there are two functions that provide recursive binary splitting. Here we provide the output of using the `rpart()` function with MPG as the dependent variable.

```
rpart(MPG ~ Cylinders + Displacement + Horsepower + Weight +
Acceleration +  Year + Origin, data=Auto)

n= 392
```

```
node), split, n, deviance, yval
     * denotes terminal node

1) root 392 23818.9900 23.44592
  2) Displacement>=190.5 170   2210.1880 16.66000
    4) Horsepower>=127 96      457.0662 14.51875 *
    5) Horsepower< 127 74      741.9541 19.43784 *
  3) Displacement< 190.5 222   7785.9020 28.64234
    6) Horsepower>=70.5 151    3347.5600 26.28013
     12) Year< 78.5 94   1222.0920 24.12021
        24) Weight>=2305 55     413.6684 22.28545 *
        25) Weight< 2305 39     362.1677 26.70769 *
     13) Year>=78.5 57    963.7389 29.84211
        26) Weight>=2580 33     224.9988 27.46061 *
        27) Weight< 2580 24     294.2333 33.11667 *
    7) Horsepower< 70.5 71    1803.7790 33.66620
     14) Year< 77.5 28    280.2500 29.75000 *
     15) Year>=77.5 43    814.4786 36.21628 *
```

The root node starts with 392 observations and total sum of squares (deviance) equal to 23,818.99. The mean of the response is given by yval. In this case the average MPG over all 392 observations is 23.45. The first split is according to the predictor Displacement and the point of the split is 190.5. The second and third nodes constitute 170 observations with Displacement greater than or equal to 190.5 and 222 observations with Displacement less than 190.5, respectively. In node 2 the deviance is 2210.19 and in node 3 the deviance is 7785.9. The numerical value of yval (mean of MPG) in the nodes are 16.7 and 28.6, respectively. The sum of deviances in nodes 2 and 3 is 9990, which is considerably less than 23,820.

The 170 observations in node 2 are now split into nodes 4 and 5 according to the variable Horsepower and the split point is Horsepower 127 and contains 96 and 74 observations respectively. These are also terminal nodes and will not be split any further. The average MPG performance is 14.5 and 19.4, respectively. Node 3 is split into two nodes according to the same variable Horsepower but the split point is Horsepower $\geq$ 70.5. Neither of these is a terminal node and each is split further based on Year and Weight. Note that at each stage the child nodes of node $x$ are numbered $2x$ and $2x + 1$.

The final constructed tree is given in Figure 10.1. The final tree has eight terminal nodes. The number of observations in each terminal node is given along with the expected MPG performance of the cars. Each of the terminal nodes corresponds to one set of rules. For example, the terminal node 15 corresponds to the following rule: A car with Displacement < 190, Horsepower < 70 and Year $\geq$ 78 is expected to have MPG 36. On the other hand, if a car has Displacement $\geq$ 190 and Horsepower $\geq$ 127, then its expected MPG is 15.

The split for regression tree, i.e., for continuous response, is through ANOVA, i.e., the objective is to reduce the sum of squares due to error. At

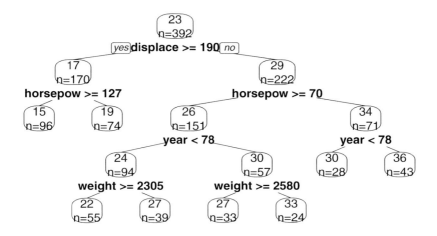

**FIGURE 10.1**
Regression tree for Auto Data.

every stage of the split, the reduction in the sum of squares is computed and the default stopping rule is when the reduction is less than 1%. A summary of `rpart()` provides the following table.

| | CP | nsplit | rel error | xerror | xstd |
|---|---|---|---|---|---|
| 1 | 0.58033113 | 0 | 1.0000000 | 1.0024038 | 0.06153852 |
| 2 | 0.11060764 | 1 | 0.4196689 | 0.4610299 | 0.04212332 |
| 3 | 0.04877326 | 2 | 0.3090612 | 0.3985313 | 0.03978300 |
| 4 | 0.04245216 | 3 | 0.2602880 | 0.3347437 | 0.03628806 |
| 5 | 0.02976827 | 4 | 0.2178358 | 0.2776215 | 0.03059567 |
| 6 | 0.01873528 | 5 | 0.1880675 | 0.2539922 | 0.02912450 |
| 7 | 0.01866186 | 6 | 0.1693323 | 0.2435878 | 0.02757468 |
| 8 | 0.01000000 | 7 | 0.1506704 | 0.2355478 | 0.02745307 |

At the root node, ($nsplit = 0$) the error is 100%. After the first split relative error is 42% and the improvement is 58%. At the second split relative error is 30% and the improvement is 11% and so on, till the 8th split where improvement is only 1%. No further split is done. This table is known as the CP-Table. It provides a brief summary of the overall performance of the tree. CP column shows the values of the *complexity parameter*. The column xerror provides cross-validated error rate and their standard errors are given in the last column. Functions of these two right-most columns will be crucial in determining an optimal tree. It suffices to say at this point that an optimal tree corresponds to minimum error. ‖

## Using Gini Index: : 300 Data Points from German Credit Data

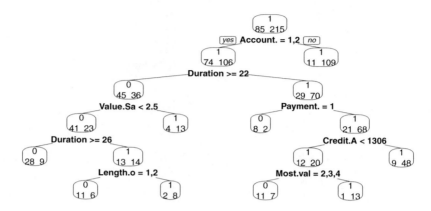

**FIGURE 10.2**
Impurity: Example based on Gini Index.

**Example 10.2.** (German Credit Data) A random subset of 300 observations from German Credit Data is used for application of classification tree. Two possible impurity measures that may be used are and entropy. Given below is the output obtained using Gini measure of impurity.

```
n= 300

node), split, n, loss, yval, (yprob)
     * denotes terminal node

1) root 300 85 1 (0.28 0.72)
   2) Account.Balance=1,2 180 74 1 (0.41 0.59)
     4) Duration.of.Credit.Month>=22.5 81 36 0 (0.56 0.44)
       8) Value.Savings.Stocks< 2.5 64 23 0 (0.64 0.36)
         16) Duration.of.Credit.Month>=25.5 37  9 0 (0.7675676 0.24) *
         17) Duration.of.Credit.Month< 25.5 27 13 1 (0.48 0.52)
           34) Length.of.Current.Employment=1,2 17  6 0 (0.65 0.35) *
           35) Length.of.Current.Employment=3,4 10  2 1 (0.20 0.80) *
       9) Value.Savings.Stocks>=2.5 17  4 1 (0.23 0.77) *
     5) Duration.of.Credit.Month< 22.5 99 29 1 (0.30 0.70)
       10) Payment.Status.of.Previous.Credit=1 10  2 0 (0.80 0.20) *
       11) Payment.Status.of.Previous.Credit=2,3 89 21 1 (0.23 0.77)
         22) Credit.Amount< 1306.5 32 12 1 (0.375 0.625)
           44) Most.Valuable.Available.Asset=2,3,4 18  7 0 (0.61 0.39) *
           45) Most.Valuable.Available.Asset=1 14  1 1 (0.071 0.929) *
         23) Credit.Amount>=1306.5 57  9 1 (0.158 0.842) *
   3) Account.Balance=3,4 120 11 1 (0.092 0.908) *
```

The root node shows that the total sample size is $n = 300$, total loss or impurity value is 85 and modal value of the class is 1. Among the 300 observations 72% of the values were equal to 1. Hence that is the modal value of the root node. Along with the modal value probability of each classification is also provided – 28% have 0 classification (non-creditworthy) and 72% have 1 classification (creditworthy). The first split is based on the classes of the variable Account.Balance. For Account.Balance levels 3 and 4, a terminal node is reached with class size ($n$) 120, modal value 1 and sample mix of 0 and 1 in that node to be in the ratio 9.2% : 90.8%.

The node given by Account.Balance levels 1 and 2 is successively split by Duration.of.Credit.Month, Value.Savings.Stocks, Duration.of.Credit.Month for a second time and Length.of.Current.Employment. The other arm is split through Payment.Status, Credit.Amount and Most.Valuable.Available.Asset. The total number of terminal nodes is 9. The tree is shown in Figure 10.2.

If the impurity measure used is Entropy, the tree is somewhat different. Number of terminal nodes is 12. Even though the first few predictors chosen for split are the same, overall a different set of predictors contribute towards the construction of the tree, which is shown in Figure 10.3.

||

**Using Entropy: 300 Data Points from German Credit Data**

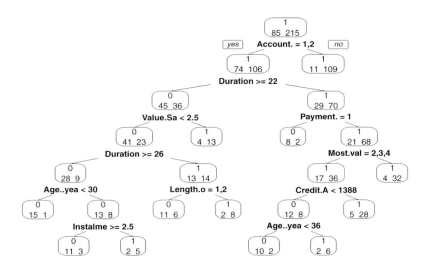

**FIGURE 10.3**
Impurity: Example based on entropy.

**Example 10.3.** (GT Compressor Decay) These data are abstracted from the UCI Machine Learning Dataset where simulated data is used to predict GT compressor decay based on a 16 feature vector. The description of the data is available at http://archive.ics.uci.edu/ml/datasets/Condition+Based+Mainte nance+of+Naval+Propulsion+Plants. We have indicated the attributes by $V_1 - V_{16}$. The dataset contains 11,934 observations and the full grown tree is shown in Figure 10.4.

|    | CP         | nsplit | rel error | xerror    | xstd        |
|----|------------|--------|-----------|-----------|-------------|
| 1  | 0.06089845 | 0      | 1.0000000 | 1.0002394 | 0.008186245 |
| 2  | 0.04316114 | 5      | 0.6744440 | 0.6743112 | 0.007067518 |
| 3  | 0.03515895 | 7      | 0.5881217 | 0.6052757 | 0.006608107 |
| 4  | 0.02492951 | 11     | 0.4474860 | 0.4525693 | 0.005629586 |
| 5  | 0.02474877 | 12     | 0.4225564 | 0.4344953 | 0.005482378 |
| 6  | 0.02072880 | 13     | 0.3978077 | 0.4154774 | 0.005262323 |
| 7  | 0.01910049 | 14     | 0.3770789 | 0.3865944 | 0.004747911 |
| 8  | 0.01631191 | 15     | 0.3579784 | 0.3733890 | 0.004711099 |
| 9  | 0.01194749 | 17     | 0.3253546 | 0.3310780 | 0.004090413 |
| 10 | 0.01144023 | 20     | 0.2895121 | 0.3059862 | 0.003914049 |
| 11 | 0.01064368 | 21     | 0.2780719 | 0.2868705 | 0.003750632 |
| 12 | 0.01061704 | 22     | 0.2674282 | 0.2750692 | 0.003657006 |
| 13 | 0.01050813 | 23     | 0.2568112 | 0.2732213 | 0.003634116 |
| 14 | 0.01000000 | 24     | 0.2463030 | 0.2654324 | 0.003566186 |

It is clear that toward the end the complexity declines at a very slow rate. The terminal nodes contain a small number of observations with a small difference in the values of response. In such cases one may not be interested in growing the tree to the fullest, but in pruning the tree to make it more applicable. Too large a tree may lead to overfitting, which will not lead to good predictability. ‖

## 10.3 Pruning a Tree

A complex tree may work well for the data for which it is developed, but may not produce good prediction. A smaller tree with a fewer number of splits might lead to a better interpretation at the cost of a small bias. This is achieved by stopping the splits at a high threshold level. If the marginal improvement in complexity is negligible, then the tree is not split any further.

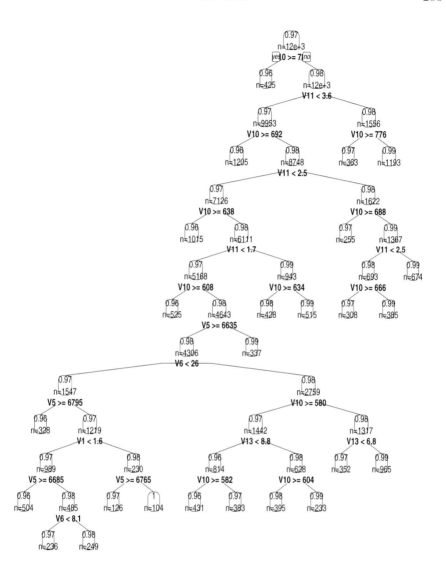

**FIGURE 10.4**
Example of a full grown tree.

However, simply stopping the splitting at a convenient point may not be a
good approach because it would not give one the chance to see if later on

a significant reduction in complexity is achieved. Hence a cross-validation approach is taken to achieve optimal pruning. The minimum value of xerror from the CP table may be used to determine the optimum threshold for cutting the tree.

In both the above examples xerror function showed monotonic decrease. Hence marginal rate of decrease of CP is taken to determine the pruning level of the tree in the second example. The pruned tree is displayed in Figure 10.5.

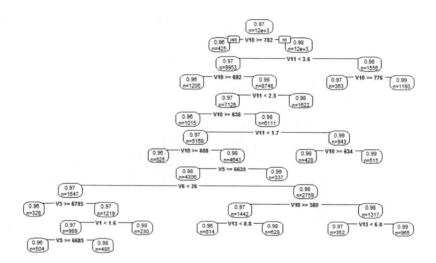

**FIGURE 10.5**
Example of pruning.

## 10.4 Aggregation Method: Bagging

Instead of depending on only one tree, the aggregation procedures make multiple passes over the data. The same analytical procedure is run and a result based on average or majority occurrence is accepted as the final result. When it comes to decision trees, multiple trees or decision rules are constructed, and the final output is averaged over each single tree. This technique corrects for overfitting and puts forth a flexible prediction model.

The central limit theorem shows that, if a sample of size $n$ is taken from a population with mean $\mu$ and variance $\sigma^2$, then the mean of the sample

## Bootstrap Sample 1

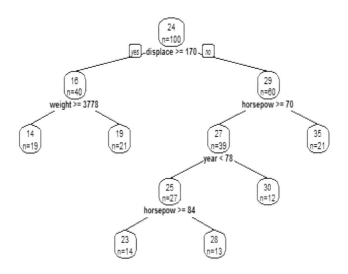

**FIGURE 10.6**
Bootstrap sample 1.

mean $\bar{x}$ is equal to the population mean $\mu$, but the variance of $\bar{x}$ is $\sigma^2/n$. As sample size increases, the variance of the sample mean decreases without any change in its expectation, so that a large sample is expected to provide a more accurate estimate of the population mean. A bootstrap procedure is built in a similar way. Since it is not possible to have multiple samples from the same population, bootstrap samples are taken from the sample already collected. From a sample of size $n$, $B$ bootstrap samples of size $m (m < n)$ are collected and sample statistics of interest are computed from the bootstrap samples. Each bootstrap sample is treated as a random sample from the actual sample, which in this case acts as the population. The sample average of bootstrap statistics and the sample variance of bootstrap statistics are used to estimate the population statistic and population variance. Bootstrap works because multiple passes at the data reduce variance but without any impact on the bias, if it exists.

Bagging is bootstrap aggregation of classification and regression trees. For the sample available, bootstrap samples are taken and on each bootstrap sample a tree is built. Each tree will be different, with different terminal nodes. While aggregating all these trees, the average value of the response, averaging over predicted values given by each bootstrap tree, is taken as the final pre-

**Bootstrap Sample 2**

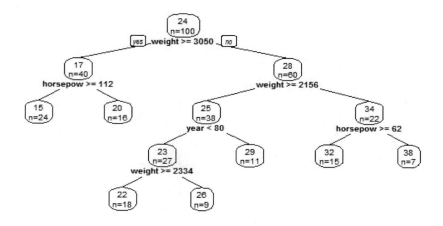

**FIGURE 10.7**
Bootstrap sample 2.

**Bootstrap Sample 3**

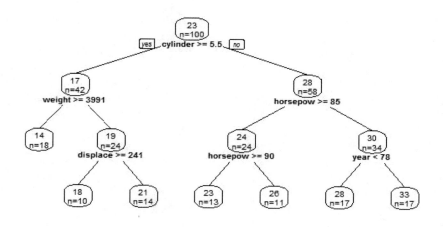

**FIGURE 10.8**
Bootstrap sample 3.

dicted value of the response in the case of a regression tree. For a classification tree, the majority vote for classification of each observation is taken to be the final predicted value. Pruning is not done to the trees, but the trees are grown to the fullest extent. However, bagging does not produce a tree-like structure, and hence the simple interpretability associated with a tree is lost.

**Example 10.4.** (Auto Data) As an illustration three bootstrap samples of size 100 are taken from the Auto Data. The regression tree as grown on the full sample is shown in Figure 10.1. The bootstrap trees are displayed in Figures 10.6–10.8.

Note that each tree may have a different number of terminal nodes and a different combination of splits. Each tree produces a different set of rules for prediction and may produce a different MPG prediction for each level combination of covariates. The final prediction will be given by a simple average of the different predicted values. In the case of classification, for each bagged tree, each observation is either correctly or incorrectly classified. Overall the difference between the proportion of correct and incorrect classification is known as the margin for that observation. Ideally, each observation should have a large margin for a bagging procedure to be stable and to provide predictions with lower variance. For every bootstrap sample of size $B < n$, $(n - B)$ observations are not included in the bootstrap sample. These observations are known as the out-of-bag observations. For each bootstrap tree, these out-of-bag observations are treated as test samples. Prediction error is based only on the out-of-bag observations. ‖

## 10.5 Random Forest

Bagging is a special case of a versatile algorithm called Random Forest (RF). In bagging a random sample of observations is split to construct each tree. In RF, at each split for each tree, a random sample of predictors is used. There are three tuning parameters in RF:

1. Number of trees to grow

2. Number of predictors to sample at each split: A rule of thumb is to choose $m \approx \sqrt{p}$, where $p$ is the number of predictors

3. Size of terminal nodes

RF decorrelates the trees. Consider a situation where one predictor is stronger than the rest. As a result, this variable will almost always be the top split in the case of bagged trees. This will cause the bagged trees to be similar or correlated. Since in RF only a random subset of predictors is considered at each split, at many splits this strong predictor will not even be considered. This will make the resultant trees look different from each other.

RF works because of variance reduction. None of the individual trees is optimum, but on the aggregate, because of a possibly different set of predictors being used at every split, RF gives good prediction. Each tree is independent because a random sample is used to construct each tree. Each split is also independent because a subset of predictors is used to determine the optimum split. RFs are able to handle a large number of predictors since at every split a reduced number of predictors is considered.

## 10.6    Variable Importance

An overall summary of importance of each predictor is measured using the sum of squares criterion for regression trees and using the Gini index or the entropy criterion for classification trees. The average decrease in the sum of squares initiated by a split due to a variable, average taken over all trees, measures the importance of a variable in the random forest. A similar interpretation may be given using Gini or other measures of impurity.

**Example 10.5.** (German Credit Data) A subset of 300 data points was randomly abstracted from German Credit data. The random forest procedure is applied to this dataset.

```
Call:
randomForest(formula = Creditability ~ ., data = Train300,
ntree = 200, importance = T, proximity = T)
            Type of random forest: classification
                  Number of trees: 200
No. of variables tried at each split: 4

      OOB estimate of  error rate: 22%
Confusion matrix:
    0   1 class.error
0 31  54  0.63529412
1 12 203  0.05581395
```

The number of trees considered is 200 and for each split 4 variables are chosen randomly. The out-of-bag (OOB) estimate of error is 22%, whereas the misclassification rate of Creditability = 1 is 5% and the same for Creditability = 0 is 63%. The error rate estimate primarily depends on the bootstrap sample size. The misclassification rate is a function of the number of trees built as well as the number of variables used in building the tree. However, the error rate is quite robust in its dependence on the number of variables used in the

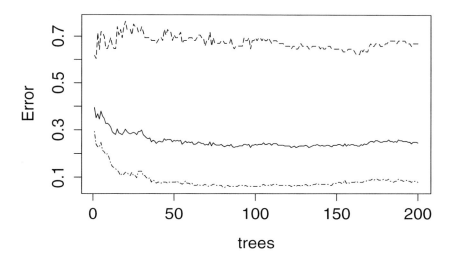

**FIGURE 10.9**
Estimate of error rate.

split. Figure 10.9 is a plot of the OOB error rate for different bootstrap sample sizes. The solid line gives the overall error rate, the dashed line above gives the error rate of misclassification of class 0, the dot-dashed line below gives the misclassification rate of class 1. Not only is the misclassification rate for class 0 high, it shows high volatility also. Whereas the overall and class 1 error rates stabilize for bootstrap sample size 50, even at sample size 200 class 0 error rates may not have stabilized.

The dependence of error rate on the number of variables in the split is minimal. Table 10.1 contains the OOB error rates for different numbers of predictors used in the split for bootstrap sample size 200. Recall that, even though the number of predictors used in each split is the same, the actual variables used in each split may be different.

In the data there are 19 predictors. Using the rule of thumb $m = \sqrt{p} \approx 4$, a low value of misclassification error for class 0 is achieved. The choice of $m = 6$ would also have been good since at this level a marginally lower OOB error rate is achieved with a lower misclassification rate for Class 0; but the misclassification rate for Class 1 has gone up.

Figure 10.10 provides a graphical ranking of the predictors according to their contribution toward reducing the overall impurity of the trees created.

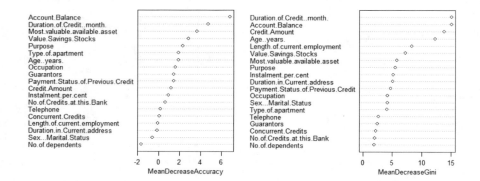

**FIGURE 10.10**
Variable importance.

It is clear from the figure that the impurity measure used has an impact on the importance of the predictors.

‖

## 10.7    Decision Tree and Interaction among Predictors

A decision tree is capable of detecting the importance of predictors in a somewhat automatic fashion. The terminal or leaf nodes result from a differential

**TABLE 10.1**
Error rates for a random forest model

| No of Predictors Used in Split | OOB Estimate of Error Rate | Class Error Rate | |
|---|---|---|---|
| | | Class 0 | Class 1 |
| 1 | 27.7% | 97% | 0% |
| 2 | 25% | 79% | 4% |
| 3 | 27% | 73% | 9% |
| 4 | 23% | 63% | 7% |
| 6 | 22.7% | 56% | 9% |
| 8 | 23.7% | 60% | 9% |
| 10 | 22.3% | 53% | 10% |

## Compressive Strength: Regression Tree

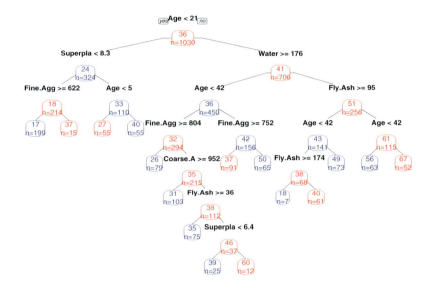

**FIGURE 10.11**
Example of selecting interaction through tree building.

dependence structure among the predictors. Interaction among predictors is created due to the fact that one predictor impacts the response in a different way, depending on the levels of a second predictor. If the number of predictors is large, then finding important interactions among predictors is a cumbersome task. Given computational capability, it is possible to consider all interactions in a regression model, but that procedure is not an efficient one. Further, inclusion of redundant interaction effects might lead to overfitting, and the predictive power of the model will decrease.

Decision trees provide a way of identifying important interactionsdecision trees. Starting from the top-most split and following the rules to the terminal nodes, the path followed may be incorporated in the model as a complicated interaction effect. Keeping matters simple and restricting ourselves to the two-factor interaction only, important relationships can be incorporated in a multiple linear regression model by utilizing the knowledge gained from the tree.

Recall the multiple linear regression model constructed on the concrete strength data. The compressive strength of concrete was the response, and several predictors like Blast furnace slag, Fly ash, Water etc. were the predictors. Multiple linear regression without any interaction produces an $R^2$ of 54%.

A regression tree was constructed on the same data and left to run its full course. See Figure 10.11. No pruning or bagging was done. From the tree it seems that there may be several important interactions. In fact, in a complex model one may think of including one interaction corresponding to each rule. But that might lead to overfitting, as this will follow the constructed tree very closely and is not guaranteed to do well in a test dataset. Therefore only a few of the more important interactions are to be included in the model. With Concrete data, it seems the two interactions Age and Superplasticize and Age and Water should be included in the model. The regression results after including two interaction effects are given in the following. As is clear from these results, both interactions are highly significant and the resultant $R^2$ jumps to 62%.

```
Call:
lm(formula = Strength ~ Blast.Furnace.Slag + Fly.Ash + Water +
Superplasticize + Coarse.Aggregate + Fine.Aggregate + Age +
Age * Superplasticize + Age * Water, data = Concrete_Data)

Coefficients:

                    Estimate  Std. Error t value  Pr(>|t|)
(Intercept)        3.039e+02  1.326e+01   22.917  < 2e-16  ***
Blast.Furnace.Slag -2.810e-02 4.592e-03   -6.121 1.33e-09  ***
Fly.Ash            -7.266e-02 5.943e-03  -12.226  < 2e-16  ***
Water              -4.757e-01 3.094e-02  -15.373  < 2e-16  ***
Superplasticize    -2.506e-01 1.124e-01   -2.230    0.026  *
Coarse.Aggregate   -9.269e-02 5.863e-03  -15.809  < 2e-16  ***
Fine.Aggregate     -1.191e-01 5.508e-03  -21.622  < 2e-16  ***
Age                 3.494e-01 6.098e-02    5.730 1.32e-08  ***
Superplasticize:Age 1.578e-02 1.777e-03    8.883  < 2e-16  ***
Water:Age          -1.285e-03 2.916e-04   -4.407 1.16e-05  ***
---
Signif.codes:  0 '***' 0.001 '**' 0.01 '*' 0.05 '.' 0.1 ' ' 1
Residual standard error: 10.28 on 1020 degrees of freedom
Multiple R-squared:  0.6245,Adjusted R-squared:  0.6212
F-statistic: 188.5 on 9 and 1020 DF,  p-value: < 2.2e-16
```

## 10.8　Suggested Further Reading

The classic reference on this topic is Breiman et al. (1984). The algorithm CART was developed by the authors and in this book they discuss the technique in great detail. However, the focus of the book is mathematical, rather than applied. For application of the decision tree techniques, James et al. (2014), Ledolter (2013) and Kuhn and Johnson (2013) are more suitable. Moreover, these books also show R applications.

# 11

## Data Mining and Multivariate Methods

Data mining implies extraction of knowledge embedded in the data. In this age of large databases and data warehouses, where almost automatically, or with very little guidance, data is being captured continuously, preconceived notions about data structure hardly exist. Data mining is a process of exploration and analysis of large amounts of data in order to discover meaningful patterns. Data mining can also be considered as a business process for maximizing the value of data collected for the business. Data is always showing patterns; it is important to know which patterns are non-random and which among them are actionable. Data captured is always past data and it is valuable only when it can be used to understand its future behavior. The most important assumption is that the future will be similar to the past so that knowledge from the past is usable in the future. In order to do that, pattern recognition in past data is important. However, business is a continuous process and the system will be evolving continuously. To understand the changing nature of the business, continuous monitoring of the patterns in the data and recognition of their slow but steady evolution are important.

Patterns may or may not represent any underlying rule. Statistical learning is said to belong to two classes – supervised and unsupervised. All model-building technologies fall under supervised learning. Unsupervised learning does not have any response associated with the predictors. The aim here is to understand the relationship among all the variables. In a way, unsupervised learning is more challenging. The first approach to data mining is data visualization, which has been dealt with in some detail in Chapter 4. Once patterns are empirically recognized, more formal techniques are required to extract actionable insight from the data. Data is almost always multivariate, having a number of attributes, discrete or continuous, associated with every observation. Any data mining or pattern recognition technique must deal with multivariate observations. In this chapter we will talk about several important multivariate techniques. Dimension reduction is a very important consideration with multivariate data, since not all attributes contribute equally significantly toward information contained in the observation.

Consider an example in marketing analytics and customer segmentation. The customer base of a business comprises different individuals with different needs and behavior patterns. As it is wrong to assume that there is an average customer and make offerings to that individual only, so it is equally impossible to cater to every individual need. Hence there is a requirement to identify similar clusters of customers and cater to the collective need of each cluster. The purchase and behavior patterns of customers involve a very large number of variables. Identifying an informative subset of those variables and using them to form homogeneous customer clusters is an important objective of the marketing strategy of a business. Similarly, assignation of a newly acquired customer to a preexisting cluster is an important aspect of analytics.

In this chapter we will discuss a few of the most important data mining techniques which have significant application in business analytics. Several supervised learning techniques have already been discussed, e.g., regression and decision trees. Data visualization, which is the first step of any pattern recognition technique, has also been considered earlier. In this chapter we concentrate on other multivariate methods.

## 11.1   Principal Component Analysis

Businesses usually maintain data in several isolated data marts. Customer data is preserved in customer relationship management (CRM) systems, sales data is preserved in sales databases, marketing research data may be preserved in a third database. Combining all the available data on a unique key may make the data voluminous, not necessarily on the number of observations, but on the attributes or items of information that were collected on each observation. Typically, pattern recognition algorithms do not work well if the number of attributes is too large. Too many attributes tend to mask the signals or patterns. Further, a large number of attributes would show correlations among themselves which may or may not signify any causal dependence. We have already dealt with the problem of the variation inflation factor in regression. Since the predictors were highly correlated among themselves, they were not effective in predicting the response. Similar issues occur in other applications also. This is known as the *Curse of Dimensionality* and must be avoided at all costs. Reducing the dimension of multivariate observations is one way of avoiding this. Principal Component Analysis (PCA) is one of the most common techniques of dimension reduction.

To put it in a nutshell, principal component analysis identifies a few linear combinations of the attributes so that they are orthogonal to each other and are responsible for a high proportion of variability. Once these are identified, further analysis is carried out using the linear combinations only, instead of the full set of attributes. So the problem reduces to finding linear combinations of attributes with the above properties. But before we proceed to finding the linear combinations, it is important to spend time on understanding the concept of total variability in a multivariate context.

**Example 11.1.** (Wine Data) Eleven chemical attributes are associated with each variety of white wine. Our goal is to predict Quality based on these chemical properties. The codes for these attributes, used in Table 11.1 below, are as defined in Table 3.2 of Section 3.4.3. Several of these attributes may be highly correlated, and putting all predictors in the model without preprocessing is not recommended. Table 11.1 is the pairwise correlation matrix of all the attributes. Simply by looking at the correlation matrix it is not possible to eliminate an appropriate number of variables so that the remaining variables may be used to develop a working application.

An application is useful if it is able to explain a large amount of variability contained in the data. The main objective of PCA is to reduce the number of variables but to retain as much variability present in the data as possible.

**TABLE 11.1**
Pairwise correlation coefficients of white wine characteristics

|        | FA    | VA    | CA    | RA    | Chl   | FSD   | TSD   | Dens  | pH   | Sulph |
|--------|-------|-------|-------|-------|-------|-------|-------|-------|------|-------|
| VA     | −0.02 |       |       |       |       |       |       |       |      |       |
| CA     | 0.29  | −0.15 |       |       |       |       |       |       |      |       |
| RA     | 0.09  | 0.06  | 0.09  |       |       |       |       |       |      |       |
| Chl    | 0.02  | 0.07  | 0.11  | 0.09  |       |       |       |       |      |       |
| FSD    | −0.05 | −0.10 | 0.09  | 0.30  | 0.10  |       |       |       |      |       |
| TSD    | 0.09  | 0.09  | 0.12  | **0.40** | 0.20 | **0.62** |    |       |      |       |
| Dens   | 0.27  | 0.03  | 0.15  | **0.84** | 0.26 | 0.29 | **0.53** |   |      |       |
| pH     | **−0.43** | −0.03 | −0.16 | −0.19 | −0.09 | 0.00 | 0.00 | −0.09 |   |       |
| Sulph  | −0.02 | −0.04 | 0.06  | −0.03 | 0.02  | 0.06  | 0.13  | 0.07  | 0.16 |       |
| Alc    | −0.12 | 0.07  | −0.08 | **−0.45** | −0.36 | −0.25 | **−0.45** | **−0.78** | 0.12 | −0.02 |

‖

## 11.1.1 Total Variation in a Multivariate Dataset

Consider a random variable $X$ having a distribution, not necessarily normal, with mean $\mu$ and variance $\sigma^2$. $X$ being univariate, $\sigma^2$ quantifies the dispersion in $X$. Suppose now two variables $X_1$ and $X_2$ are being considered together, with means $\mu_1$ and $\mu_2$ and variances $\sigma_1^2$ and $\sigma_2^2$. However, $X_1$ and $X_2$ are not independent, and hence the covariance between $X_1$ and $X_2$, denoted by $\sigma_{12}$, needs to be considered also. The dispersion matrix or the variance-covariance matrix may be written as

$$\Sigma = \begin{bmatrix} \sigma_1^2 & \sigma_{12} \\ \sigma_{12} & \sigma_2^2 \end{bmatrix}.$$

Generalizing this concept even further, if $p$ variables are considered together, then the dispersion matrix will look like

$$\Sigma = \begin{bmatrix} \sigma_1^2 & \cdots & \sigma_{1p} \\ \vdots & \ddots & \vdots \\ \sigma_{1p} & \cdots & \sigma_p^2 \end{bmatrix}.$$

On the main diagonal of the matrix will lie the variances, and the off-diagonals will contain the covariances. Note that this is a square matrix, i.e., the numbers of rows and columns are the same. The total variation contained in $\Sigma$ may then be defined as the sum of the diagonal elements. This is called the trace of the dispersion matrix and is given by

$$\text{trace}(\Sigma) = \sum_{i=1}^{p} \sigma_i^2.$$

**Example 11.1.** (Continued) Noting that there are 11 variables in the White Wine data, the main diagonal of the dispersion matrix will contain the variances as shown below.

| FA | VA | CA | RA | Chl | FSD |
|---|---|---|---|---|---|
| 0.71 | 0.01 | 0.01 | 25.70 | 0.0004 | 289.34 |
| TSD | Dens | pH | Sulph | Alc | |
| 1806.25 | 0.0004 | 0.02 | 0.01 | 1.51 | |

Since this is a sample dispersion matrix, $S$ is used to denote the estimate of $\Sigma$. Hence $\text{trace}(S) = 2123.57$, which is the sum of all the elements in the diagonal. ‖

## 11.1.2 Construction of Principal Components

The principal components $Y_i$, $i = 1, \ldots, p$ are defined as linear combinations of the original variables having the form

$$\begin{aligned} Y_1 &= e_{11}X_1 + e_{12}X_2 + \cdots + e_{1p}X_p, \\ Y_2 &= e_{21}X_1 + e_{22}X_2 + \cdots + e_{2p}X_p, \\ &\cdots \\ Y_p &= e_{p1}X_1 + e_{p2}X_2 + \cdots + e_{pp}X_p. \end{aligned}$$

To make a distinction, we will denote the variables by $X$ and the principal components (PCs) by $Y$. Note that the number of PCs is the same as the number of variables. Further, each PC is a linear combination of all the variables. The coefficients $e_{ij}$ may be any real number including zero. The equations for PCs may be thought of as regression equations without the intercept. As soon as the coefficients are estimated, the PCs are known.

We begin with some basic linear algebra. To keep matters simple, let us only define the eigenvalues and eigenvectors of a square matrix. A square matrix of order $p$, i.e., a square matrix having $p$ rows and $p$ columns, will have $p$ eigenvalues, which are the $p$ solutions of the equation

$$|A - \lambda I| = 0,$$

where $I$ is the identity matrix of order $p$. A square matrix is an identity matrix if all its diagonal elements are 1 and off-diagonal elements are 0. The symbol $|\cdots|$ denotes the determinant of a matrix. The $p$ eigenvalues are denoted as $\lambda_1, \lambda_2, \ldots, \lambda_p$. The corresponding eigenvectors $e_i, i = 1, 2, \ldots, p$ are obtained by solving the equations

$$(A - \lambda_i I)e_i = 0, \quad i = 1, 2, \ldots, p.$$

Even if the mathematics of eigenvalues and eigenvectors is not well understood, it is important that their properties are. The $p$ eigenvalues may not all be distinct, i.e., it is possible to have $\lambda_i = \lambda_j$ for one or more pairs of $i$ and $j$, $i \neq j$. The eigenvectors are orthogonal to each other, i.e., product of two eigenvectors will be identically 0. Even if two eigenvalues are equal in measure, the corresponding eigenvectors are orthogonal. Another important property of eigenvectors that is often exploited is that the sum of squares of the components of an eigenvector is equal to 1.

The principal components are so defined as to have $Var(Y_i) = \lambda_i$, and the coefficients $e_{ij}$ are elements of eigenvectors. The PCs are ranked so that $\lambda_1 \geq \lambda_2 \geq \cdots \geq \lambda_p$. The PC with the largest variance is called PC1, the component with the next largest variance is known as PC2 and so on. Note also that

$$\text{trace}(\Sigma) = \sum_{i=1}^{p} \sigma_i^2 = \sum_{i=1}^{p} \lambda_i.$$

Thus the total variation of the variables $X_1, X_2 \ldots, X_p$ equals the sum of the variances of $Y_1, \ldots, Y_p$, even though none of the pairs $\sigma_i, \lambda_j$ needs to be equal. In addition, $Cov(Y_i, Y_j) = 0$, so that all the PCs are uncorrelated.

However, merely replacing the variables by the PCs does not lead to dimension reduction, even though it solves the problem of high dependence. The hierarchical nature of variances of PCs comes into play here. The proportion of total variance explained by the $i$-th PC is

$$\frac{\lambda_i}{\lambda_1 + \lambda_2 + \cdots + \lambda_p}.$$

Instead of using all $p$ PCs, the top $k$ PCs are used, so that the ratio

$$\frac{\lambda_1 + \lambda_2 + \cdots + \lambda_k}{\lambda_1 + \lambda_2 + \cdots + \lambda_p}$$

approximately equals 1.

Typically, the first few PCs are enough to bring the ratio close to 1. To apply PCA it is recommended that instead of working with eigenvalues and eigenvectors of $\Sigma$, the dispersion matrix of original variables, scaled versions are used. The reason is that, if the variances of two variables are widely different, then their contribution toward constructing PCs also varies. If the variables are scaled, all variances are identically 1 and the trace is equal to $p$, the number of variables. In fact, the dispersion matrix for the scaled variables is equivalent to their correlation matrix.

**Example 11.1.** (Continued) To obtain principal components for the White Wine Data, scaling of the variables is recommended since the maximum variance is 1806, observed for Total.Sulfur.Dioxide (total sulphur dioxide) and the minimum is 0.0004 observed for the variable Chlorides.

```
> WWPred <- White.Wine[,1:11]
> PCWW <- prcomp(WWPred, scale=T)
> summary(PCWW)
```

```
Importance of components
                        PC1  PC2  PC3  PC4  PC5  PC6  PC7  PC8  PC9 PC10 PC11
Standard deviation     1.80 1.26 1.11 1.01 0.99 0.97 0.85 0.77 0.64 0.54 0.14
Proportion of Variance 0.29 0.14 0.11 0.09 0.09 0.09 0.07 0.05 0.04 0.03 0.00
Cumulative Proportion  0.29 0.44 0.55 0.64 0.73 0.81 0.88 0.93 0.97 1.00 1.00
```

Summary of PCA gives the above three important information about the procedure. Since there are 11 variables, the number of PCs is also 11. The standard deviations are the square roots of the variances of the PCs, $\sqrt{\lambda_i}$. Note that they are given in decreasing order since the variances of PC1 is the highest and the last PC has the lowest variance. It also shows the proportion of total variance each of the PCs explain and the cumulative proportion. This corresponds to $\frac{\lambda_i}{\lambda_1+\lambda_2+\cdots+\lambda_{11}}$, $i = 1, 2, \ldots, 11$ and its cumulative values. Note that PC1 explains 29% of total variance, PC2 explains 14% of total variance and so on whereas PC8 onwards explains 5% or even less. The first 6 PCs explain more than 80% of the total variance; the first 8 PCs explain more than 90%.

```
> print(PCWW)
```

```
Standard deviations:
1.80 1.26 1.11 1.01 0.99 0.97 0.85 0.77 0.64 0.54 0.14

Rotation:
                     PC1   PC2   PC3   PC4    PC5   PC6   PC7   PC8   PC9  PC10  PC11
Fixed.Acidity        0.16 -0.59 -0.12 -0.02 -0.25 -0.10  0.20  0.59 -0.33 -0.13 -0.17
Volatile.Acidity     0.01  0.05  0.59 -0.27 -0.64  0.12 -0.27  0.03  0.15 -0.22 -0.02
Citric.Acid          0.14 -0.35 -0.50 -0.15 -0.05  0.13 -0.71 -0.15  0.20 -0.04 -0.01
Residual.Sugar       0.43  0.01  0.21  0.27 -0.01 -0.29 -0.21 -0.39 -0.41  0.09 -0.49
Chlorides            0.21 -0.01  0.10 -0.71  0.33  0.40  0.08 -0.10 -0.39  0.05 -0.03
Free.Sulfur.Dioxide  0.30 -0.28  0.31 -0.18  0.49  0.17 -0.08 -0.14 -0.57  0.03
Total.Sulfur.Dioxide 0.41  0.24 -0.12  0.06 -0.29  0.28  0.07  0.25  0.15  0.71 -0.04
Density              0.51  0.01  0.13  0.02  0.08 -0.33 -0.11  0.07 -0.09 -0.07  0.76
pH                  -0.13  0.58 -0.13 -0.10  0.12 -0.19 -0.43  0.53 -0.26 -0.11 -0.14
Sulphates            0.04  0.22 -0.43 -0.44 -0.40 -0.48  0.31 -0.27  0.01 -0.06 -0.04
Alcohol             -0.44 -0.04 -0.11  0.14 -0.34  0.14 -0.13 -0.20 -0.62  0.27  0.36
```

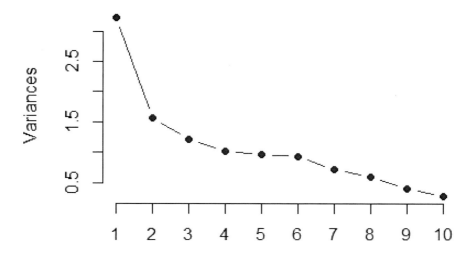

**FIGURE 11.1**
An example of a screeplot.

The coefficients of the linear combinations are given in the table above. Mathematically, the $i$-th row is the eigenvector corresponding to the $i$-th eigenvalue $\lambda_i$. Writing out the linear combinations explicitly, we get

$$
\begin{aligned}
\text{PC1} \;=\; & 0.16 \times \text{FA} + 0.01 \times \text{VA} + 0.14 \times \text{CA} + 0.43 \times \text{RS} + 0.21 \times \text{Ch} + 0.30 \times \text{FSD} \\
+\; & 0.41 \times \text{TSD} + 0.51 \times \text{Dens} - 0.13 \times \text{pH} + 0.04 \times \text{Sulph} - 0.44 \times \text{Alc} \\
\text{PC2} \;=\; & -0.59 \times \text{FA} + 0.05 \times \text{VA} - 0.35 \times \text{CA} + 0.01 \times \text{RS} - 0.01 \times \text{Ch} + 0.29 \times \text{FSD} \\
+\; & 0.24 \times \text{TSD} + 0.01 \times \text{Dens} + 0.58 \times \text{pH} + 0.22 \times \text{Sulph} - 0.04 \times \text{Alc}
\end{aligned}
$$

and so on. The coefficients are the elements of the eigenvectors of the correlation matrix. Note that the variables are in their standardized form. ‖

How many PCs should be considered for further analysis? If the goal is dimension reduction, then using all 11 PCs may not make sense, even though they are now orthogonal and covariance among them will be zero. The number of PCs to be used is based on a graphical technique known as a Screeplot.

```
> plot(PCWW, type="line", col=c("dark blue"), main="", pch=19)
> mtext("Screeplot of White Wine", side=1, line=3, cex=0.8)
```

From Figure 11.1 it seems that, contribution of PC7 onwards is negligible. While there is no hard and fast rule regarding how many PCs to choose, if the screeplot shows a pronounced 'elbow', or a sharp turn after which the rest of the curve flattens out, then PCs up to the elbow are considered. In this plot there is no such elbow but after PC6 individual contributions to the total variance are small. However, it is rare for the chosen number of PCs to explain less than 80% of the total variance. Another consideration for choosing the number of PCs is based on the corresponding eigenvalues – they are expected to be at least 1. In this example the lowest eigenvalue is 0.94.

Generally each PC is a combination of all the variables; hence interpretation is not always easy. To identify which PC relate to which of the variables, pairwise correlations between PCs and variables need to be investigated. Higher correlation between pairs indicate relatively high dependence of PC on those variables. The threshold above which correlations may be considered high is subjective, but one may use 70% as an empirical value.

**Example 11.1.** (Continued) The table below shows pairwise correlation between PC1 – PC6 and the original 11 variables.

| | PC1 | PC2 | PC3 | PC4 | PC5 | PC6 |
|---|---|---|---|---|---|---|
| Fixed.Acidity | 0.28 | -0.74 | -0.13 | -0.02 | -0.25 | -0.10 |
| Volatile.Acidity | 0.01 | 0.06 | 0.65 | -0.28 | -0.63 | 0.12 |
| Citric.Acid | 0.26 | -0.43 | -0.56 | -0.15 | -0.05 | 0.13 |
| Residual.Sugar | 0.77 | 0.01 | 0.24 | 0.28 | -0.01 | -0.28 |
| Chlorides | 0.38 | -0.01 | 0.11 | -0.72 | 0.32 | 0.38 |
| Free.Sulfur.Dioxide | 0.54 | 0.36 | -0.31 | 0.31 | -0.17 | 0.48 |
| Total.Sulfur.Dioxide | 0.73 | 0.31 | -0.14 | 0.06 | -0.29 | 0.27 |
| Density | 0.92 | 0.01 | 0.14 | 0.02 | 0.08 | -0.32 |
| pH | -0.23 | 0.73 | -0.14 | -0.10 | 0.12 | -0.19 |
| Sulphates | 0.08 | 0.28 | -0.48 | -0.45 | -0.40 | -0.47 |
| Alcohol | -0.78 | -0.04 | -0.12 | 0.14 | -0.33 | 0.13 |

If 70% is the threshold for high correlation, PC1 increases with Density, Residual Sugar and Total Sulphur Dioxide, but decreases with Alcohol percentage. PC2 increases if Fixed Acidity decreases and pH increases. PC4 increases if Chlorides in wines decreases. The other 3 PCs do not show high correlation with any variable. However, if the threshold is 60%, then PC3 may be interpreted to be positively related with Volatile Acidity while PC5 is negatively related. The PCs have zero correlation among themselves.

Just as each observation in the dataset has a value for each attribute $(X_i)$, each observation also has a value for each PC. These values are called the PC scores. These scores are computed from the scaled attribute values and the linear coefficients $e_{ij}$. In the table below the scores corresponding to the first 5 observations of the White Wine data are shown.

| | PC1 | PC2 | PC3 | PC4 | PC5 | PC6 |
|---|---|---|---|---|---|---|
| [1,] | 3.68 | -0.55 | 0.93 | 1.14 | 0.29 | -0.89 |
| [2,] | -0.64 | 0.43 | 0.36 | -1.00 | 0.71 | -0.48 |
| [3,] | 0.16 | -1.19 | 0.02 | -0.27 | 0.37 | -0.50 |
| [4,] | 1.46 | 0.10 | 0.00 | 0.42 | 0.47 | 0.77 |
| [5,] | 1.46 | 0.10 | 0.00 | 0.42 | 0.47 | 0.77 |

PC scores may be useful in locating data concentration and identification of outliers. ‖

Principal Component Analysis may also be considered as a possible data preparation step for regression. If number of explanatory variables is too large, PCA may be applied at the initial stage and instead of the variables, the PCs enter the model as predictors. Another useful application of PCA is Factor Analysis, which is explained in the next section.

## 11.2 Factor Analysis

Factor analysis (FA) aims to identify a small number of latent variables that explain most of the variance in a larger pool of observed variables. There are three types of factor analyses, namely, exploratory factor analysis, confirmatory factor analysis and structural equation modeling. The first two techniques are identical from a statistical point of view. Exploratory factor analysis is used to reveal the number of factors and the variables that are associated with specific factors. When we conduct a confirmatory factor analysis, we have clear expectations regarding the factor structure and we want to test if the expected structure is indeed present. Structural equation modeling is outside the purview of our discussion. Here we primarily deal with exploratory factor analysis through principal component analysis.

FA is, in some sense, an inversion of principal components. In PCA new variables are created through linear combinations of the observed variables, whereas in FA the observed variables are modeled as linear functions of the latent factors. But in both PCA and FA, dimensions of the data are reduced. In PCA interpretation of principal components is often not very clean. A particular variable may, on occasion, contribute significantly to more than one of the components. Ideally, we would like each variable to contribute significantly to only one component. A technique called factor rotation is employed to improve interpretation.

Consider again $p$ variables in the dataset $X_1, X_2, \ldots, X_p$ and suppose that the factors $f_1, f_2, \ldots, f_m$, $m < p$, are able to explain all the observed attributes. Hence it can be written as

$$
\begin{aligned}
X_1 &= \mu_1 + l_{11}f_1 + l_{12}f_2 + \cdots + l_{1m}f_m + \epsilon_1 \\
X_2 &= \mu_2 + l_{21}f_1 + l_{22}f_2 + \cdots + l_{2m}f_m + \epsilon_2 \\
&\cdots \quad \cdots \\
X_p &= \mu_p + l_{p1}f_1 + l_{p2}f_2 + \cdots + l_{pm}f_m + \epsilon_p.
\end{aligned}
$$

The coefficient $l_{ij}$ is called the loading of $i$-th variable on the $j$-th factor and the matrix

$$
L = \begin{bmatrix}
l_{11} & l_{12} & \cdots & l_{1m} \\
l_{21} & l_{22} & \cdots & l_{2m} \\
\cdots & \cdots & \cdots & \cdots \\
l_{p1} & l_{p2} & \cdots & l_{pm}
\end{bmatrix}
$$

is known as the matrix of factor loadings. The model is similar to a regression model where the variables are the responses and the latent factors are the predictors. It is also assumed that $E(\epsilon_i) = 0$, for all $i = 1, 2, \ldots, p$ and $E(f_j) = 0$ for all $j = 1, \ldots, m$. Hence $E(X_i) = \mu_i$, for $i = 1, \ldots, p$. It is also assumed that $Var(f_j) = 1$, $Var(\epsilon_i) = \psi_i$, all $f_j$ are uncorrelated, all $\epsilon_i$ are uncorrelated and $f_j$ and $\epsilon_i$ are also uncorrelated, for all $j = 1, \ldots, m$ and $i = 1, 2, \ldots, p$. Under this model the variance of attribute $i$ is

$$
Var(X_i) = \sigma_i^2 = \sum_{j=1}^{m} l_{ij}^2 + \psi_i.
$$

The term $h_i^2 = \sum_{j=1}^{m} l_{ij}^2$ is called the communality for variable $X_i$. The larger the communality, the better the model performs for that variable. The term $\psi_i$, $i = 1, \ldots, p$ is known as the specific variance.

To estimate the factor loadings, a knowledge of orthogonal matrices and spectral decomposition of a square matrix is necessary. Avoiding the deduction, we simply note that the estimated factor loadings are given by

$$
\hat{l}_{ij} = \hat{e}_{ij} \sqrt{\hat{\lambda}_i}
$$

where $\hat{\lambda}_i$ is the $i$-th eigenvalue of the sample covariance matrix $S$ and $\hat{e}_{ij}$ is the $j$-th element of the associated eigenvector. These are the elements of the matrix of factor loadings.

Communalities may be thought of as multiple $R^2$ values for regression models predicting the variables from the latent factors. The communality for a given variable can be interpreted as the proportion of variation in that variable explained by all the factors collectively. An FA model can be assessed by the communalities. Values closer to 1 indicate that the factors are able to explain the underlying structure of the variables. The total communality value is identically equal to the sum of the eigenvalues used in the model, i.e.,

$$
\sum_{i=1}^{p} \hat{h}_i^2 = \sum_{j=1}^{m} \hat{\lambda}_j.
$$

The specific variances can be estimated by taking the difference of sample variance of the variables and the communalities. Note that, for the scaled variables, the variance of $X_i$ is 1, for all $i$.

**Example 11.2.** (Wine Data) In exploratory factor analysis, the number of factors to extract is a decision that rests with the analyst. Often several different choices are tried out and the minimum number of factors that reasonably estimates the variability is proposed.

```
> FAWW <- factanal(WWPred, 3, rotation="varimax")
> print(FAWW, digits=2, cutoff=.6, sort=TRUE)

Call:
factanal(x = WWPred, factors = 3, rotation = "varimax")

Uniquenesses:
      Fixed.Acidity      Volatile.Acidity        Citric.Acid
               0.00                  0.99               0.91
      Residual.Sugar             Chlorides  Free.Sulfur.Dioxide
               0.00                  0.86               0.89
Total.Sulfur.Dioxide               Density                 pH
               0.70                  0.00               0.72
          Sulphates               Alcohol
               0.96                  0.20

Loadings:
                     Factor1 Factor2 Factor3
Residual.Sugar         0.98
Density                0.78
Fixed.Acidity                  0.99
Alcohol                                -0.78
Volatile.Acidity
Citric.Acid
Chlorides
Free.Sulfur.Dioxide
Total.Sulfur.Dioxide
pH
Sulphates


                     Factor1 Factor2 Factor3
SS loadings            2.03    1.42    1.29
Proportion Var         0.18    0.13    0.12
Cumulative Var         0.18    0.31    0.43
```

To test the hypothesis that three factors are sufficient, the chi-square statistic is 5471.48 on 25 degrees of freedom. The p-value is 0.

With three factors, only 43% of the variability is explained.

```
> FAWW <- factanal(SWWPred, 6)
> print(FAWW, digits=2, curpoff = 0.6, sort=TRUE)

Call:
factanal(x = SWWPred, factors = 6)

Uniquenesses:
```

| Fixed.Acidity | Volatile.Acidity | Citric.Acid |
|---|---|---|
| 0.00 | 0.00 | 0.88 |
| Residual.Sugar | Chloride | Free.Sulfur.Dioxide |
| 0.00 | 0.84 | 0.44 |
| Total.Sulfur.Dioxide | Density | pH |
| 0.19 | 0.00 | 0.43 |
| sulphates | Alcohol | |
| 0.88 | 0.00 | |

Loadings:

| | Factor1 | Factor2 | Factor3 | Factor4 | Factor5 | Factor6 |
|---|---|---|---|---|---|---|
| Residual.Sugar | 0.95 | 0.13 | 0.23 | | | -0.14 |
| Density | 0.76 | 0.56 | 0.19 | 0.24 | | |
| Alcohol | -0.30 | -0.93 | -0.15 | | 0.11 | |
| Free.Sulfur.Dioxide | 0.13 | | 0.72 | | | |
| Total.Sulfur.Dioxide | 0.20 | 0.31 | 0.78 | 0.14 | 0.11 | 0.17 |
| Fixed.Acidity | | | | 0.98 | | -0.16 |
| Volatile.Acidity | | | | -0.11 | 0.99 | |
| pH | | | | -0.32 | | 0.67 |
| Citric.Acid | | | | 0.31 | -0.12 | |
| Chlorides | | 0.38 | | | | |
| Sulphates | | | | | | 0.34 |

| | Factor1 | Factor2 | Factor3 | Factor4 | Factor5 | Factor6 |
|---|---|---|---|---|---|---|
| SS loadings | 1.64 | 1.45 | 1.28 | 1.25 | 1.03 | 0.66 |
| Proportion Var | 0.15 | 0.13 | 0.12 | 0.11 | 0.09 | 0.06 |
| Cumulative Var | 0.15 | 0.28 | 0.40 | 0.51 | 0.60 | 0.66 |

```
Test of the hypothesis that 6 factors are sufficient.
The chi square statistic is 347.4 on 4 degrees of freedom.
The p-value is 6.4e-74
```

With 6 predictors in the model 66% of the variability is explained, but the null hypothesis that 6 factors are sufficient for the analysis, is rejected. However higher than 6 factors are not possible for this data.

```
> FAWW <- factanal(SWWPred, 7)
Error in factanal(SWWPred, 7) : 7 factors are too many for 11 variables
```

For White Wine Data, no set of factors is able to capture the latent structure. Uniquenesses given in the output is the amount of variance in the variable not explained by the latent factors. Communalities can be computed from the uniqueness values given. Since cut-off threshold is fixed at 60%, only loadings above that level are shown.                                                                    ‖

Factor rotation is done to improve the interpretability of factors. There are several possible rotation mechanism of which Varimax is one. It suffices to say that R uses Varimax rotation by default. We will not go into any further depth regarding the different aspects of factor rotation.

## 11.3 Classification Problems

Classification problems arise when two or more groups (or classes) are known a priori and one or more new observations are required to be put into one class or the other. Consider the German Credit Data. Every bank needs to make a decision whether a new applicant can be sanctioned for a loan or not. This is a two-class problem where each applicant is classified as a good or bad risk depending on the applicant's profile matching with that of already existing observations. Similarly, whether or not a bank fails can also be considered as a classification problem. In healthcare analytics, profiling the risk of all the subjects in the database may be considered a multiple-class classification problem where the classes already identified may be extremely high risk, moderately high risk, high-normal risk, low risk, etc., for every disease. Correct classification would lead to proper insurance premiums charged and optimum utilization of healthcare facilities. One way of assigning each observation to a class is through binary logistic regression, which has already been considered in Chapter 9. An extension of binary logistic regression to multiple classes exists but its usage is limited. To categorize observations to one of the several classes, a discriminant analysis procedure is applied.

## 11.4 Discriminant Analysis

In Section 5.6, the Bayes' theorem was introduced. The Bayes' theorem deals with posterior probability, i.e., in a cause–effect relationship, after considering the effect, probability is assigned to different causes. The discriminant analysis procedure is dependent on the application of the Bayes' theorem.

Suppose there are $g$ classes, each class representing a population. One may consider airline passengers belonging to $g = 3$ populations – the economy class passengers, the business class passengers and the first class passengers. A multivariate set of attributes is observed on each unit. Suppose there are $n$ units in the dataset and $n_j$ of them are coming from the $j$-th population, $j = 1, 2, \ldots, g$. The objective of discriminant analysis is to estimate the posterior probability that the $i$-th observation is coming from the $j$-th population. Based on the observations whose class memberships are known, a discriminant rule is developed. To a new observation whose class membership is not known, the discriminant rule is applied and the observation is assigned to that class whose posterior probability is the highest.

Notationally, let the $j$-th population be denoted by $\pi_j$, $j = 1, \ldots, g$. Let $p_j = Pr(\pi_j)$ be the probability that a randomly selected observation comes from the $j$-th population. Each population is characterized by a probability distribution. Even if in each case the form of the distribution is normal, each distribution will have a different set of parameters $(\mu_j, \sigma_j)$. Let $f(x_i, \pi_j)$ be the probability rule that the $i$-th observation is coming from the $j$-th population. Without going into the details of the likelihood function, let us simply say that the posterior probability that an observation is a member of the population $\pi_k$ is

$$Pr(\pi_k | x_i) = p_k f(x_i | \pi_k) / (\sum_{j=1}^{g} p_k f(x_j | \pi_k)).$$

The classification rule will assign $x_i$ to that population for which the posterior probability is the greatest. Discriminant analysis develops the classification rule with known group membership. Hence a dataset is required which contains $n$ observations all of whose group membership is completely known. In case of the German Credit Data all observations were given a value 1 or 0 depending on whether they were creditworthy or not. For the White Wine Data, each wine had a known value on Quality attribute. The prior probability of a population is simply the proportion of observations in the dataset belonging to the same population. If $n_j$ is the number of observations in the $j$-th class, then

$$\hat{p}_j = n_j / n.$$

Note that $\hat{p}_1 + \hat{p}_2 + \cdots + \hat{p}_g = 1$. Alternatively, if for any reason, the sample proportions are not to be used, $\hat{p}_j$ may also be taken as $1/g$. This would be used if, notwithstanding the sample values, there is reason to believe that in the totality all populations are represented in equal proportion.

Recall that each observation comes from a multivariate population and hence a variance-covariance matrix needs to be considered. From each group of observations estimates of population mean and variance-covariance matrix are computed. It is also important to test whether the variance-covariance matrix can be assumed to be identical. Formal hypothesis test is to be done as $H_0$ : $\sigma_1 = \sigma_2 = \cdots = \sigma_g$ against the alternative $H_A$: At least one $\sigma_j$ is different from the rest. If the null hypothesis is rejected, then quadratic discriminant analysis is applicable; if the null hypothesis is not rejected then linear discriminant analysis is used. Note that we perform DA only if the populations are different, and usually a difference in populations is manifested through their means. If the population means are not different, then there is no question of making any discrimination among them, because all classes will then be identical.

Based on the sample mean and variance estimates and estimated prior probability, the linear or quadratic discriminant functions are computed. Each observation is assigned a discriminant score. A discriminant score is also computed for a new observation, and this is classified into the group corresponding to the largest discriminant score.

It is important to note that the populations must have a multivariate normal distribution. Discriminant scores are dependent on the means of the population, and it is difficult to define a concept of mean if a discrete random variable with several levels is used as a predictor.

**Example 11.3.** (German Credit Data) To apply DA to German Credit Data only the continuous variables are used. The function `lda()` is available in the R Library MASS. The creditworthy and non-creditworthy observations are coded 1 and 0, respectively.

```
> ldafit<- lda(Creditability ~ Value.Savings.Stocks +
  Length.of.Current.Employment + Duration.of.Credit.Month +
  Credit.Amount + Age, data = German.Credit)
> ldafit
Call:
  lda(Creditability ~ Value.Savings.Stocks
  + Length.of.Current.Employment + Duration.of.Credit.Month
  + Credit.Amount + Age, data = German.Credit)

Prior probabilities of groups:
  0   1
0.3 0.7

Group means:
      Value.Savings.Stocks  Length.of.Current.Employment
0             1.67                        3.17
1             2.29                        3.47

      Duration.of.Credit.Month   Credit.Amount        Age
0                24.86              3938.127         33.96
1                19.21              2985.443         36.22

Coefficients of linear discriminants:
LD1
Value.Savings.Stocks                0.38
Length.of.Current.Employment        0.26
Duration.of.Credit.Month           -0.06
Credit.Amount                      -4.916199e-05
Age                                 0.01
```

To use the discriminant rule to predict class membership of each observation in the data, a prediction using the discriminant function is done.

```
> ldapred<- predict(ldafit, German.Credit)
> table(ldapred$class, German.Credit$Creditability)

      0    1
  0  67   41
  1 233  659
```

It appears that $(67 + 659) = 726$ observations are correctly classified. Hence accuracy of LDA is 72.6%.

LDA does not seem to be providing predicted values with high accuracy. A quadratic discriminant analysis is also done but that does not improve the result. The function for QDA in R is qda() in the R Library MASS. However, results from QDA application are not shown. The table comparing actual classification with predicted classification is known as the confusion table. The form of the confusion table is given below.

| Predicted Class | Actual Class | | | |
|---|---|---|---|---|
| | 1 | 2 | ... | $g$ |
| 1 | $n_{11}$ | $n_{12}$ | ... | $n_{1g}$ |
| 2 | $n_{21}$ | $n_{22}$ | ... | $n_{2g}$ |
| ... | ... | ... | ... | ... |
| $g$ | $n_{g1}$ | $n_{g2}$ | ... | $n_{gg}$ |

The observations on the main diagonal, $n_{ii}$, $i = 1, \ldots, g$ are classified correctly; all the other off-diagonal observations are not. Ideally, the perfect classification will have only non-zero observations on the main diagonal and will have zero observations on the off-diagonal cells. However, such an ideal classification is not possible to find. One measure of accuracy is the proportion of observations on the main diagonal of the confusion matrix.

## 11.4.1   K Nearest Neighbor (KNN) Algorithm

This is an algorithmic procedure to classify a new observation into one of the known classes. Like any classification method, we start with a sample, sometimes called a training sample, of known class membership. A new observation is classified according to its maximum similarity to the observations belonging to a certain class. However, taking a decision regarding which data points the new observation is most similar to is an open question. The K Nearest Neighbor algorithm makes a decision based on the number of observations within a certain radius. It does not take into account all observations belonging to a certain class. The algorithm compares the new observation with observations closest to it, according to some criterion, and assigns it class membership accordingly. The most important tuning parameter in this algorithm is $k$, the number of neighbors considered to make a decision. Once the positive integer $k$ is specified, then the $k$ nearest observations to the new item is considered. The most common class among them is assigned to the new item. KNN can be a computation-intensive algorithm if the sample size is large, since the distance between the new item and each of the observations considered needs to be computed and sorted. Variables need to be standardized if the attributes are to play a similar role and variances are of different orders. However, whether to apply normalization is rather subjective. The complexity is determined by the parameter $k$. If $k$ is large, the boundaries of classification are smoother.

## 11.5 Clustering Problem

Market segmentation is one of the fundamental strategic decisions for any business. Whenever a business has a number of offerings for its customers, it would like to match its offering to customer needs so that its products become more and more acceptable. Since each customer has different needs and is willing to pay a different price, customization of products is required. However, it is not possible to customize products to each customer's needs. Hence businesses divide their customers into seemingly homogeneous groups and customize their products according to the group's demands. Clustering is the technique to divide a population into homogeneous groups depending on the values of attributes. There are two clustering procedures – hierarchical and K-means clustering. In the hierarchical clustering procedures, single units join successively to form agglomerative clusters and the sizes of the clusters increase, ending with a single cluster where all units combine. In K-means clustering procedure, the number of clusters is predefined. The procedure starts with K initial clusters and at every iteration one unit is shifted from one cluster to another so that the clusters become more similar within themselves but more dissimilar among themselves. The clustering procedure depends on a concept of distance between a pair of observations.

### 11.5.1 Similarity and Dissimilarity

Let us consider only two units in a sample, denoted by $X_1$ and $X_2$. Let us also assume on each of these units $m$ attributes have been measured. For the Wine Data, $m = 11$, if only the physicochemical properties of the wines are considered. On a few of these attributes, two units may register similar values on a few others the units may register quite different values. The distance between two units is an overall measure to determine how close, or otherwise, these units are. There are many different types of distances. We will list here only the most common ones. The Euclidean distance is the most frequently used distance measure.

$$d(X_1, X_2) = \sqrt{\sum_{j=1}^{m}(X_{1j} - X_{2j})^2}.$$

This is the square root of the sum of the squared differences between the measurements for each variable. Minkowski's distance is an extension of the Euclidean distance.

$$d(X_1, X_2) = \left[\sum_{j=1}^{m}|X_{1j} - X_{2j}|^p\right]^{1/p}.$$

Another commonly used distance is known as the Mahalanobis distance, which depends on normality assumption and the multivariate mean and dispersion matrix.

Any distance measure must satisfy a few properties. Most of these properties are common sense, such as the distance from $X_1$ to $X_2$ is the same as the distance from $X_2$ to $X_1$ (symmetry). Distance is a positive quantity. If two items are identical, then the distance between them is 0, and that is the only time the distance between two items can be 0. Distances also satisfy the triangle equality.

If a dataset has $n$ observations, an $n \times n$ distance matrix can be computed to accommodate all pairwise distances. This matrix, like a correlation matrix, is symmetric, but unlike the correlation matrix, contains 0 on the main diagonal. Note that distance is a relative quantity. Whether two items in a dataset are similar or not depends on their relative position vis-a-vis all other items. There is no upper bound regarding what value a distance measure can take.

**Example 11.4.** (Auto Data) A subset of nine cars is identified and their values on different attributes are shown in the table below.

| Id | Make | Model | MPG | Cylinders | Displacement | Horsepower | Weight | Acceleration |
|---|---|---|---|---|---|---|---|---|
| 1 | Cadillac | Eldorado | 23 | 8 | 350 | 125 | 3900 | 17.4 |
| 2 | Cadillac | Seville | 16.5 | 8 | 350 | 180 | 4380 | 12.1 |
| 3 | Mercedes | 300d | 25.4 | 5 | 183 | 77 | 3530 | 20.1 |
| 4 | Mercedes | 240d | 30 | 4 | 146 | 67 | 3250 | 21.8 |
| 5 | Mercedes | 280s | 16.5 | 6 | 168 | 120 | 3820 | 16.7 |
| 6 | Subaru | subaru | 26 | 4 | 108 | 93 | 2391 | 15.5 |
| 7 | Subaru | subaru | 32.3 | 4 | 97 | 67 | 2065 | 17.8 |
| 8 | Subaru | subaru dl | 30 | 4 | 97 | 67 | 1985 | 16.4 |
| 9 | Subaru | subaru dl | 33.8 | 4 | 97 | 67 | 2145 | 18 |

Examples of a few alternative distances are shown below.

```
> DE <- dist(Comp3[,2:8], "euclidean")

> round(DE, digits=1)
        1      2      3      4      5      6      7      8
2   483.2
3   408.8  872.4
4   683.8 1153.9  282.7
5   199.0  591.9  293.7  573.1
6  1528.6 2005.6 1141.6  860.3 1430.5
7  1853.3 2331.6 1467.6 1186.0 1757.3  327.4
8  1932.5 2411.0 1547.4 1266.0 1837.2  407.0   80.1
9  1774.1 2252.2 1387.7 1106.1 1677.4  247.8   80.0  160.1

> DM <- dist(Comp3[,2:8], "minkowski", p=3)
> round(DM, digits=1)
```

```
        1       2       3       4       5       6       7       8
2   480.2
3   381.3   852.6
4   656.8 1132.6   280.2
5   187.0   566.6   290.3   570.2
6 1511.1 1990.2 1139.1   859.0 1429.0
7 1836.6 2316.1 1465.1 1185.0 1755.1   326.1
8 1916.5 2396.0 1545.1 1265.0 1835.1   406.0    80.0
9 1756.8 2236.2 1385.1 1105.0 1675.1   246.1    80.0   160.0

> DH <- dist(Comp3[,2:8], "manhattan")
> round(DH, digits=1)
        1       2       3       4       5       6       7       8
2   549.8
3   593.1 1142.9
4   928.4 1478.2   335.3
5   279.2   808.6   364.3   669.6
6 1796.9 2336.9 1241.2   939.3 1530.7
7 2161.7 2711.5 1573.2 1241.3 1902.9   378.6
8 2240.0 2783.8 1652.3 1322.4 1975.8   450.9    87.7
9 2082.4 2632.2 1493.5 1161.6 1823.6   299.3    82.7   168.4

> DX <- dist(Comp3[,2:8], "maximum")
> round(DX, digits=1)
     1     2     3     4     5     6     7     8
2   480
3   370   850
4   650  1130   280
5   182   560   290   570
6  1509  1989  1139   859  1429
7  1835  2315  1465  1185  1755   326
8  1915  2395  1545  1265  1835   406    80
9  1755  2235  1385  1105  1675   246    80   160
```

Pairwise distance is given in the appropriate cells of the matrix. Depending on the distance calculated, the numerical values change. It so happens that the relative positions of the cars according to distances do not show marked change. This is an expectation, which holds in general, but there is no guarantee that it will hold always. The overall pattern may remain the same. For example, Cadillac and Mercedes are similar among themselves, even though Cadillac models may be closer to Mercedes than another Cadillac, but they all are very different from Subaru.

## 11.5.2 Hierarchical Clustering

Hierarchical clustering is an agglomerative clustering technique. The aim of the procedure is to put together the units which are similar to each other. The process starts with $n$ clusters of size 1 and ends with one cluster of size $n$. Reaching the endpoint is not the principal goal of the procedure, but the process itself throws light on the similarity among the units. There are several methods of creating clusters, by using different linkages among the units that are already included in the clusters. Linkages are functions of the distances, but they are more than pairwise distance. Linkage is defined for clusters, and, depending on the linkage, larger clusters are formed from smaller ones. In *single linkage clustering* the distance between two clusters is defined to be the minimum distance between any single unit in the first cluster and any single unit in the second cluster. At each stage of the process the two clusters that have the smallest single linkage distance are combined to form a single cluster.

In *complete linkage clustering* the distance between two clusters is defined to be the maximum distance between any single unit in the first cluster and any single unit in the second cluster. At each stage of the process the two clusters that have the smallest complete linkage distance are combined to form a single cluster.

In *average linkage clustering* the distance between two clusters is defined to be the average distance between units in the first cluster and units in the second cluster. At each stage of the process the two clusters that have the smallest average linkage distance are combined to form a single cluster.

In the *centroid method*, the distance between two clusters is the distance between the two mean vectors of the clusters. At each stage of the process two clusters that have the smallest centroid distance are combined to form a single cluster.

Ward's method does not directly define a measure of distance between two points or clusters. It is an ANOVA-based approach. At each stage, those two clusters merge, which provides the smallest increase in the combined error sum of squares.

Hierarchical clustering produces a tree-like structure called a dendrogram, which depicts the history of clustering. The distance is shown on the $Y$-axis and on the $X$-axis all sample units are placed. The dendrogram shows at which height (or distance) two or more clusters combine to form a new cluster.

**Example 11.3.** (Continued) The nine models of cars are used as an illustration to see how different methods of clustering try to form agglomerative clusters. Single linkage and complete linkage are used and the distance matrix used in both cases is the Euclidean distance (see Figure 11.2).

```
> f1 <- hclust(DE, method="single")
> plot(f1, main="Single Linkage Clustering", sub="", xlab="")

> f2 <- hclust(DE, method="complete")
> plot(f2, main="Complete Linkage Clustering", sub="", xlab="")
```

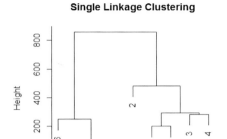
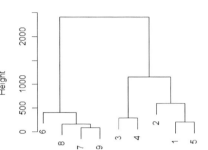

**FIGURE 11.2**
Hierarchical clustering: Example 1.

Hierarchical clustering starts with all nine units representing nine clusters. The lines show at what level which two units or clusters combine. With single linkage units 7, 8 and 9 combine at the lowermost level; with complete linkage units 7 and 9 combine first and then unit 8 combines with the two-member cluster. Continuing with single linkage method, units 1 and 5 combine, units 3 and 4 combine; then these two-member clusters combine. Next unit 2 combines with this four-member cluster. Continuing with complete linkage, units 1 and 5 combine; units 3 and 4 combine; then unit 2 combines with the cluster containing units 1 and 5; next the two-member cluster with units 3 and 4 combines with the three-member cluster.

It is clear that primarily there are two distinct conglomerations. One comprises units 1–5 and the other comprises units 6–9. The latter are all Subaru and the former are the luxury cars. However, the procedure stops only when all nine units form a single cluster. Because the two conglomerations are so different, the final agglomeration occurs at a comparatively higher distance. ∥

**Example 11.4.** (Auto Data with 50 cars) We have applied single linkage and the centroid method with to arrive at two different systems of clusters. In each case clusters are identified by rectangles drawn around them. It seems that, depending on which linkage method is applied, not only the allocation changes, but the number of clusters also may vary to a significant extent. See Figures 11.3 and 11.4.

```
> AutoClust1<- hclust(DistMat,, method="single")
> plot(AutoClust1, main="Euclidean Distance, Single Linkage",sub="",
  xlab="")
> rect.hclust(AutoClust1, k=10, border="red")
> AutoClust2<- hclust(DistMat,, method="centroid")
```

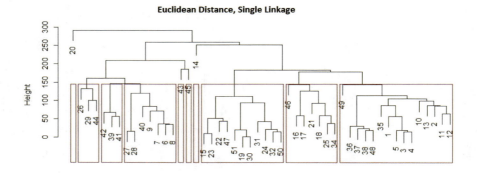

**FIGURE 11.3**
Hierarchical clustering: Example 2.

```
> plot(AutoClust2, main="Euclidean Distance, Centroid Method",sub="",
  xlab="")
> rect.hclust(AutoClust2, k=5, border="red")
```

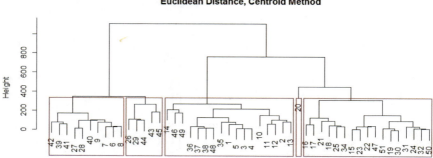

**FIGURE 11.4**
Hierarchical clustering: Example 3.

## 11.5.3   K-Means Clustering

Here the number of clusters is fixed, but membership in the clusters is not. It is a non-hierarchical procedure that starts with a partition of all $n$ units in the sample into K groups. For each group, the mean is calculated. Then each unit is shifted to that group, where the distance between the unit and group mean is the least. After every shift, the group means are recomputed. Note that, after every shift, only two group means are required to be recomputed, since at every step one unit moves from one group to another. This continues until there are no more shifts. The grouping may depend on the initial partitioning. It is not guaranteed that, starting from any partition, the final group allocations will be identical.

There are a few arguments for not fixing the number of clusters beforehand. If no information regarding the population is known, then no initial guess for $K$ is available. Even if the groups in the population are well defined, the sampling procedure may lead to insignificant representation of one or more smaller groups. In that case, insisting on having the same number of clusters as expected in the population may not lead to sensible clustering. The presence of outliers is always a potential problem, especially when the number of clusters is fixed. An outlier would tend to work as a seed, leading to spurious clusters being formed.

Hierarchical clustering procedures have other problems, as documented in previous sections. It is not always clear where to cut the dendrogram tree and form the clusters. Indeed, the aim of an agglomerative procedure is not to come up with a final recommended number of clusters. However, it is possible to apply a hierarchical clustering procedure on the data once to see the relative position of the units. Based on the results, one can try out alternative K values.

**Example 11.5.** (Auto Data: all observations) Suppose that all 392 cars are to be classified into three clusters according to their origin. Note that 1 indicates US, 2 European and 3 Japanese cars. Based on the other attributes, the units are classified into these three clusters.

```
> kfit<- kmeans(Auto[,2:8],3)
> Autoclust<- data.frame(Auto, kfit$cluster)
> with(Autoclust, table(origin, kfit.cluster))
      kfit.cluster
origin  1  2  3
     1 89 96 60
     2  1 17 50
     3  0  9 70
```

The table in the R output shows that Japanese cars are remarkably consistent and only 9 out of 79 have been classified into a different group. However, most of the European and a large proportion of American cars are also classified into the same cluster as Japanese cars. In an ideal situation, the main diagonal should contain most of the observations.  ‖

## 11.6   Suggested Further Reading

Multivariate statistics is based on fundamentals of mathematical knowledge and the best books, therefore, require an advanced level of statistical training. One of the best books in this field is Johnson and Wichern (2007). For a comparatively easier read, one may consider Tabachnick and Fidell (2013) and Grimm and Yarnold (2000). For excellent sources of R application with multivariate statistics, one may consider Schumacker (2015) and Everitt and Hothorn (2011). James et al. (2014) and Kuhn and Johnson (2013) also have applications of multivariate techniques in data mining along with R codes.

# 12

## Modeling Time Series Data for Forecasting

All businesses and organizations do forecasting. From roadside vendors and peddlers to major airlines and hotels as well as drug manufacturers and governments, all try to forecast for the next several years to plan and optimize their output, maximize profits and minimize losses, plan for manpower resources and overall smoothly run their businesses. Of course, forecasting methods are not the same for small retailers and pharmaceuticals running drug trials all over the world. In fact, in many cases, even sizable organizations do not employ statistical methods for forecasting, but go by their gut feeling or at best employ naive methods.

A good forecasting method takes into account the systematic changes in the data. There are many different methods of forecasting. Forecasting involves predicting the values of one or more important variables, or responses, for the future. Regression techniques can also be used as a forecasting method, if it may be assumed that the predictors are all known in the future. However, possibly the most used method of forecasting is time series analysis, where one or more variables are predicted for the future, depending on their past performances. In this chapter we will deal with univariate time series only, where a single variable is studied and forecast for the future.

Forecasting will never be perfect; the challenge is to make a business forecast model better than the competition. Better forecasts result in better customer service and lower costs, as well as better relationships with suppliers and customers. The forecast can, and must, make sense based on the big picture, economic outlook, market share, and so on. The best way to improve forecast accuracy is to focus on reducing the error of forecasting. Bias is the worst kind of forecast error. The aim of all forecasters is to make accurate forecasts with less cost and effort. To achieve the best possible predictions, one has to learn about the present scenario, make realistic assumptions about the future and change assumptions and study their impact on future values. Forecasts are to be benchmarked against naive method forecasts as well as industry standards.

## 12.1  Characteristics and Components of Time Series Data

A time series is defined as a set of quantitative observations arranged in chronological order. Time series data is collected over a sequence of discrete time points covering a predefined range. Thus time is treated as a discrete variable. To measure the growth of a country, the gross national product is measured annually. Profits and losses of companies are measured every quarter. Sales volumes are measured monthly. Stock market movements are followed daily. All of the above examples represent time series data. Even minute-by-minute movements of stock value can be modeled using time series. The two most important characteristics of time series data are that all observations must be collected at the same interval and observations must be contiguous. In the same time series, yearly and monthly observations must not be combined. Neither should there be any gap in the range of the time points for which the data are collected.

In Figure 12.1 we present several examples of time series data collected at various time intervals – yearly, quarterly, monthly and weekly. Time series data can also be collected daily or hourly or at any other precisely measurable intervals.

Naturally, chronologically observed variables are not independent. Hence standard procedures for an independent set of observations will not work

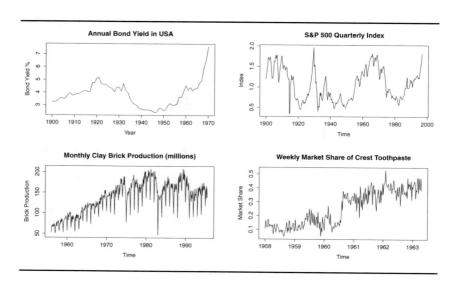

**FIGURE 12.1**
Examples of several typical time series.

for time series. It is also clear from the above figures that, over time, the observations show increasing or decreasing definitive movements or cyclical movements that are repeated yearly. These distinct patterns are characteristics of a time series.

The major components of a time series are trend, seasonality, cyclical variation and irregular variation.

Trend is the long-term increasing or decreasing movement in a time series. Every series shows a tendency to either move in a certain direction or stagnate over time. The series Annual Bond Yield shows a definite increasing trend after 1950. The series has an increasing trend, whereas the S&P 500 Quarterly series does not show any trend over the years considered. The Weekly Market Share proportion of Crest Toothpaste shows an upward movement all through the six years under consideration.

Seasonality indicates fluctuations within a year which are not random in nature. The series on monthly clay brick production shows similar movement within a year, regardless of whether there is a trend or not. The repetitive fluctuation within a year is known as seasonal movement of a time series, since this regular pattern is associated with seasons within a year. Seasonal patterns are relevant for time series with intervals less than a year.

Cyclicality of a time series is associated with ups and downs associated with a business across years. Schematically, a business may go through a sequence of starting up, prosperity, decline and, finally, recovery. Such long-term movements are known as cyclical patterns. This may also be noticed in retail business where demands for particular items may show come-back after regular intervals. The study of cyclical patterns is more difficult than the other two, and we will omit this component from our discussion.

In addition, there are irregular or random components of a time series which are caused by sudden shocks to the series. These movements are not repetitive and do not show any pattern. This component is similar to the error component in a regression model and is also known as the noise.

If $Y(t)$ is the value of a time series at time $t$, then $Y(t)$ may be expressed as a sum or as a product of the three components.

- Additive time series: $Y(t) = T(t) + S(t) + I(t)$

- Multiplicative time series: $Y(t) = T(t) \times S(t) \times I(t)$

where $T(t)$ is the trend value of the series at time $t$, $S(t)$ is the seasonality index at time $t$ and $I(t)$ is the value of the irregular or error component associated with the observation at the $t$-th time point. The multiplicative model is based on the assumption that the component's trend and seasonality are not independent, but show an interaction effect. Typically, this is noticed when seasonal fluctuations increase as trend values increase. The time series in Figure 12.2 indicates that the total number of international airlines passengers is increasing over the years and with every passing year the number of passengers across months is showing higher fluctuations.

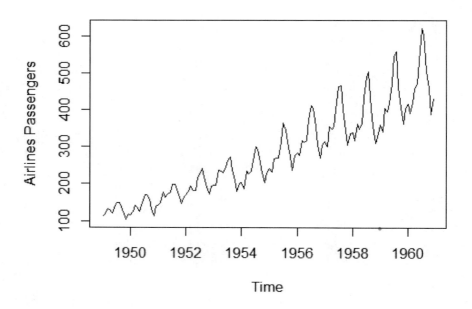

**FIGURE 12.2**
International airlines passengers.

In comparison to that, the sales of champagne seem to be constant over the years, as shown in Figure 12.3. The seasonality effect, however, shows higher fluctuations in recent years. The series on air traffic is a possible case for multiplicative time series, while the series of champagne sales would possibly be better modeled as an additive series.

### 12.1.1    Time Series Analysis Techniques

Two distinct features of time series are clear from the above discussion. Each observation in a time series may comprise trend, seasonality and error components. Each observation in a time series depends on the previous one or more observations. Depending on these two characteristics, two different techniques may be employed to analyze time series. The main goal in analyzing a time series is to obtain an understanding of the underlying forces and structure that produced the observed data and to fit a model and proceed to forecasting, monitoring or even feedback and feed-forward control. The classical time series analysis procedure decomposes the series into its constituent components.

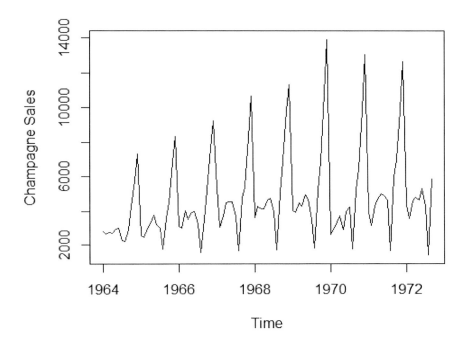

**FIGURE 12.3**
Monthly champagne sales (millions).

The regression approach models each observation as a function of previous observations.

Whatever may be the ultimate analysis technique, the important questions to consider when first looking at a time series relate to the existence of trend, seasonality, outliers and constant variance. If the time series includes a trend component, then, on the average, the measurements tend to increase (or decrease) over time. If, on the other hand, the time series includes a seasonal component, then there is a regularly repeating pattern of highs and lows related to calendar time such as seasons, quarters, months, days of the week and so on. We have already considered these questions for the Airline Passengers Data and for the Champagne Sales Data. The former shows both trend and seasonality, while the latter shows only a seasonal effect.

The next issue to consider is whether there are outliers in the data. Just like in regression, outlier detection in time series is important, albeit difficult. In regression, outliers are way off the fitted line. In time series data, outliers are suddenly different from the other data points. In the S&P 500 quarterly series there may be an outlier for 1929 Q3. For the Clay Brick Production series,

January 1981 may be an outlier. Outliers are sometimes incorporated in the history of the time series. Consider the example of passenger traffic through an airport, which suddenly closed down on a certain day due to heavy fog or snowfall. That day passenger volume will be extraordinarily low, if not exactly zero. This observation is definitely an outlier and must be treated as such from the beginning.

Another important factor to take into account is whether the variance remains constant over time. This is also a difficult assumption to test. Variation of the irregular component, which is the random error component, causes fluctuations in the time series. At the same time, fluctuations are caused by seasonality also. Typically, daily movement of stock exchanges or such high volatile data shows signs of non-constant variance. Special models (ARCH or GARCH) have been developed for those series, which are out of the purview of our discussion here.

Finally, one would like to check whether there are any abrupt changes in either the level of the series or the variance. This is a very real problem and has a serious implication in the course of time series analysis. Consider a production process, and let the output of that process be the series of interest. If the process is modernized at some point of time and as a result has become more efficient, a sudden change in the level of output is expected. If this information is not available to the analyst, and if the information before the modernization is used to model the series, the forecasts may not have acceptable accuracy. The weekly market share of Crest toothpaste seems to show a clean break in the middle part of 1980. However, it is difficult to detect exactly at which point the break occurred. Similarly it is possible that around 1972–1973 there is a change in the nature of the Clay Brick Production series.

## 12.2   Time Series Decomposition

Decomposition procedures identify the trend and seasonal factors in a time series. More extensive decompositions might also include long-run cycles, holiday effects, day of week effects and so on. Here we consider only trend and seasonal decompositions. One of the main objectives for decomposition of a time series is to estimate seasonal effects that can be used to create and present seasonally adjusted values. A seasonally adjusted value removes the seasonal effect from a value so that trends can be seen more clearly. For instance, the number of international airline passengers has increased every year from 1949 to 1960. At the same time, in certain months the passenger volume is low compared to the yearly mean, and in certain months the volume is relatively high. To investigate the trend, the monthly variation in the passenger volume must be adjusted.

A more difficult problem is presented by the monthly Clay Brick Production series. Passenger volume shows an increasing trend over the whole range of the series, but brick production increases steadily until 1971–1972, after which it shows an oscillating movement. To study trend in this context, the seasonal adjustment must be done to begin with.

## 12.2.1   Trend Estimation in a Time Series: Moving Average

The first step in decomposition of a time series is to estimate the trend. Although there are many simplistic and sophisticated procedures to estimate trend, they may be divided into two main categories:

(i) Smoothing the series using moving averages or exponential smoothing procedures;

(ii) Modeling the trend using a regression approach.

The second step is to de-trend the time series either by subtracting the trend component (for additive series) or by dividing by the trend component (for multiplicative series). Seasonal factors are estimated using the de-trended series. For monthly data, an effect for each month of the year is estimated; for quarterly data, an effect for each quarter is estimated. If the data is weekly, then 52 effects, one for each week, are estimated. The simplest method for estimating these effects is to average the de-trended values for a specific season. For instance, to get a seasonal effect for January, we average the de-trended values for all Januarys in the series and so on. Median values, instead of mean, may also be used. The seasonal effects are usually adjusted so that they average to 0 for an additive decomposition or average to 1 for a multiplicative decomposition.

The moving average (MA) procedure for estimating trend is also known as linear filtering. The MA at each point in time determines (possibly weighted) averages of observed values that surround a particular time. This smooths out the irregular roughness to show a clearer signal. The number of observations included in computing the average is known as the period of the MA. For seasonal data, the MA smooths out seasonality to identify the trend. If the length of the MA period is odd, the MA is automatically centered. If the length is even, to center the MA, a second averaging is required. To smooth out seasonal data, the length of the period must be equal to the number of seasonality effects. Table 12.1 presents the centered MA values for a part of the quarterly S&P 500 Index series.

The first value of an uncentered MA is the average of all four quarters of 1900. The second value is the average of 1900 Q2–Q4 and 1901 Q1 and so on. However, these values do not correspond to any quarter, and hence the first value in the centered MA column was calculated as the average of these two uncentered MA values. The value 1.37 corresponds to 1900 Q3. Because the period considered is 4, there will not be any MA values for the first 2 and the last 2 quarters. The centered MA values obtained in the last column may

**TABLE 12.1**
Illustration of MA with the quarterly S&P 500 Index series

| Year | Quarter | S&P Index | Uncentered MA | Centered MA |
|------|---------|-----------|---------------|-------------|
| 1900 | Q1 | 1.198938 | | |
| 1900 | Q2 | 1.299910 | | |
| | | | 1.33 | |
| 1900 | Q3 | 1.348435 | | 1.37 |
| | | | 1.41 | |
| 1900 | Q4 | 1.455093 | | 1.47 |
| | | | 1.52 | |
| 1901 | Q1 | 1.552336 | | 1.56 |
| | | | 1.61 | |
| 1901 | Q2 | 1.725968 | | 1.64 |
| | | | 1.66 | |
| 1901 | Q3 | 1.694321 | | |
| 1901 | Q4 | 1.681869 | | |

be considered as trend values corresponding to the appropriate quarters. The form of the trend under quarterly seasonality is given by

$$T(t) = \frac{1}{8}Y(t-2) + \frac{1}{4}Y(t-1) + \frac{1}{4}Y(t) + \frac{1}{4}Y(t+1) + \frac{1}{8}Y(t+2)$$

where $T(t)$ is the value of the trend at time point $t$, $Y(t)$ is the value of the series at time point $t$, $t-i$ denotes $i$ time points before $t$ and $t+i$ denotes $i$ time points after $t$. It is clear that the two values at the end get lower weight compared to the middle values, due to centering. For monthly data, the MA estimate of the trend is

$$T(t) = \frac{1}{24}Y(t-6) + \frac{1}{12}Y(t-5) + \frac{1}{12}Y(t-4) + \cdots + \frac{1}{12}Y(t+4)$$
$$+ \frac{1}{12}Y(t+5) + \frac{1}{24}Y(t+6).$$

For seasonal data, the period of the MA is fixed and is equal to the period of seasonality. For non-seasonal data, an appropriate period for the MA needs to be determined. Determination of the period of the MA is not easy and there is no mathematical formula or rule of thumb that works in all situations. If the period of the MA is relatively short, then the trend line will not be smooth enough. If the period is too long, then the trend line may be too flat. The objective of finding MA is to smooth the rough edges of a time series to see what trend or pattern might emerge. Only an optimum MA period is able to capture the trend pattern. It is also to be noted that the longest possible span

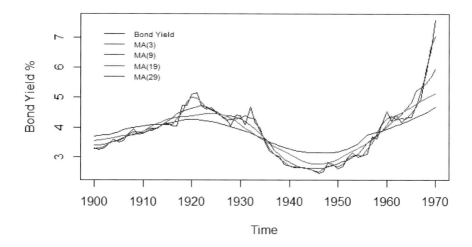

**FIGURE 12.4**
Bond Yield: Comparison of moving averages.

is the whole length of the series with an average taken over the entire set of values. Here the MA is identical to the series mean. At the other extreme, the shortest possible period is 1, where each $T(t) = Y(t)$. None of them will provide any insight into the trend estimation problem.

**Example 12.1.** (Trend Estimation for Annual Bond Yield) Figure 12.4 illustrates the effect of smoothing on the original series by several MAs of different periods. For the sake of convenience we have considered only odd periods. The red line denotes MA(3); it follows the original series most closely. MA(29) is denoted by the dark blue line, which almost completely disregards the ups and downs of the original series and is too flat. The other MAs are in between, among which possibly the blue line, MA(9) is the most preferable. It ignores the short-term random fluctuations but follows the long-term pattern closely.

However, it is difficult to determine whether MA(9) or MA(11) would be the "best" just by looking at the curves. In practice, moving averages will look similar if the periods are close, particularly if the periods are at least moderately large. When the moving average method is used for forecasting, some objective criterion should be employed to determine the optimum period. ‖

**Example 12.2.** (Trend Estimation for a Seasonal Series) For a seasonal series the period of the MA must be equal to the period of seasonality. In Figure 12.5 the 12-point MA has been overlaid on the original series of Clay

Brick Production. This appears to smooth out the series nicely. However, here the trend is not a monotonic function of time. ‖

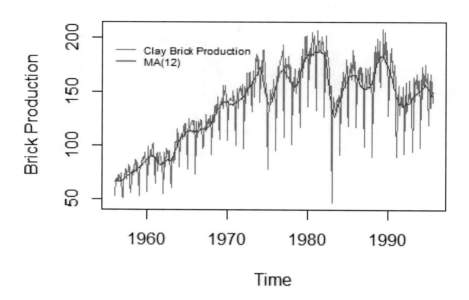

**FIGURE 12.5**
Clay Brick Production: Moving average.

## 12.2.2    Decomposition of a Time Series

The decomposition process on a time series identifies and separates out the three components from the original series, namely, trend, seasonality and irregular components. R generally uses the MA method to smooth the trend, though other smoothing methods are also possible. Once the trend component is isolated, either by subtracting or by dividing as the case may be, the average values of each season will provide the seasonal values.

**Example 12.3.** It has already been suggested that the International Air Passengers series is best modeled with a multiplicative seasonality. Figure 12.6 shows the components of the International Air Passengers series. The monthly seasonality components are expanded in Figure 12.7.

    The trend is practically linear. We fit a line by least squares linear regression to the MA trend values. The fitted line is given by

$$\hat{T}(t) = 84.65 + 2.667t$$

## Decomposition of multiplicative time series

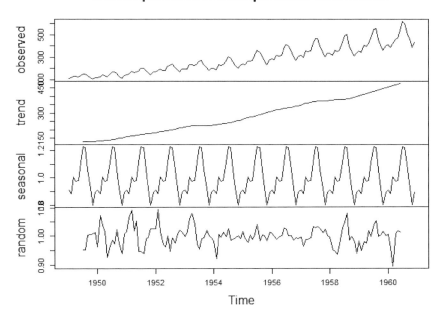

**FIGURE 12.6**
Decomposition of international air passenger series.

where $t$ denotes the time point and $\hat{T}(t)$ denotes the estimated trend value for that component. Monthly seasonality values are given in Table 12.2.

Since the seasonality is multiplicative, the sum of seasonality indices is 12, i.e., the average value is 1. Seasonality implies that in some months passenger volume is less than the average but in some other months it is more than the average. March and September passenger volumes seem to be close to the average. January, February, April and May volumes are lower, whereas June–Aug volumes are higher. This is expected, as in summer airlines operate

**TABLE 12.2**
Monthly seasonality indices

| Jan | Feb | Mar | Apr | May | Jun | Jul | Aug | Sep | Oct | Nov | Dec |
|------|------|------|------|------|------|------|------|------|------|-----|-----|
| 0.91 | 0.88 | 1.01 | 0.98 | 0.98 | 1.11 | 1.23 | 1.22 | 1.06 | 0.92 | 0.8 | 0.9 |

## Multiplicative Seasonality of Air Passengers

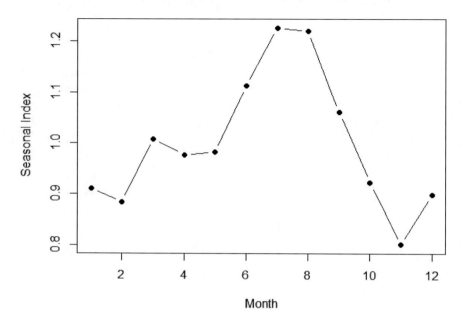

**FIGURE 12.7**
Seasonality component.

on close to full capacity. However, after September the volume goes down and in winter months volume remains low. After decomposition, the irregular component is expected to be random, known formally as the white noise. However, in the present example the irregular component has not reached that stage yet. To get further insight into the series, the irregular component can be modeled using autoregression or some other methods. More on autoregression will be discussed later.  ‖

### 12.2.3   Single Exponential Smoothing

One criticism of the MA method is that, to produce a smoothed series, only a small number of observations within the window around the observation of interest is used; all the other observations are ignored. Further, all observations in the window are weighted equally. One can consider an alternative method of smoothing where, theoretically, all past observations are used. However, the weights associated with the past observations are reduced exponentially with their distance from the current observation. Depending on the nature of

**TABLE 12.3**
Step by step illustration of single exponential smoothing; $\bar{\alpha} = 1 - \alpha$

| Time Period | Obs | Forecast | | |
|---|---|---|---|---|
| 0 | $Y(0)$ | No forecast is available for this period | | |
| 1 | $Y(1)$ | $\widehat{Y}(1)$ | $=$ | $Y(0)$ |
| 2 | $Y(2)$ | $\widehat{Y}(2)$ | $=$ | $\alpha Y(1) + \bar{\alpha}\widehat{Y}(1)$ |
| | | | $=$ | $\alpha Y(1) + \bar{\alpha}\widehat{Y}(1)$ |
| 3 | $Y(3)$ | $\widehat{Y}(3)$ | $=$ | $\alpha Y(2) + \bar{\alpha}\widehat{Y}(2)$ |
| | | | $=$ | $\alpha Y(2) + \bar{\alpha}[\alpha Y(1) + \bar{\alpha}Y(0)]$ |
| | | | $=$ | $\alpha Y(2) + \alpha\bar{\alpha}Y(1) + \bar{\alpha}^2 Y(0)$ |
| 4 | $Y(4)$ | $\widehat{Y}(4)$ | $=$ | $\alpha Y(3) + \bar{\alpha}\widehat{Y}(3)$ |
| | | | $=$ | $\alpha Y(3) + \alpha\bar{\alpha}[\alpha Y(2) + \alpha\bar{\alpha}Y(1) + \bar{\alpha}^2 Y(0)]$ |
| | | | $=$ | $\alpha Y(3) + \alpha\bar{\alpha}Y(2) + \alpha\bar{\alpha}^2 Y(1) + \bar{\alpha}^3 Y(0)$ |

the time series, exponential smoothing may have up to three parameters. The degree of smoothing is determined by the choice of these parameters.

Consider first the simplest case where the exponential smoothing process has a single parameter. This is used to smooth the level of the series, which is similar to the intercept parameter of a regression. It assumes there is no trend or seasonality in the series. Such a series is an example of a *stationary time series*, more about which will be described later along with autoregressive series. The smooth value of the series at time point $(t + 1)$ is expressed as a weighted average of the observed value at $t$ and forecasted value at $t$

$$\widehat{Y}(t + 1) = \alpha Y(t) + (1 - \alpha)\widehat{Y}(t), 0 < \alpha < 1$$

where $\alpha$ is the smoothing constant. Typically, $\alpha < 0.5$ with default value 0.2 is used by many software. With a relatively small value of $\alpha$, smoothing is more extensive as past observations get lower weight. With a relatively large value of $\alpha$, smoothing is less extensive as past values get lower weight. Table 12.3 clearly explains the smoothing procedure.

The degree of smoothing is controlled by the tuning parameter $\alpha$. When $\alpha > 0.5$, the observed value $Y(t - 1)$ in the previous time point gets higher weight in forecasting of the current value $\widehat{Y}(t)$. If $\alpha < 0.5$, the observed value $Y(t - 1)$ in the previous time point gets lower weight. For $\alpha = 0.5$, the weights are the same. The choice of $\alpha$ is determined by the degree of faith which the analyst puts in the observed value compared to the forecast value. It is clear that, as the series progresses, the observations in the distant past contribute less and less to the current forecast. Practically speaking, contributions of observations that are three or more periods away from the current value are often negligible.

**Example 12.4.** Figure 12.8 illustrates single exponential smoothing for the

**S&P 500 Index: Single Exponential Smoothing**

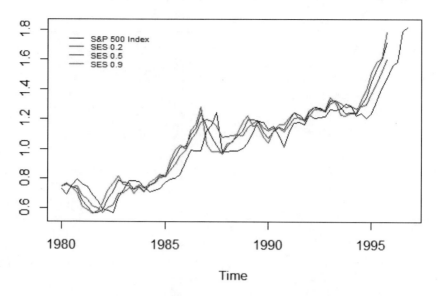

FIGURE 12.8
Single exponential smoothing.

S&P 500 Quarterly Index from 1980 to 2000 for three different tuning parameters.                                                                                          ‖

## 12.2.4   Double Exponential Smoothing

Double exponential smoothing is used when there is a trend but no seasonality. Essentially the method creates a forecast by combining exponentially smoothed estimates of the trend (slope of a straight line) and the level. Two different weights, or smoothing parameters, are used to update these two components at each time.

$$\alpha: \quad \text{smooths level; } 0 < \alpha < 1,$$
$$\beta: \quad \text{smooths trend; } 0 < \beta < 1.$$

The smoothed level is more or less equivalent to a simple exponential smoothing of the data values. The Smoothed trend is more or less equivalent to a simple exponential smoothing of the first differences. It has been mentioned that single exponential smoothing works well for stationary series. If a series has only trend and no seasonality, then the first differenced series is stationary. Hence the current level estimate is

$$L(t) = \alpha Y(t) + (1 - \alpha)(L(t-1) + T(t-1))$$

and the trend estimate is

$$T(t) = \beta(L(t) - L(t-1)) + (1 - \beta)T(t-1),$$

where $L(t)$ is the level of the series at time $t$, $T(t)$ is the trend of the series at time $t$ and $Y(t)$ is the observed value of the series at time $t$. Figure 12.9 shows various options for the $(\alpha, \beta)$ combination for the S&P 500 Index.

## 12.2.5 Triple Exponential Smoothing (Holt–Winters Model)

Triple exponential smoothing is used when the data contains both trend and seasonality. The method creates a forecast by combining exponentially smoothed estimates of the level, the trend and the seasonality. Hence three different weights, or smoothing parameters, are used to update these compo-

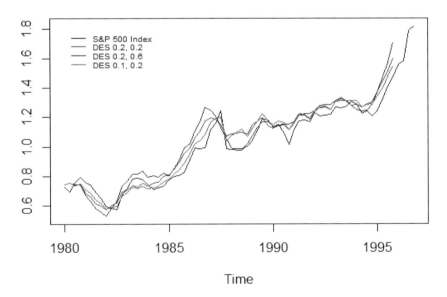

**FIGURE 12.9**
Double exponential smoothing.

nents at each time.

$$\alpha: \quad \text{smooths level; } 0 < \alpha < 1$$
$$\beta: \quad \text{smooths trend; } 0 < \beta < 1,$$
$$\gamma: \quad \text{smooths seasonality; } 0 < \gamma < 1.$$

There are two main Holt–Winters models, depending on the type of seasonality. For multiplicative seasonality, the current level estimate is

$$L(t) = \alpha[Y(t)/S(t-s)] + (1-\alpha)(L(t-1) + T(t-1)),$$

the trend estimate is

$$T(t) = \beta(L(t) - L(t-1)) + (1-\beta)T(t-1),$$

and the seasonality estimate is

$$S(t) = \gamma[Y(t)/L(t)] + (1-\gamma)S(t-s),$$

where $S(t)$ is the seasonal component with the period of seasonality being $s$. For additive seasonality, the above system of equations may be modified suitably.

**Example 12.5.** (Forecast Using Holt–Winters Model) Exponential smoothing models are typically used to forecast one period ahead. Single and double exponential smoothing models should not be used to forecast for more than one month or one quarter ahead. The Holt–Winters model, being a seasonal one, can be used to predict for the future year. As an example, the forecast of international air passengers using Holt–Winters model for the last year (1960) has been considered and compared with the observed values for accuracy. Passenger volumes from 1949 to 1959 have been used to fit the model. Fitting of a model includes determination of the time series model parameters, i.e., the values of $\alpha$, $\beta$ and $\gamma$ in this case. Comparing different sets of values of $\alpha$, $\beta$ and $\gamma$, it was found that $\alpha = 0.2$, $\beta = 0.35$ and $\gamma = 0.65$ give an acceptable fit. Time series model fitting may require many iterations and testing for optimum values of the parameters. At this stage such an elaborate exercise has not been performed. Figure 12.10 shows the actual and fitted value of passenger volume. Visual inspection indicates the fits are close. Predictions for 1960 Jan–Dec passenger volume and actuals are shown in Figure 12.11. This figure compares the observed and forecast passenger volume for one year only to emphasize the seasonal movement across months.

Table 12.4 shows the forecast passenger volume compared to the actual monthly passenger volume. The average absolute error is 4%, which is computed by taking the average of the absolute differences between the observed and forecast passenger volume as a percentage of actual passenger volume. ‖

So far we have discussed the decomposition method of time series analysis

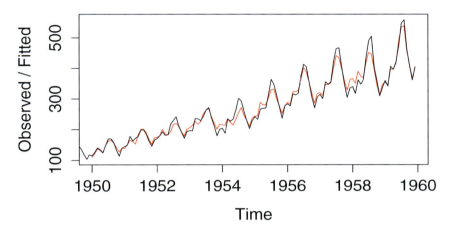

**FIGURE 12.10**
Holt–Winters model for international air passengers.

where different components of the series are identified, smoothed and then extrapolated. This approach is very intuitive and appealing to understand the nature of the series, especially when seasonality is considered. However, typically such an approach is not applicable to extrapolate into more than one period in the future. There are modifications to the decomposition method which are used to predict into multiple periods in the future, but those modifications are not always mathematically sound.

In the next section the regression-based alternative approach of analyzing and forecasting time series is considered.

**TABLE 12.4**
Actual and forecast passenger volume

| 1960 | Jan | Feb | Mar | Apr | May | Jun |
|---|---|---|---|---|---|---|
| Actual | 417 | 391 | 419 | 461 | 472 | 535 |
| Forecast | 420.26 | 406.67 | 467.62 | 461.76 | 484.91 | 547.06 |
| – | Jul | Aug | Sep | Oct | Nov | Dec |
| Actual | 622 | 606 | 508 | 461 | 390 | 432 |
| Forecast | 614.21 | 620.99 | 528.14 | 475.57 | 431.94 | 470.92 |

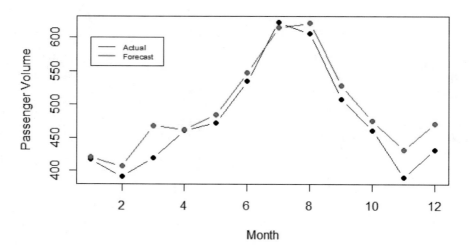

**FIGURE 12.11**
1960 passenger volume: actual and forecast

## 12.3    Autoregression Models

An alternative approach to modeling time series is to formulate the current observation as a function of observations for previous one or more time periods. Before autoregressive models are introduced properly, the concept of a weakly *stationary time series* is to be understood. At the level of our discussion, weak stationarity will suffice, and we will not require the properties of a strongly stationary series. A time series $Y(t)$ is weakly stationary if $E(Y(t))$ is same for all $t$, $Var(Y(t))$ is same for all $t$, $Cov(Y(t), Y(t - h))$ is same for all $t$, $t = 0, 1, 2, \ldots$ and $h$ denotes the lag from the current time. The last condition ensures that autocorrelation of a particular lag is the same across the whole series. The autocorrelation function (ACF) at a lag $h$ is defined as

$$\text{ACF}(h) = Cov(Y(t), Y(t - h)) / (sd(Y(t)) \times sd(Y(t - h)))$$

where $Y(t)$ and $Y(t-h)$ is the same series but running at a lag of $h$; $sd(Y(t))$ represents the standard deviation of the $Y(t)$ series. The ACF is always between $-1$ and $+1$. ACF(0) is the autocorrelation of the series $Y(t)$ with itself and its value is identically equal to 1. Autocorrelation makes sense only if a series is stationary. Note that the presence of trend and seasonality in a series violates the assumptions of stationarity. Intuitively, stationarity indicates that any segment of the time series under consideration is like any other segment of the same size.

**TABLE 12.5**

Example of lag series up to order 3

| | | Bond Yield | | |
|---|---|---|---|---|
| Year | Original Series | Lag 1 Series | Lag 2 Series | Lag 3 Series |
| 1900 | 3.3 | — | — | — |
| 1901 | 3.25 | 3.3 | — | — |
| 1902 | 3.3 | 3.25 | 3.3 | — |
| 1903 | 3.45 | 3.3 | 3.25 | 3.3 |
| 1904 | 3.6 | 3.45 | 3.3 | 3.25 |
| 1905 | 3.5 | 3.6 | 3.45 | 3.3 |
| 1906 | 3.55 | 3.5 | 3.6 | 3.45 |
| 1907 | 3.8 | 3.55 | 3.5 | 3.6 |
| 1908 | 3.95 | 3.8 | 3.55 | 3.5 |
| 1909 | 3.77 | 3.95 | 3.8 | 3.55 |
| 1910 | 3.8 | 3.77 | 3.95 | 3.8 |

**Example 12.6.** In Table 12.5 the first few observations from the Bond Yield series are shown along with lag series of different lags. Note that, for each unit increment in the lag order, the observations are shifted by one year. If the lag 1 series is denoted by $L_1$, then $L_1(t+1) = Y(t)$. If the lag 2 series is denoted by $L_2$, then $L_2(t+2) = L_1(t+1) = Y(t)$ for all $t$, except where observations are not available.

Autocorrelations are the correlation coefficients of the original and the lag series. ACF(1) is the correlation between $Y(t)$ and $L_1(t)$. It is computed over all pairs of observations that are available. Figure 12.12 shows autocorrelation of the Bond Yield series up to lag 50. It is extremely unlikely that any autoregression model will use lags of that order. This is to show how autocorrelation follows the natural course of a series. Autocorrelation patterns are useful in identifying parameters of ARIMA (autoregressive integrated moving average) models which are discussed later. The dashed lines on either side of the ACF $= 0$ line represent the thresholds of significance. Correlation values contained within the lines are deemed non-significant, i.e., not significantly different from 0. ‖

Another related concept is *partial autocorrelation*. The partial correlation is the correlation between two variables adjusting for another set of variables, which may have an effect on both. Suppose $Y$ denotes the response and $X_1$, $X_2$, and $X_3$ are predictors. The partial correlation between $Y$ and $X_3$ is the correlation between these two variables *after* adjusting for the effect that $(X_1, X_2)$ may have on both of them. Formally, the partial correlation between $Y$ and $X_3$ correlates the residuals from the regressions predicting $Y$ from $X_1$ and $X_2$, and $X_3$ from $X_1$ and $X_2$. The partial autocorrelation function at lag $h$, referred to as PACF($h$), is the correlation between $Y(t)$ and $Y(t-h)$, having adjusted for the intermediate lag variables $Y(t-1), Y(t-2), \ldots, Y(t-h+1)$.

## Autocorrelation of Bond Yield

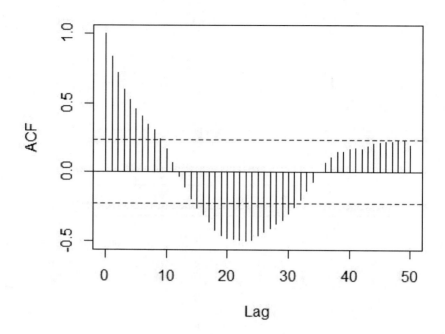

**FIGURE 12.12**
Example of autocorrelation.

As PACF($h$) is a correlation coefficient for each $h$, the PACF function is bounded between $-1$ and $+1$. Also from the definition it is clear that PACF(0) $= 1$ and PACF(1) $=$ ACF(1). The partial autocorrelations of the Bond Yield series are given in Figure 12.13. All the PACF values except for PACF(1) are within the dashed lines and hence non-significant. This indicates that the continued significance of ACF values up to very high lags, as seen in Figure 12.12, is an artifact of the dependence of $Y(t)$ on the intermediate values.

## 12.3.1    Introducing ARIMA($p, d, q$) Models

ARIMA models are the most general class of models for forecasting a time series. The full expansion of the name is Autoregressive (AR) Integrated (I) Moving Average (MA). A simpler form of this model is called the ARMA. ARIMA models depend on stationarity. A time series is stationary if its statis-

## Partial Autocorrelation of Bond Yield

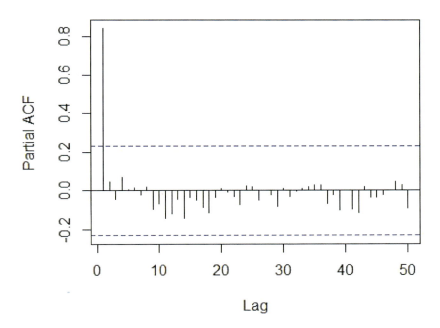

**FIGURE 12.13**
Example of partial autocorrelation.

tical properties are all constant over time, i.e., $E(Y(t)) = \mu$ and $Y(t) = \mu + \omega(t)$, where $\omega(t) \sim$ i.i.d $N(0, \sigma^2)$. A stationary series has no trend and its variation around the mean is constant over time. We have noted that, for a stationary series, autocorrelations of all orders remain constant over time but may depend on the order. A stationary time series is a combination of signal and white noise. An ARIMA model identifies and separates the signal from the noise and extrapolates into the future to obtain forecasts. The ARIMA forecasting equation for a stationary time series is a linear equation in which the predictors consist of lags of the dependent variable and/or lags of the forecast errors.

ARIMA models are characterized by three different parts, each of which contributes one parameter of the ARIMA model. The current observation is considered to be a function of the past $p$ observations. Each of the past observations, in turn, is a random variable and is associated with an error term. The current observation, in addition to the past observations, may be a function of past $q$ error terms. If the series is stationary, no differencing is required and an ARMA$(p, q)$ model is obtained. Notationally,

$$
\begin{aligned}
Y(t) \; = \; & \beta_1 Y(t-1) + \beta_2 Y(t-2) + \beta_3 Y(t-3) + \cdots + \beta_p Y(t-p) \\
& + \omega(t) + \alpha_1 \omega(t-1) + \alpha_2 \omega(t-2) + \cdots + \alpha_q \omega(t-q),
\end{aligned}
$$

where $\beta_1, \beta_2, \ldots, \beta_p$ are autoregressive parameters of various orders, $\omega(t)$ is white noise (i.i.d. random variables with with mean 0, variance $\sigma^2$) and $\alpha_1, \alpha_2, \ldots, \alpha_q$ are moving average parameters, $|\alpha_1| < 1$. Theoretically, $(p, q)$ may take any value, but usually values higher than 2 are not preferred in practical situations. When the series is not stationary, then, instead of directly modeling $Y(t)$, the first difference $\Delta^{(1)} Y(t) = Y(t) - Y(t-1)$ or the second difference $\Delta^{(2)} Y(t) = [Y(t) - Y(t-1)] - [Y(t-1) - Y(t-2)] = Y(t) - 2Y(t-1) + Y(t-2)$ of the series is modeled. Higher-order differencing may also be considered, but practically speaking, it is almost never done. The appropriate order of differencing is represented by the parameter $d$. When the differenced series is modeled, it is called an ARIMA model. If the left hand side of the equation contains the differenced value, appropriate adjustments are also required on the right hand side.

## 12.3.2    Special Cases of ARIMA$(p, d, q)$ Models

The realm of ARIMA models is practically all encompassing. The most difficult part of fitting an ARIMA model is to identify its order, i.e., the values of $p$, $d$ and $q$. The ACF and PACF are the only tools of identification unless any specific domain knowledge is applicable. Theoretically, it is easy to show how particular patterns of ACF and/or PACF lead to appropriate identification of $p$ and $q$, but in practice the patterns are so mixed up that often the order is determined by trial and error. Before ARIMA models are studied in their entire generality, a few common and simple ARIMA models are introduced.

### 12.3.2.1    Random Walk Model: ARIMA$(0, 1, 0)$

If a time series is not stationary, then the simplest model is an AR$(1)$ model, which, when the coefficient of the independent variable is 1, reduces to a random walk model

$$
\widehat{Y}(t) = \delta + Y(t-1), \tag{12.1}
$$

which is equivalent to $\Delta^{(1)} Y(t) = Y(t) - Y(t-1) = \delta$.

### 12.3.2.2    ARIMA$(1, 1, 0)$

This is the model where the errors of a random walk model are autocorrelated. To counteract the lack of independence, a differenced predictor is added to the model, so that

$$
\widehat{Y}(t) - Y(t-1) = \delta + \phi_1 (Y(t-1) - Y(t-2))
$$

or

$$\widehat{Y}(t) = \delta + Y(t-1) + \phi_1(Y(t-1) - Y(t-2)).$$

### 12.3.2.3 ARIMA$(0,1,1)$

Another possible corrective measure to adjust for autocorrelated error terms of a random walk model is to introduce a moving average term. In that case Equation (12.1) can be expressed as

$$\widehat{Y}(t) = \delta + \widehat{Y}(t-1) + \alpha_1\omega(t-1).$$

Replacing $\omega(t-1)$ by $Y(t-1) - \widehat{Y}(t-1)$, the equation reduces to

$$\widehat{Y}(t) = \delta + Y(t-1) - (1-\alpha_1)\omega(t-1).$$

A special case of ARIMA$(0, 1, 1)$ without the constant $\delta$ is equivalent to the single exponential smoothing model with the smoothing parameter equal to $(1-\alpha_1)$.

### 12.3.2.4 ARIMA$(0,2,1)$ and ARIMA$(0,2,2)$ Models without Constant

Both these models correspond to double exponential smoothing where the level and the trend components are being dealt with separately. The ARIMA$(0,2,2)$ model without a constant can be written as

$$\widehat{Y}(t) = 2Y(t-1) - Y(t-2) - (1-\alpha_1)\omega(t-1) - (1-\alpha_2)\omega(t-2).$$

The importance of these models will be clear when examples are used to fit ARIMA models. At this point we change gear and take a critical look at one of the most important properties of time series, namely, stationarity.

## 12.3.3 Stationary Time Series

A stationary time series is defined as $Y(t) = \mu + \omega(t)$, where $\omega(t) \sim$ i.i.d. $N(0, \sigma^2)$. A first-order *autoregressive* process AR(1) is defined as $Y(t) = \mu + \phi Y(t-1) + \omega(t)$. An alternative formulation of AR(1) is

$$
\begin{aligned}
Y(1) &= \mu + \phi Y(0) + \omega(1) \\
Y(2) &= \mu + \phi Y(1) + \omega(2) \\
&= \mu + \phi[\mu + \phi Y(0) + \omega(1)] + \omega(2) \\
&= \omega(2) + \phi\omega(1) + \phi^2 Y(0) + (1+\phi)\mu.
\end{aligned}
$$

After successive substitution, we get the general form of $Y(t)$ as

$$Y(t) = \omega(t) + K\mu + \phi\omega(t-1) + \phi^2\omega(t-2) + \phi^3\omega(t-3) + \cdots + \phi^t\omega(0) + \phi^t Y(0)$$

where $\phi$ is any positive number and $K$ is a function of $\phi$. A requirement for the AR(1) series to be stationary is to have $|\phi| < 1$. The mean and variance of AR(1) series are $E(Y(t)) = \mu/(1-\phi)$ and $Var(Y(t)) = \sigma^2/(1-\phi^2)$. Correlations between values at $h$ time periods apart are given by $ACF(h) = \phi^h$. If $|\phi| < 1$, then the observations farther apart would be less and less dependent. As a result, the ACF will converge to 0 with increasing lag.

The parameter $\phi$ may be interpreted as a shock factor in the series through which the dependence of current observations on previous observations is formulated. If $\phi > 1$, the effect of any random shock (event) propagates through the system and exerts an increasingly larger influence on the system. This assumption is not intuitively appealing. In a macro-economic set-up, the effect of economic depression or any other sudden event like a natural calamity or military operation has an immediate impact on the economy of a country. However, as time passes such particular events start losing their potency to control the economy. If $\phi = 1$, the impact of such shocks stay in the system; it is neither amplified nor does it die away. However, when $\phi < 1$, shocks to the system gradually die away. This is the most common scenario in any time series where events of the recent past have greater influence on the current status of the series than events in the remote past. Autocorrelations approaching zero also bear witness to that.

The AR(1) process is said to have a unit root if $\phi = 1$. An AR(1) process having one unit root can be made stationary by differencing only once. If a series contains more than one unit root, the series needs to be differenced more than once. If a non-stationary series must be differenced $d$ times to become stationary, then it is called an integrated series of order $d$, represented as $I(d)$. Notationally,

$$Y(t) \sim I(d) \implies \Delta^{(d)}Y(t) \sim I(0),$$

where $\Delta^{(d)}$ represents the $d$-th order of differencing. Many common economic and financial time series possess one unit root. Two examples of unit root series are given in Figure 12.14.

It is clear that after differencing both the Bond Yield and Clay Brick series once, the autocorrelation pattern shows significant changes. However, whether the first difference converts the series into a stationary one cannot be checked only visually. There are formal tests of significance to check for stationarity. One such test is the augmented Dickey–Fuller test for stationarity. This test checks whether the time series is non-stationary using an autoregressive model. The null hypothesis states $H_0$: Unit root exists for the series versus the alternative $H_A$: There is no unit root for the series. If the null hypothesis is rejected, then the series is stationary. In simple notation, the time series model may be written as $Y(t) = \phi Y(t-1) + \omega(t)$ and after taking the first difference $\Delta Y(t) = \psi Y(t-1) + \omega(t)$. Hence the null model is a random walk model with $\psi = 0$.

We will only consider the one-sided alternative $\psi < 0$, which indicates a stationary AR(1) process. For integrated models of higher order, the augmented Dickey–Fuller (ADF) test is used, which incorporates or augments $p$ lags of the dependent variable in the model.

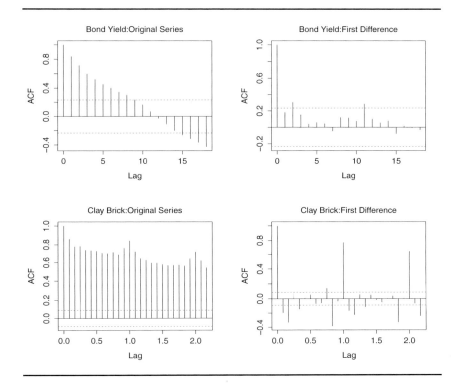

**FIGURE 12.14**
Examples of unit root series.

```
> adf.test(BondYield)

        Augmented Dickey-Fuller Test

data:  BondYield
Dickey-Fuller = 0.6607, Lag order = 4, p-value = 0.99
alternative hypothesis: stationary
```

The series Bond Yield does not reject the null hypothesis of non-stationarity. Even taking the first difference does not introduce stationarity here.

```
> adf.test(dBY1)

        Augmented Dickey-Fuller Test

data:  dBY1
```

```
Dickey-Fuller = -1.3917, Lag order = 4, p-value = 0.8226
alternative hypothesis: stationary
```

However, in the series of monthly Clay Brick Production, the first differenced series is a stationary series but the original series is not. Note that the original series shows strong 12-month seasonality, hence after every cycle of 12 months is completed a spike in autocorrelation is noticed. The ADF test ignores seasonal autocorrelation, more of which is discussed along with seasonal ARIMA models later. The lag order in the ADF test is a parameter that R enters as a default if no value is specified. For the Bond Yield series, lags are considered until the 4th order but for the Clay Brick series lags up to the 7th order are considered.

```
> adf.test(ClayBricks)

Augmented Dickey-Fuller Test

data:  ClayBricks
Dickey-Fuller = -3.2284, Lag order = 7, p-value = 0.0831
alternative hypothesis: stationary

> adf.test(dClBr1)

Augmented Dickey-Fuller Test

data:  dClBr1
Dickey-Fuller = -12.4751, Lag order = 7, p-value = 0.01
alternative hypothesis: stationary
```

In the case of the original series, the null hypothesis was not rejected at the 5% level, but it is rejected for the first differenced series. So it can be concluded that the Clay Brick series is a unit root series, and, after taking first difference, the series becomes stationary. An alternative test for the unit root is also available and is called the Phillips–Peron test. There is not much difference in the two methods and they will almost always give similar results. The main criticism against both tests is that both have low power if the series is stationary but has a root close to 1.

### 12.3.4 Identification of ARIMA$(p, d, q)$ Parameters

Non-seasonal ARIMA models are identified by the number of autoregression terms $p$ and the number of moving average terms $q$ in the model. The order of difference is important to convert any series into a stationary series. It must be stated at the outset that the three parameters of ARIMA models are not estimated by solving any closed form equations, nor are they approximated by using any numerical method. To determine a working ARIMA model requires looking at and trying out several sets of combinations of $p$, $d$ and $q$ and putting forward a model which gives acceptable fit over the course of the time series. The autocorrelations and partial autocorrelations of different orders provide indications for the possible values of $p$ and $q$. Determination of order of differencing is a different process and will be discussed later. For now it is assumed that the series has already been made stationary.

The plot of PACF helps in identifying AR terms and the plot of ACF helps in identifying MA terms, albeit tentatively. Recall that PACF is the correlation between $Y(t)$ and $Y(t-h)$ after adjusting for all lower-order lags. If $Y(t)$ and $Y(t-1)$ have significant correlation, then $Y(t-1)$ and $Y(t-2)$ will also have significant correlation, and hence $Y(t)$ and $Y(t-2)$ will also have significant correlation, since both are correlated to $Y(t-1)$. If the effect of $Y(t-1)$ is eliminated from both $Y(t)$ and $Y(t-2)$ and the correlation vanishes, then it indicates first-order dependence, but not of any higher order. All higher-order correlations are satisfactorily explained by the lag 1 correlation.

EuStockMarkets is a time series dataset available in the R datasets library. It contains the daily closing prices of major European stock indices from 1991 to 1999, e.g., Germany DAX (Ibis), Switzerland SMI, France CAC and UK FTSE. The data is sampled in business time, i.e., weekends and holidays are omitted.

Since this is daily data, lags of very high order are considered (see Figure 12.15). Even after 100 lags, the autocorrelation remains extremely high. However, when partial autocorrelation is considered, only PACF(1) is significant, but all the others drop off sharply. This indicates the presence of a single autoregression term in the series. The order of the lag beyond which the PACF has a sharp fall is the number of autoregression terms in the model. For AR models, ACF tapers to zero. Theoretically, it can be shown that, if the model is AR(1), then $\mathrm{ACF}(h) = [\rho(1)]^h$. For an AR(2) model, autocorrelation converges to 0 sinusoidally.

An indication of the number of MA terms to be included in the model is found from the number of significant ACF terms in a stationary series. Another indication of the presence of an MA term is a negative lag 1 autocorrelation in a differenced series.

The order of differencing also may be determined through ACF and PACF. Heuristically, it is quite clear that, if a series shows positive autocorrelation up to a high order, a higher order of differencing is required. The correct order of differencing is the lowest number of differencing that converts a series into a stationary series. Differencing, however, tends to introduce negative correlations. If the original series shows positive autocorrelations to a very high order, it is possible that taking the first difference introduces negative correlation. If that happens, no more differencing is necessary, since that may further drive the autocorrelations into the negative zone. Consider the time series based on daily returns of the Google stock from 20 Aug 2004 to 13 Sep 2006. This data is available in R library TSA (see Figure 12.16).

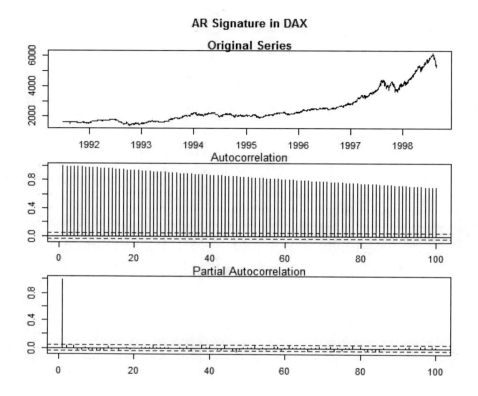

**FIGURE 12.15**
Examples of AR series.

This is a stationary series as interpreted from the result of Augmented Dickey–Fuller test.

```
> adf.test(google)

Augmented Dickey-Fuller Test

data:  google
Dickey-Fuller = -7.982, Lag order = 8, p-value = 0.01
alternative hypothesis: stationary
```

Figure 12.17 illustrates how the autocorrelation function of a stationary series, the first difference of a stationary series and the second difference of a stationary series behave. In the original series, none of the autocorrelations

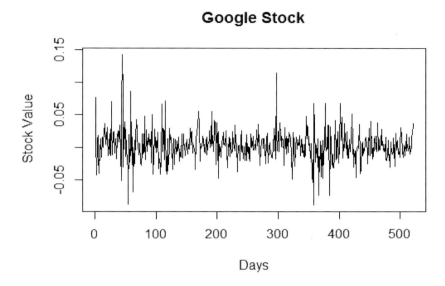

**FIGURE 12.16**
Example of a stationary series.

are significant, except very high order ones. However, artificial significance in autocorrelation is introduced after taking the difference of this series.

Consider now the example of the first difference of the series DAX. Recall that, in the original series, all autocorrelations to a very high order were significant.

The first autocorrelation of the first differenced series has become so small that it does not even show up in Figure 12.18; its value is 0.001, which is practically zero. If the lag 1 autocorrelation is extremely small or even negative, then no more differencing is required. In fact, a large negative value of ACF(1) indicates over-differencing.

**Example 12.7.** (Market Share of Crest Toothpaste) The weekly market share of Crest toothpaste is available for 6 years from 1958 to 1963. A close inspection of the series indicates that, in the middle of the third year, market share improved week over week and it seems as if there is a break in the series. Since it is a realistic assumption that future performance is dependent more on the recent past compared to the distant past, it is a good practice to consider only the later part of a series if there is a distinct change in the pattern. We will concentrate on the market share from the middle of 1960 to 1963.

Figure 12.19 shows significant autocorrelation up to high orders; the first

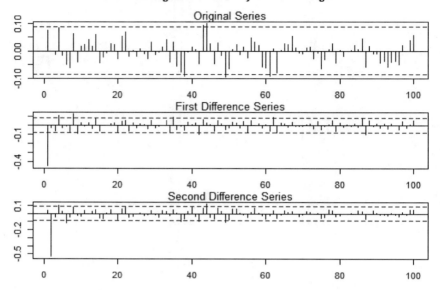

**FIGURE 12.17**
The ACF in a differenced stationary series.

few of them are not eliminated even when partial autocorrelations are considered. However, the augmented Dickey–Fuller test rejects the null hypothesis of non-stationarity with a p-value of 4%.

```
> adf.test(ToothpasteL)

Augmented Dickey-Fuller Test

data:  ToothpasteL
Dickey-Fuller = -3.5183, Lag order = 5, p-value = 0.04329
alternative hypothesis: stationary
```

Therefore it seems one or more AR terms will be needed in the series. Figure 12.20 shows the autocorrelations of the first and second differences of the toothpaste series. In both cases the lag 1 ACF is significantly negative. This may indicate the presence of at least one MA term. ‖

In any model building exercise, the simplicity of the model is an important issue, and time series is not an exception. In general, it may not be advisable to include high-order AR and MA terms in a single model. It is possible that

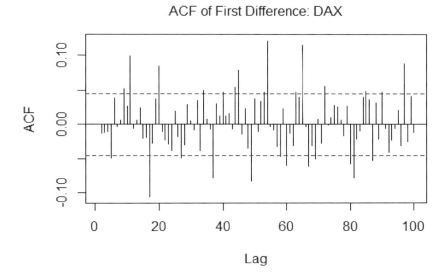

**FIGURE 12.18**
The ACF in a differenced non-stationary series.

AR and MA terms in a model cancel each others effects. Instead of multiple AR and multiple MA terms in a model, it may be better to use a fewer number of AR or a fewer number of MA terms in the model.

### 12.3.4.1 Seasonal ARIMA Models

Seasonal ARIMA models are more complex models with adjustment for seasonality, which is taken care of by performing seasonal differencing. Just as first differencing eliminates linear trend and second differencing eliminates quadratic trend, seasonal difference will eliminate additive seasonality, but not multiplicative seasonality. High-order seasonal differencing is not used in practice. If there is monthly seasonality in the data, then a difference of $Y(t)$ and $Y(t-12)$ defines first-order seasonal difference. If there is quarterly seasonality, then the first-order seasonal difference is given by $Y(t) - Y(t-4)$.

The most general form of the seasonal ARIMA model is ARIMA$(p, d, q) \times$ ARIMA$(P, D, Q)$ where $P$, $D$ and $Q$ define the seasonal AR components, the seasonal difference and the seasonal MA terms, respectively, in addition to the ARIMA$(p, d, q)$ model.

**Example 12.8.** (International Air Passengers) The original series (Figure 12.2) shows pronounced trend and seasonality.

The ACF has very high positive values to large orders and then it assumes significantly high negative values (see Figure 12.21). It is also clear that, at every multiple of 12, the ACF swings higher than the neighboring values. This is a typical pattern of ACF when the series contains a pronounced seasonal effect.

It is interesting to see the effect of differencing on this series. The first difference has effectively removed the linear trend and only the seasonal pattern remains. It is also clear that the seasonal fluctuations are more pronounced in recent times than in the past. The differenced series is not stationary, with the ADF test having a p-value of 16%. For a seasonal series, the lag order of the ADF test, the parameter $k$, is important to control and must always have a number higher than the seasonal periodicity. The first seasonal differenced series has been able to remove the seasonality but the series is still not stationary, the p-value of the ADF test being 25%. The last panel is the first seasonal differenced series, of the first differenced series, which now has completely removed the trend and seasonality. This is a stationary series with the p-value of the ADF test being 4%.

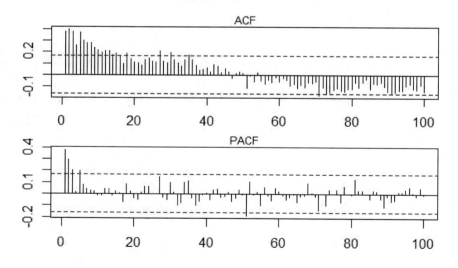

**FIGURE 12.19**
The ACF and the PACF of a non-stationary series.

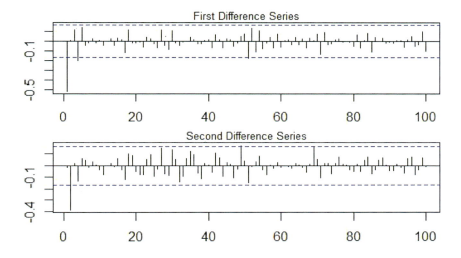

**Autocorrelation of Crest Toothpaste Market Share**

**FIGURE 12.20**
The ACF and the PACF of a differenced non-stationary series.

It is interesting to note the changes in ACF after differencing the series in various orders. The first differenced series shows high ACF values at 12-month intervals, showing periodicity. The first seasonal differenced series shows an oscillating pattern in ACF, possibly indicating one or more MA terms in the series. But the twice differenced series shows a random pattern. ‖

## 12.3.5  Fitting of ARIMA Models

To fit an ARIMA model, the first step is to make the data stationary. Non-seasonal and seasonal differencing of various orders attempt to make the series stationary. Other methods that might make a series stationary are transforming the series to stabilize its variance, e.g., the logarithmic transformation, or, in the case of price, by deflating them using price indices. Patterns of ACF and PACF are studied and several working models with AR and MA terms are considered. A close inspection of residual diagnostics helps to determine whether the residuals are white noise or whether they show any pattern. Patterns in ACF and PACF of the residual series, or lack thereof, indicate the

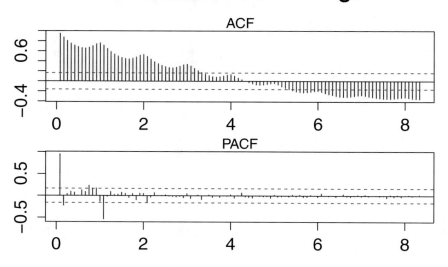

**FIGURE 12.21**
The ACF and the PACF of a seasonal ARIMA model.

nature of the residuals. In the most general form, an ARIMA model states

$$\widehat{y}(t) \;=\; \beta_1 y(t-1) + \beta_2 y(t-2) + \beta_3 y(t-3) + \ldots + \beta_p y(t-p)$$
$$+\omega(t) + \alpha_1 \omega(t-1) + \alpha_2 \omega(t-2) + \ldots + \alpha_q \omega(t-q),$$

where

$$y(t) \;=\; Y(t) \text{ if order of differencing is zero,}$$
$$y(t) \;=\; Y(t) - Y(t-1) \text{ if order of differencing is one,}$$
$$y(t) \;=\; Y(t) - 2Y(t-1) + Y(t-2) \text{ if order of differencing is two.}$$

Very rarely, differencing is done more than twice. But differencing may also involve both seasonal and non-seasonal differencing and then the above correspondence may be more complicated. The final fitted model is in terms of the original observations, not in terms of the differenced ones. As a rule, it is not recommended to have $p + q > 2$. Also, practitioners prefer to use either only AR or only MA terms in the model to keep matters simple.

    The simplest way to fit an ARIMA model in R is to use the `auto.arima()` function in the `forecast` library of R. This function finds an optimum ARIMA model among specified options for all parameters. The criteria used are

Akaike's information criterion (AIC) and Bayes information criterion (BIC), and the model that minimizes one or the other is recommended. AIC and BIC are both calculated based on the likelihood function of the observations (see Section 8.8.1). To fit a model according to AR and MA signature patterns, the function `Arima()` is used. As in the case of regression, residuals are examined for probable patterns, and, if a pattern is detected, then an alternative model needs to be proposed. There are tests available to see if the model residuals represent white noise or not.

**Example 12.9.** (DAX series)

```
> auto.arima(DAX)
Series: DAX
ARIMA(2,2,1)

Coefficients:
          ar1       ar2      ma1
      -0.0088   -0.0253   -0.9899
s.e.   0.0233    0.0233    0.0035

sigma^2 estimated as 1065:  log likelihood=-9112.25
AIC=18218.04    AICc=18218.06    BIC=18240.15
```

The function `auto.arima()` chooses an ARIMA(2, 2, 1) model for the DAX series. It recommends that the series be differenced twice to achieve stationarity and two AR terms and one MA term are included. The function also gives values of the AR and MA coefficients along with their standard errors. According to the discussions in the previous sections, the conclusion reached was that one differencing is enough for this series as the ACF of the first differenced series is almost negligible. Hence an alternative model was also fit through `Arima()`.

```
> Arima(DAX1, order=c(1,1,0), include.drift=T)
Series: DAX1
ARIMA(1,1,0) with drift

Coefficients:
          ar1     drift
       0.0008   2.0697
s.e.   0.0233   0.7543

sigma^2 estimated as 1056:  log likelihood=-9103.87
AIC=18213.75    AICc=18213.76    BIC=18230.33
```

An inspection of the autocorrelations (not presented here) reveal that the autocorrelations of lower orders are mostly very small (in magnitude). However, it cannot be claimed, at least not just by a cursory inspection of the autocorrelation values, that all correlations up to a certain order are insignificant. Hence a statistical test is required that simultaneously tests for overall significance.

To test whether the residuals are white noise and no other pattern is hidden among them, a special test is applied where groups of autocorrelations are examined together to see whether they are all non-significant. The Box–Ljung test defines $H_0$: The data are distributed independently versus the alternative hypothesis $H_A$: The data are not independent. The test statistic is

$$Q = n(n+2) \sum_{k=1}^{h} \frac{\rho_k^2}{(n-k)},$$

which follows a $\chi^2$ distribution. In the above statistics, $n$ denotes the sample size or size of the series, $\rho_k$ is the autocorrelation at the $k$-th lag and it simultaneously tests whether all autocorrelations up to lag $h$ are non-significant. When this statistic is used to test the randomness of an ARIMA model its degrees of freedom takes into account the parameters of the ARIMA model.

```
> Box.test(DAX$residuals, lag=10, type="Ljung")

Box-Ljung test

data:  DAX$residuals
X-squared = 14.7362, df = 10, p-value = 0.142

> Box.test(DAX$residuals, lag=15, type="Ljung")

Box-Ljung test

data:  DAX$residuals
X-squared = 35.2352, df = 15, p-value = 0.002276
```

The above results indicate that correlations up to the tenth lag may be independent but not beyond that. Autocorrelations of the 15th lag and higher order are not independent. But in practice it will be a rare case when the residuals are truly independent to a very large order. Hence we may conclude that the working independence of residuals has been established. The ARIMA$(1,1,0)$ model may therefore be proposed for the DAX data.    ‖

**Example 12.10.** (Google Stock) Recall that a time series plot, autocorrelations as well as the augmented Dickey–Fuller test indicates that the Google stock series is stationary. However, the `auto.arima()` function fails to recognize the stationarity, even though the Box–Ljung test shows a high p-value for a high lag order.

```
> auto.arima(google)
Series: google
ARIMA(2,0,2) with non-zero mean

Coefficients:
          ar1      ar2     ma1     ma2   intercept
      -0.4542  -0.8042  0.5324  0.8203     0.0027
s.e.   0.1755   0.1045  0.1752  0.1033     0.0011

sigma^2 estimated as 0.000559:  log likelihood=1211.69
AIC=-2411.38    AICc=-2411.21    BIC=-2385.84

> Box.test(google, lag=50, type="Ljung")

Box-Ljung test

data:  google
X-squared = 59.1318, df = 50, p-value = 0.1765
```

‖

It is always important not to depend blindly on an automatic procedure. Each dataset is different, with varied nuances. A good practice is to use `auto.arima()` to obtain a starting point for fitting an ARIMA model. But the recommended model must always be confirmed by looking at autocorrelation and partial autocorrelation patterns and finally by testing whether the residuals are reduced to white noise.

**Example 12.11.** (Bond Yield) Recall that the first differencing of the series failed to make it stationary. Autocorrelation of the first differenced series showed significant autocorrelations of lower order. Several models were considered and the following results were obtained.

```
> Arima(BondYield, order=c(0,1,1))
Series: BondYield
ARIMA(0,1,1)

Coefficients:
         ma1
      0.1471
s.e.  0.0999

sigma^2 estimated as 0.09022:  log likelihood=-15.14
AIC=34.29    AICc=34.47    BIC=38.78

Box-Ljung test P-value = 8%
```

```
> Arima(BondYield, order=c(0,1,2))
Series: BondYield
ARIMA(0,1,2)

Coefficients:
         ma1      ma2
      0.1098   0.3426
s.e.  0.1171   0.1088

sigma^2 estimated as 0.08072:  log likelihood=-11.37
AIC=28.73    AICc=29.1    BIC=35.48

Box-Ljung test P-value = 49%


> Arima(BondYield, order=c(0,2,1))
Series: BondYield
ARIMA(0,2,1)

Coefficients:
         ma1
      -0.7638
s.e.   0.0973

sigma^2 estimated as 0.07902:  log likelihood=-10.78
AIC=25.56    AICc=25.75    BIC=30.03

Box-Ljung test P-value = 21%
```

It seems that ARIMA$(0,2,1)$ fits the data reasonably well given that it has the lowest AIC value and the residuals have also reached stationarity.  ‖

**Example 12.12.** (International Air Passengers) It has been established that this is a multiplicative seasonal series. The first difference of the first seasonally differenced series is stationary.

```
> auto.arima(IntPax)
Series: IntPax
ARIMA(0,1,1)(0,1,0)[12]

Coefficients:
          ma1
       -0.3184
s.e.    0.0877
```

```
sigma^2 estimated as 137.3:   log likelihood=-508.32
AIC=1020.64    AICc=1020.73    BIC=1026.39


> Box.test(I1$residuals, lag=20, type="Ljung")


Box-Ljung test


data:  I1$residuals
X-squared = 24.2365, df = 20, p-value = 0.2322
```

The model recommended by `auto.arima()` includes an MA term only. Formally, the model can be written as $\text{ARIMA}(0,1,1) \times (0,1,0)$, seasonality at 12 months. The residuals do not significantly deviate from white noise. ‖

The fitting of a time series model is as much an art as a science. These models are complex and sometimes it is difficult to satisfy all assumptions that are required for model fitting.

## 12.4    Forecasting Time Series Data

The main objective of studying time series data is to forecast the series for the future. Unlike the coefficient of multiple determination in regression model building, there is no absolute measure of time series model fit. The criteria mentioned above (AIC, AICc, BIC, etc.) are measures of the relative performance of a model. Among the candidate models, the one that shows the least values of these statistics is the recommended model. But that does not give any indication of how close to the actual observations the fitted observations lie. It is possible that none of the candidate models describes the data well and the model chosen is merely the closest available. Recall also that, as in regression, the best fitted model may not always have the best predictive power.

It is important that the accuracy of a forecast is determined using genuine forecasts, i.e., the accuracy is examined comparing the forecasts with actual observations. However, in the case of time series, actual observations are in the future and are not currently available! Usually, therefore a portion of the current series is separated out as a validation sample and the model is fitted to the rest of the data. The validation part of the time series is always the most recent part of the series; random separation of training and test samples is not possible for obvious reasons. The size of the test data need not be very large but it must cover at least the forecast horizon, i.e., for how far ahead in the future the forecasting model is going to be used. There may be other restrictions on the training and test sample depending on the total length of the series, the seasonality and the parameters of the fitted ARIMA model.

**TABLE 12.6**
Definitions of accuracy measures

| Accuracy Measure | Statistic |
|---|---|
| Mean Absolute Error (MAE) | $\mathrm{Mean}(|e(t)|)$ |
| Root Mean Squared Error (RMSE) | $\sqrt{\mathrm{Mean}(e^2(t))}$ |
| Mean Absolute Percentage Error (MAPE) | $\mathrm{Mean}\,(|e(t)/Y(t)|) \times 100$ |
| Median Absolute Percentage Error (MdAPE) | $\mathrm{Median}\,(|e(t)/(Y(t)|) \times 100$ |
| Mean Absolute Scaled Error (MASE) | Has a different format for seasonal and non-seasonal series |

Note that time series patterns are essentially nonlinear, i.e., they show upward and downward movements at irregular intervals. Depending on whether the training and test samples separate out close to the point of inflection, the predictions may look worse than they actually are. In time series all forecasts are one-step-ahead forecasts. The immediately previous value is used in forecasting the current value. Hence the separation of training sample (the older values) and the test samples (the recent values) is important. To reduce its effect, often multiple test samples are used in a series to determine forecast accuracy. In this scheme, each test sample contains one value less than the previous test sample; the chronological first value of the test sample in the previous stage is moved to the training sample in the next stage. Having multiple validation samples also works for a short time series. This method is known as the rolling forecasting validation technique.

Suppose the length of the test sample is $T$. Let the difference between the actual observation $Y(t)$ and the forecast observation $\hat{Y}(t)$ be denoted by $e(t)$. Some of the most common measures of accuracy are given in Table 12.6.

Mean absolute error (MAE) and root mean squared error are scale dependent. They have the same scale and unit as the original data series. RMSE is possibly the more commonly applied of these two. For comparing different forecast methods on the same dataset, both have wide application. MAPE and MdAPE are scale independent since both are expressed as a percentage of the original observations. These are to compare performance of different forecasts on the same series as well as forecasting methods on different series. MAPE has the disadvantage of being artificially high if the original observation is very small; if it is zero, then MAPE is infinity. This drawback can be avoided by using the median value of the percent errors instead of the arithmetic average.

Hyndman and Koehler (2006) suggested a measure of forecast accuracy

**TABLE 12.7**
Comparison of Holt–Winters and ARIMA forecasts

| 1960 | Jan | Feb | Mar | Apr | May | Jun |
|---|---|---|---|---|---|---|
| Actual | 417 | 391 | 419 | 461 | 472 | 535 |
| HW Forecast | 420.26 | 406.67 | 467.62 | 461.76 | 484.91 | 547.06 |
| ARIMA Forecast | 424.29 | 406.29 | 470.29 | 460.29 | 484.29 | 536.29 |
| – | Jul | Aug | Sep | Oct | Nov | Dec |
| Actual | 622 | 606 | 508 | 461 | 390 | 432 |
| HW Forecast | 614.21 | 620.99 | 528.14 | 475.57 | 431.94 | 470.92 |
| ARIMA Forecast | 612.29 | 623.29 | 527.29 | 471.29 | 426.29 | 469.29 |

normalized by closeness of fitted and observed series for the training data. They defined two quantities:

$$Q_{NS} \text{ for non-seasonal series} = 1/(N-1) \sum_{j=2}^{N} |Y(t) - Y(t-1)|$$

$$Q_S \text{ for seasonal series } = 1/(N-m) \sum_{j=m+1}^{N} |Y(t) - Y(t-m)|.$$

The Mean Absolute Scaled Error is the average of $|e(t)/Q|$. The normalizing factor $Q$ is computed by taking a naive one-step-ahead forecast and comparing it with the observed value. But other forecasts can also be accommodated in computing $Q$. This factor also takes into account the volatility of the series. If a series is highly volatile, it is expected that forecasting errors are large and a normalizing factor is useful for comparison.

**Example 12.13.** (Forecast Comparison – International Air Passengers) The series of International Airlines Passengers (1949 Jan–1960 Dec) is a seasonal series with increasing trend and multiplicative seasonality. The two possible models are the Holt–Winters exponential model and the seasonal ARIMA. To compare the performance of these two models, the training set is taken to be the series from 1949 Jan–1959 Dec and the test set is defined to be 1960 Jan– Dec. Both models for this data have been developed in the previous sections of this chapter and the same parameters are used here for forecasting. The forecast values are given in Figure 12.22 along with a prediction limit of 95%. Table 12.7 shows the actual observations for 1960 Jan–Dec passenger volume and two different forecasts.

A comparison of the accuracy measures is presented in Table 12.8. To compute the Mean Absolute Scaled Error (MASE), $Q$ is computed from the actual observations as $Q = 30.45$. Both models seem to perform almost equally well.

# Prediction for International Air Passengers

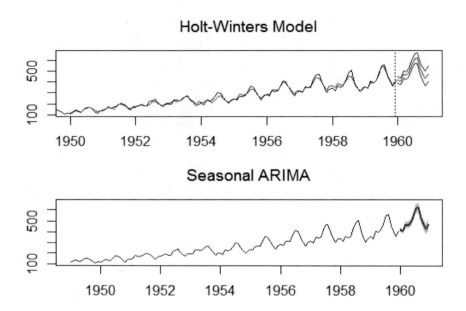

**FIGURE 12.22**
Comparative study of Holt–Winters and ARIMA forecasts.

It is interesting to note that, except for MASE, ARIMA is doing marginally better.

**TABLE 12.8**
Comparison of accuracy measures

|       | Holt–Winters Model | ARIMA Model |
| ----- | ------------------ | ----------- |
| MAE   | 19.30              | 18.19       |
| RMSE  | 24.34              | 23.54       |
| MAPE  | 4.35%              | 4.11%       |
| MdAPE | 2.95%              | 2.73%       |
| MASE  | 0.61               | 0.68        |

**Example 12.14.** (Clay Brick Production) It has been seen before that the series Clay Brick is not reduced to stationarity after taking the first difference. This is natural, as there is a 12-month seasonality present in the data. However, `auto.arima()` fails to recognize that only the first differencing does not make the series stationary.

```
> auto.arima(ClayBricks)
Series: ClayBricks
ARIMA(2,1,1)(1,0,0)[12]

Coefficients:
          ar1      ar2     ma1    sar1
      -0.9165  -0.4972  0.4426  0.818
s.e.   0.0872   0.0438  0.0977  0.025

sigma^2 estimated as 105.8:  log likelihood=-1788.08
AIC=3586.16    AICc=3586.29    BIC=3606.98

> Box.test(C1$residuals, lag=20, type="Ljung")

Box-Ljung test

data:  C1$residuals
X-squared = 69.7878, df = 20, p-value = 1.973e-07
```

The residuals show significant deviation from randomness even for relatively lower lag. Hence this model cannot be considered as a working model. Taking this model as a starting point though, a seasonal difference is introduced in the model.

```
> Arima(ClayBricks, order=c(2, 1, 1), seasonal=c(1, 1, 0))
Series: ClayBricks
ARIMA(2,1,1)(1,1,0)[12]

Coefficients:
          ar1      ar2     ma1     sar1
      -0.9855  -0.5595  0.4272  -0.3644
s.e.   0.0700   0.0421  0.0795   0.0440

sigma^2 estimated as 101.9:  log likelihood=-1728.55
AIC=3467.1    AICc=3467.23    BIC=3487.79

Box.test(C2$residuals, lag=10, type="Ljung")

Box-Ljung test
```

```
data:  C2$residuals
X-squared = 33.3147, df = 10, p-value = 0.0002411
```

Seasonal difference also failed to make the series stationary. Other model building strategies were also applied, e.g., additive and multiplicative decomposition. But the residuals coming out of decomposition were not stationary either.

So the question remains that, when standard techniques fail, what should be the best way to forecast for a time series. Time series forecasting is helped by domain knowledge. Since in most cases no other information, e.g., information on other predictors is available, forecasting of a long series is difficult. Closer inspection of Clay Brick Production indicates possible changes in the data generating mechanism, even though 12-month seasonality continues. The series can definitely be divided into two distinct parts – from the beginning of the series to 1975, when the series shows a consistent upward trend, and in the more recent years it shows a cyclic pattern. Also, for an overlong series it may not be prudent to use information from the distant long past, since series values for next year will hardly depend on what happened 30 years ago.

To arrive at an acceptable forecasting model, therefore, only 1990 January onwards values are chosen. The last year observed, 1995, has values for Jan–Aug only. The training set is defined to be 1990–1993 and performance of the model is tested on 1994 Jan–1995 Aug. Three alternative models are chosen for forecasting. The first model, $ARIMA(2,0,2) \times (1,0,0)$, is recommended by the `auto.arima()` procedure. However, it was felt that the series being seasonal, a seasonal differencing is in order. Taking the first model as a start, one seasonal difference parameter has been added. The third model is a simplification of the second model, with one less AR parameter. The information criteria and results of Box–Ljung test are given below.

| ARIMA Models | AIC | p-value for Box–Ljung Test |
|---|---|---|
| $(2,0,2) \times (1,0,0)$ | 402.17 | 0.17 |
| $(2,0,2) \times (1,1,0)$ | 298.13 | 0.61 |
| $(2,1,1) \times (1,1,0)$ | 283.63 | 0.80 |

The second and third models are definitely preferable to the first one as far as the model fitting criteria go. The following table contains the values of measures of forecasting accuracy. Note that the forecast is made for 2 years but the accuracy measures are computed over 20 observations.

| | ARIMA | | |
|---|---|---|---|
| | $(2,0,2) \times (1,0,0)$ | $(2,0,2) \times (1,1,0)$ | $(2,1,1) \times (1,1,0)$ |
| MAE | 7.54 | 8.64 | 8.32 |
| RMSE | 9.27 | 10.93 | 11.69 |
| MAPE | 5.18% | 5.88% | 5.71% |
| MdAPE | 4.23% | 4.91% | 3.65% |
| MASE | 0.49 | 0.56 | 0.54 |

The third model ARIMA$(2,1,1) \times (1,1,0)$ has the least AIC value but does not always have the lowest value of all measures of forecast accuracy. See Figure 12.23 for alternative forecasts for the Clay Brick Production series. ‖

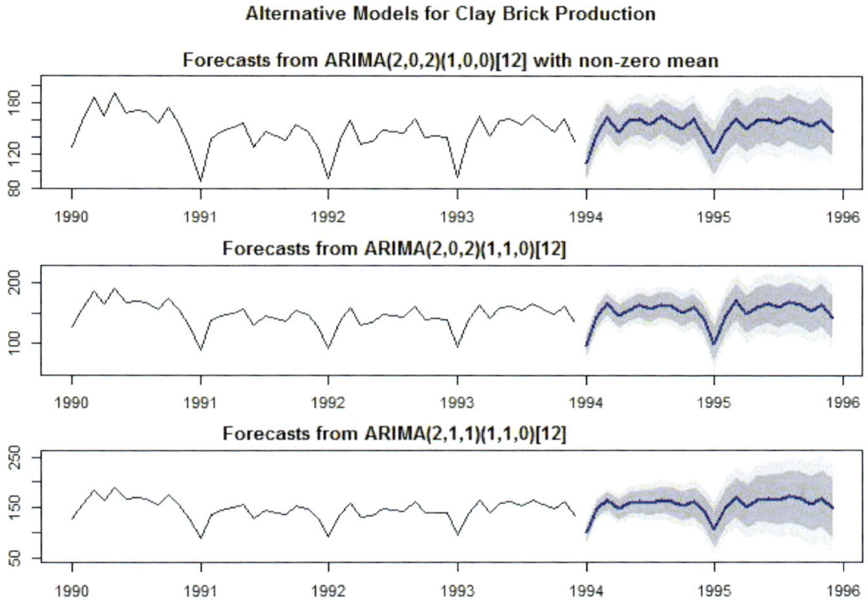

**FIGURE 12.23**
Examples of ARIMA Forecasts.

**Case Study 12.1.** (Australian Wine Sales). The data on Australian wine sales have been abstracted from the following site given by the URL https://datamarket.com/data/set/22q2/monthly-australian-wine-sales-thousands-of-litres-by-wine-makers-in-bottles-1-litre#!ds=22q2!2ekt&display=line.

The series has has six components. Monthly sales (in 1000 litres) are abstracted from 1980 Jan to 1995 July for six different types of wine, namely Dry White, Fortified, Red, Rose, Sparkling and Sweet White. This could have been taken as a multivariate time series data, but we treat each series as a univariate series. Interest lies in predicting each series accurately as well as predicting the sum total.

For 1994 July and Aug sales figures are missing for Rose wine. Time series cannot be modeled if there is any missing data somewhere in the middle. This missing observation needs to be imputed. The most natural method of imputation is replacing the missing observation by an average. It can be the average of two neighboring observations, or it can be the average of similar observations. For example, if it is a daily time series like passenger volume between Kolkata and Delhi airports, it is possible that the volume depends on the day of the week. In that case replacing a Wednesday's missing value by taking the average of Tuesday and Thursday might be incorrect; the average of passenger volumes on other Wednesdays of the same month might provide a better replacement. In the present scenario of wine sales the variations among individual values seem insignificant. Hence both missing values are replaced by average of the two closest observations.

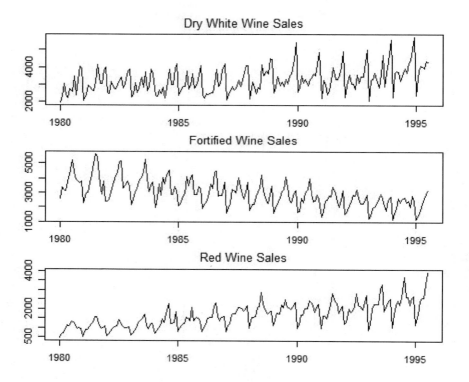

**FIGURE 12.24**
Wine sales: Example 1.

From Figures 12.24 and 12.25 is it quite clear that each type of wine has a distinct pattern. Dry white wine and sparkling wine sales do not show perceptible trends. Fortified wine and rose wine sales show downward trends while red wine sales show an upward trend. Sales of sweet white wine shows the most randomness and the early part of the series should be discounted.

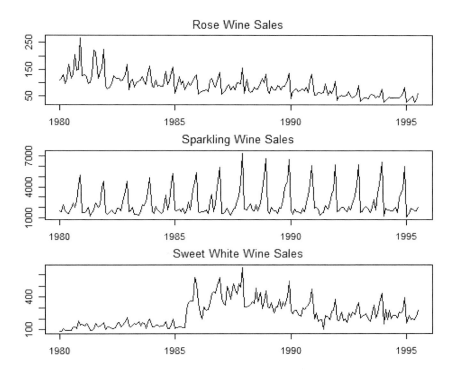

**FIGURE 12.25**
Wine sales: Example 2.

For all the analysis and model building undertaken henceforth Sweet white wine series starts from 1986 January.

All models for all series are developed using data till 1993 Dec. Data from 1994 Jan to 1995 July are used to test forecast accuracy of the proposed models. Once a model is chosen then the full series has been used to predict sales for the next two years. Autocorrelations and partial autocorrelations of the wine sales series show considerably different patterns (Figures 12.26 and 12.27). These observations are used to propose appropriate models for each sales series. The only consistent observation among sales series of all the six types of wine is that there is a 12 month periodicity.

All six series have been decomposed to understand the trend and seasonality patterns better.

The trends for Fortified, Rose and Sweet White wines are linear and downward while for Red wine the trend is linear and upward. For the other two types of wine the trend pattern may also indicate cyclicality.

Except for Fortified and Red wine, all others show higher sales towards end

**Autocorrelation for Wine Series**

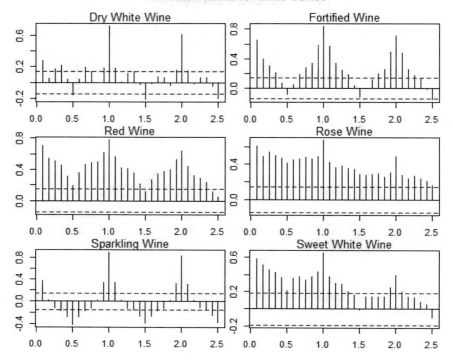

**FIGURE 12.26**
Example of autocorrelation.

of the year. This makes sense as it corresponds to Christmas and the festive season. It seems that around July – August there is another high season for wine sales (see Figure 12.28).

Armed with these observations both ARIMA and Holt–Winters models are attempted. To get an idea of Holt–Winters parameters, it is useful to note that the single exponential smoothing parameter $\alpha$ is equivalent to $1 + \theta_1$, where $\theta_1$ is the MA coefficient of ARIMA(0, 1, 1) model. The double exponential smoothing parameters $\alpha$ and $\beta$ may be obtained from the MA coefficients of ARIMA(0, 2, 2) as $\alpha = 1 + \theta_1$ and $\beta = 1 + \theta_2$. To get at least a starting point of Holt–Winters parameters MA parameters of ARIMA(0, 2, 2)× (0, 1, 1) may be used. However it is difficult to get an easy starting point for Holt–Winters model with seasonality. For final model recommendations Holt–Winters models were not found to be suitable.

To fit an appropriate ARIMA model, `auto.arima()` has been used to

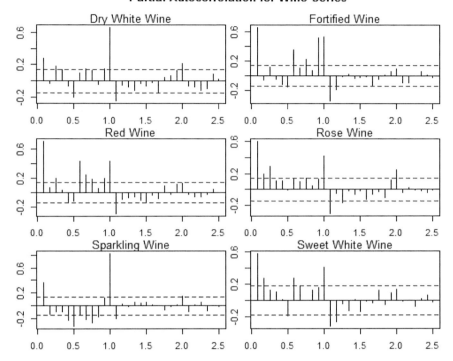

**FIGURE 12.27**
Example of partial autocorrelation.

begin with and then several attempts for simplification have been made. If a simpler model has a lower value of AIC and the residuals are stationary, then that ARIMA model has been considered as the recommended model. For each series the final ARIMA model is tested for forecast accuracy; see Table 12.9.
◈

It is clear from the above exercise that an ARIMA model which is able to reduce the residuals into a stationary series may not provide a very good forecast. But it must be remembered that the forecasts may not be perfect as far as the future is concerned. If ARIMA or any other model is not used for forecasting, planning for the future is done according to any naive method, which may include simple moving averages or estimation of trend and adjustment by seasonality. As long as a model-based forecast does considerably better than the naive forecast, it is worth using the models.

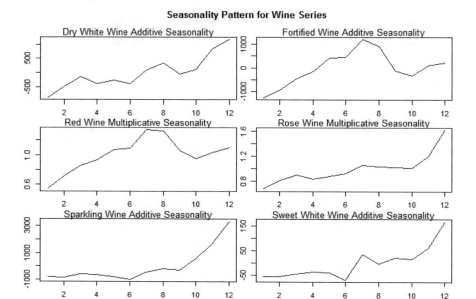

**FIGURE 12.28**
Seasonality patterns of Wine Sales series.

## 12.5   Other Time Series

ARIMA models, though useful enough to explain many different variations in time series, are naturally not able to explain all. A very important class of models are the so called ARCH and GARCH models. An ARCH (autoregressive conditionally heteroscedastic) model takes into account the changing variance in a series. Although usually an increasing variance is the case for applicability of the ARCH model, it can also be used to describe a situation in which there may be short periods of increased variation. In fact, gradually increasing variance connected to a gradually increasing mean level might be better handled by transforming the variable. ARCH models are typically used in the context of econometric and finance problems having to do with the amount that investments or stocks increase (or decrease) per time period, so there is a tendency to describe them as models for that type of variable. An ARCH model could be used for any series that has periods of increased or decreased variance. This might, for example, be a property of residuals after an ARIMA model has been fit to the data. A further extension of the ARCH model is known as the GARCH model.

**TABLE 12.9**
Comparison of forecast accuracy of wine sales series

| Wine Type | Dry White | Fortified | Red |
|---|---|---|---|
| Recommended ARIMA model | $(0,0,0) \times (0,1,1)$ with drift | $(0,1,1) \times (1,1,0)$ | $(1,0,1) \times (0,1,1)$ with drift |
| Forecast Accuracy Measures | | | |
| MAPE | 12.29% | 11.23% | 10.38% |
| MdPE | 13.58% | 11.43% | 9.60% |
| MASE | 1.68 | 0.89 | 1.36 |
| Wine Type | Rose | Sparkling | Sweet White |
| Recommended ARIMA model | $(0,1,1) \times (1,1,0)$ | $(0,0,1) \times (0,1,2)$ with drift | $(1,1,0) \times (2,0,0)$ with drift |
| Forecast Accuracy Measures | | | |
| MAPE | 28.9% | 17.66% | Forecast |
| MdPE | 27.9% | 8.39% | errors |
| MASE | 0.77 | 1.03 | too high |

Another variation is time series regression when both response and predictors are measured as a time series and the errors are assumed to follow an AR structure. The same concept can be extended to a lagged regression where lags of response should be regressed on lags of predictors. The cross-correlation function (CCF) is used to identify possible models. In the present chapter only the basics of time series have been discussed. None of the more complex time series models nor the frequency domain approach to time series have been considered. These should be the part of a more advanced treatment.

## 12.6 Suggested Further Reading

Time series is a topic that has extensive applications as well as theoretical connotations. For forecasting, time series is used almost exclusively. In this chapter we have used only basic models based on trend and seasonality. There are innumerable books on time series at various levels of difficulty. Among them, Hamilton (1994) is a comprehensive text. For time series with R applications, Shumway and Stoffer (2010) is good as a first course. At the same level, other useful books are Montgomery et al. (2008), Derryberry (2014), Hyndman and Athanasopoulos (2014) and Bisgaard and Kulahci (2011). For a more mathematical treatment of time series, Hyndman et al. (2008), Gourieroux (1997) and Tsay (2013) can be consulted.

# References

Agresti, A. and Franklin, C. (2009). *Statistics: The Art and Science of Learning from Data*, 2nd Ed. Pearson Prentice Hall, Upper Saddle River, NJ.

Akritas, M. (2014). *Statistics with R for Engineers and Scientists.* Pearson, New York.

Anderson, D. R., Sweeney, D. J. and Williams, T. A. (2014). *Statistics for Business and Economics*, 11th Ed. Cengage, New Delhi.

Bisgaard, S. and Kulahci, M. (2011). *Time Series Analysis and Forecasting by Example.* Wiley, New York.

Black, K. (2012). *Applied Business Statistics: Making Better Business Decisions.* Wiley, New York.

Borner, K. and Polley, D. E. (2014). *Visual Insights: A Practical Guide to Making Sense of Data.* The MIT Press, Cambridge, MA.

Breiman, L., Friedman, J., Stone, C. J. and Olshen, R. A. (1984). *Classification and Regression Trees.* Chapman and Hall/CRC, New York.

Burns, P. (2013). *The R Inferno.* http://www.burns-stat.com/pages/Tutor/R_inferno.pdf.

Camm, Jeffrey D., Cochran, James J., Fry, Michael J., Ohlmann, Jeffrey W. and Anderson, David R. (2013). *Essentials of Business Analytics.* Cengage, New Delhi.

Casella, G. and Berger, R. (1990). *Statistical Inference*, 2nd Ed. Duxbury, Pacific Grove, CA.

Chatterjee, S. and Hadi, A. S. (2012). *Regression Analysis by Example*, 5th Ed. Wiley, New York.

Collett, D. (1991). *Modelling Binary Data.* Chapman and Hall, New York.

Cook, D. and Weisberg, S. (1982). *Residual and Influence in Regression.* Chapman and Hall, New York.

Dalgaard, P. (2008). *Introductory Statistics with R.* Springer, New York.

Davenport, T. H. and Harris, J. G. (2007). *Competing on Analytics: The New Science of Winning.* Harvard Business School Publishing Corporation, Brighton.

Davenport, T. H., Harris, J. G. and Morison, R. (2010). *Analytics at Work: Smarter Decisions, Better Results.* Harvard Business School Publishing Corporation, Brighton.

Davenport, T. H. and Kim, J. (2013). *Keeping Up with the Quants.* Harvard Business School Publishing Corporation, Brighton.

Davino, C., Furno, M. and Vistocco, D. (2013). *Quantile Regression: Theory and Applications.* Wiley, New York.

DeGroot, M. H. and Schervish, M. J. (2012). *Probability and Statistics*, 4th Ed. Pearson, New York.

Derryberry, D. R. (2014). *Basic Data Analysis for Time Series with R.* Wiley, New York.

Dill, J., Earnshaw, R., Kasik, D., Vince, J. and Wong, P. C. (Eds.) (2012). *Expanding the Frontiers of Visual Analytics and Visualization.* Springer, New York.

Draper, N. R. and Smith, H. (1998). *Applied Regression Analysis*, 3rd Ed. Wiley, New York.

Everitt, B. S. and Hothorn, T. (2011) *An Introduction to Applied Multivariate Analysis with R.* Springer, New York.

Feller, W. (1968). *An Introduction to Probability Theory and Its Applications*, 3rd Ed. John Wiley and Sons, New York.

Freedman, D., Pisani, R. and Purves, R. (2007). *Statistics*, 4th Ed. Norton & Co., New York.

Gourieroux, C. (1997). *ARCH Models and Financial Applications.* Springer, New York.

Grimm, L. G. and Yarnold, P. R. (Eds.) (2000). *Reading and Understanding More Multivariate Statistics.* American Psychological Association, Washington, DC.

Hamilton, J. D. (1994). *Time Series Analysis.* Princeton University Press, Princeton.

Hardoon, D. R. and Shmueli, G. (2013). *Getting Started with Business Analytics: Insightful Decision-Making.* CRC Press, Boca Raton, FL.

Harrell, F. E., Jr. (2015). *Regression Modeling Strategies with Applications to Linear Models, Logistic and Ordinal Regression, and Survival Analysis.* Springer, New York.

Hosmer, D. W., Lemeshow S. and Sturdivant, R. X. (2013). *Applied Logistic Regression*, 3rd Ed. Wiley, New York.

Hothorn, T. and Everitt, B. S. (2014). *A Handbook of Statistical Analyses using R*, 3rd. Ed. CRC Press, Boca Raton, FL.

Hyndman, R. J. and Athanasopoulos, G. (2014). *Forecasting: Principles and Practice.* OText (Online Open Text) https://www.otexts.org/book/fpp.

Hyndman, R. J. and Koehler, A. (2006). Another look at measures of forecast accuracy. *International Journal of Forecasting.* 22, 679-688.

Hyndman, R. J., Koehler, A., Ord, J. K. and Snyder, R. D. (2008). *Forecasting with Exponential Smoothing: The State Space Approach.* Springer, New York.

James, G., Witten, D., Hastie, T. and Tibshirani, R. (2014). *An Introduction to Statistical Learning: With Applications in R.* Springer, New York.

Johnson, N. L., Kemp, A. and Kotz, S. (2005). *Univariate Discrete Distributions*, 3rd Ed. John Wiley and Sons, New York.

Johnson, N. L., Kotz, S. and Balakrishnan, N. (1994). *Continuous Univariate Distributions*, Volume 1, 2nd Ed. John Wiley and Sons, New York.

Johnson, N. L., Kotz, S. and Balakrishnan, N. (1995). *Continuous Univariate Distributions*, Volume 2, 2nd Ed. John Wiley and Sons, New York.

Johnson, N. L., Kotz, S. and Balakrishnan, N. (1997). *Discrete Multivariate Distributions.* John Wiley and Sons, New York.

Johnson, R. A. and Wichern, D. W. (2007). *Applied Multivariate Statistical Analysis*, 6th Ed. Prentice Hall, Upper Saddle River, NJ.

Kabacoff, R. I. (2011). *R in Action:Data Analysis and Graphics with R.* Manning, Shelter Island, New York.

Kachigan, S. K. (1991). *Multivariate Statistical Analysis: A Conceptual Introduction*, 2nd Ed. Radius Press, New York.

Keim, D., Kohlhammer, J., Ellis, G. and Mansmann, F. (Eds.) (2010). *Mastering the Information Age: Solving Problems with Visual Analytics.* Eurographics Association, Goslar, Germany.

Kirchgassner, G. and Wolters, J. (2007). *Introduction to Modern Time Series.* Springer, Berlin.

Kleinbaum, D. G., and Klein, M. (2010). *Logistic Regression.* Springer, New York.

Kotz, S., Balakrishnan, N. and Johnson, N. L. (2000). *Continuous Multivariate Distributions, Volume 1, Models and Applications,* 2nd Ed. John Wiley and Sons, New York.

Kuhn, M. and Johnson, K. (2013). *Applied Predictive Modeling.* Springer, New York.

Kundu, D. and Basu, A. (Eds.) (2004). *Statistical Computing: Existing Methods and Recent Developments.* Narosa, New Delhi.

Kutner, M. H., Nachtsheim, C. J., Neter, J. and Li, W. (2003). *Applied Linear Statistical Models,* 5th Ed. McGraw-Hill, Boston.

Ledolter, J. (2013). *Data Mining and Business Analytics with R.* Wiley, New York.

Levin, R. I. and Rubin, D. S. (2011). *Statistics for Management,* 6th Ed. Prentice Hall, New York.

Lind, D., Marchal, W. and Wathen, S. (2014). *Statistical Technique in Business and Economics,* 16th Ed. McGraw-Hill, New York.

Linoff, G. S. and Berry, M. J. A. (2011). *Data Mining Techniques: For Marketing, Sales, and Customer Relationship Management,* 3rd Ed. Wiley, New York.

Mann, P. S. (2013). *Introductory Statistics,* 8th Ed. Wiley, New York.

Montgomery, D. C., Jennings, C. L. and Kulahci, M. (2008) *Introduction to Time Series Analysis and Forecasting,* 2nd Ed. Wiley, New York.

Montgomery, D. C., Peck, E. A. and Vining, G. G. (2012). *Introduction to Linear Regression,* 5th Ed. Wiley, New York.

Moore, D. S., McCabe, G. P. and Craig, B. A. (2009). *Introduction to the Practice of Statistics.* Freeman.

Ott, R. and Longnecker, M. (2010). *An Introduction to Statistical Method and Data Analysis,* 6th Ed. Brooks/Cole, Belmont, California, Cengage Lerning.

Pal, N. and Sarkar, S. (2009). *Statistics: Concepts and Applications,* 2nd Ed. Phi Learning.

Pardoe, I. (2012). *Applied Regression Modeling,* 2nd Ed. Wiley, New York.

Peng, R. D. (2015). *R Programming for Data Science, 2nd Ed.* http://leanpub.com/rprogramming.

Ross, S. (2010). *Introduction to Probability Models*, 10th Ed. Academic Press, Amsterdam.

Rousseeuw, P. J. and Leroy, A. M. (2003). *Robust Regression and Outlier Detection*. Wiley, New York.

Rozanov, Iu. A. (1978). *Probability Theory: A Concise Course*. Dover Publications, New York.

Ryan, T. P. (2009). *Modern Regression Methods*, 2nd Ed. Wiley, New York.

Saxena, Rahul and Srinivasan, Anand (2013). *Business Analytics: A Practitioners Guide*. Springer, New York.

Schumacker, R. E. (2015). *Using R with Multivariate Statistics*. Sage.

Shumway, R. H. and Stoffer, D. S. (2010). *Time Series Analysis and Its Application*, EZ 3rd Ed. Springer, New York.

Stuart, A. and Ord, K. (2010). *Kendall's Advanced Theory of Statistics, Distribution Theory, Vol 1*. Wiley, New York.

Tabachnick, B. G. and Fidell, L. S. (2013). *Using Multivariate Statistics*, 6th Ed. Pearson.

Takezawa, K. (2005). *Introduction to Nonparametric Regression*. Wiley, New York.

Tijms, H. (2004). *Understanding Probability: Chance Rules in Everyday Life*. Cambridge University Press.

Tsay, R. S. (2013). *Multivariate Time Series Analysis: With R and Financial Applications*. Wiley, New York.

Walpole, R. E., Myers, R. H., Myers, S. L. and Ye, K. E. (2010). *Probability and Statistics for Engineers and Scientists*, 9th Ed. Pearson.

Weisberg, S. (2013). *Applied Linear Regression*, 4th Ed. Wiley, New York.

Wickham, H. (2009). *ggplot2: Elegant Graphics for Data Analysis*. Springer, New York.

# Index